Discrete Dynamical Systems, Bifurcations and Chaos in Economics

This is volume 204 in
MATHEMATICS IN SCIENCE AND ENGINEERING
Edited by C.K. Chui, *Stanford University*

A list of recent titles in this series appears at the end of this volume.

Discrete Dynamical Systems, Bifurcations and Chaos in Economics

WEI-BIN ZHANG
COLLEGE OF ASIA PASIFIC MANAGEMENT
RITSUMEIKAN ASIA PASIFIC UNIVERSITY
BEPPU-SHI, OITA-KEN
JAPAN

ELSEVIER

Amsterdam – Boston – Heidelberg – London – New York – Oxford
Paris – San Diego – San Francisco – Singapore – Sydney – Tokyo

ELSEVIER B.V.
Radarweg 29
P.O. Box 211,
1000 AE Amsterdam
The Netherlands

ELSEVIER Inc.
525 B Street, Suite 1900
San Diego, CA 92101-4495
USA

ELSEVIER Ltd
The Boulevard, Langford Lane
Kidlington, Oxford OX5 1GB
UK

ELSEVIER Ltd
84 Theobalds Road
London WC1X 8RR
UK

First edition 2006

Library of Congress Cataloging in Publication Data
A catalog record is available from the Library of Congress.

British Library Cataloguing in Publication Data
A catalogue record is available from the British Library.

ISBN-13: 978-0-444-52197-2
ISBN-10: 0-444-52197-6
ISSN: 0076-5392

Transferred to digital printing 2007

Preface

Difference equations are now used in modeling motion and change in all areas of science. In particular, applications of difference equations in economics have recently been accelerating mainly because of rapid development of nonlinear theory and computer.

Application of difference equations to economics is a vast and vibrant area. Concepts and theorems related to difference equations appear everywhere in academic journals and textbooks in economics. One can hardly approach, not to mention digest, the literature of economic analysis without "sufficient" knowledge of difference equations. Nevertheless, the subject of applications of difference equations to economics is not systematically studied. The subject is often treated as a subsidiary part of (textbooks of) mathematical economics. Due to the rapid development of difference equations and wide applications of the theory to economics, there is a need for a systematic treatment of the subject. This book provides a comprehensive study of applications of difference equations to economics.

This book is a unique blend of the theory of difference equations and its exciting applications to economics. The book provides not only a comprehensive introduction to applications of theory of linear (and linearized) difference equations to economic analysis, but also studies nonlinear dynamical systems which have been widely applied to economic analysis in recent years. It provides a comprehensive introduction to most important concepts and theorems in difference equations theory in a way that can be understood by anyone who has basic knowledge of calculus and linear algebra. In addition to traditional applications of the theory to economic dynamics, it also contains many recent developments in different fields of economics. We emphasize "skills" for application. Except conducting mathematical analysis of the economic models like most standard textbooks on mathematical economics, we use computer simulation to demonstrate motion of economic systems. A large fraction of examples in this book are simulated with Mathematica. Today, more and more researchers and educators are using computer tools to solve – once seemingly impossible to calculate even three decades ago – complicated and tedious problems.

I would like to thank Editor Andy Deelen at Elsevier for effective co-operation. I completed this book at the Ritsumeikan Asia Pacific University, Japan. I am grateful to the university's free academic environment. I take great pleasure in expressing my gratitude to my wife, Gao Xiao, who has been supportive of my efforts in writing this book in Beppu City, Japan. She also helped me to draw some of the figures in the book.

Wei-Bin Zhang

Contents

Chapter 1

Difference equations in economics

The necessity of knowledge about theory of difference equations is evident if one opens almost any current journal in any subfield of theoretical as well as applied economics. Nevertheless, there is no book which is concentrated on applications of contemporary theory of difference equations to economics. The purpose of this book is to introduce the theory of difference equations and its applications to economics.

A difference equation expresses the rate of change of the current state as a function of the current state. A simple illustration of this type of dependence is changes of the GDP (Gross Domestic Product) over years. Consider the GDP of the economy in year t as the state variable in period t, which is denoted by $x(t)$. Let us consider a case that the rate of change of the GDP is constant. Then, the motion of the GDP is described mathematically as

$$\frac{x(t+1)-x(t)}{x(t)} = g.$$

As the growth rate g is given for each year, the GDP in period t is given by solving the difference equation

$$x(t+1) = (1+g)x(t).$$

If we know a special year's GDP, $x(0)$, then the GDP in year t is given by

$$x(t) = x(0)(1+g)^t.$$

In fact, if we know any special year's GDP, then the equation predicts the GDP in any time. We can explicitly solve the above difference function because g is a constant. It is reasonable to consider that the growth rate is affected by many factors, such as the current state of the economic system, the knowledge of the economy, international environment. When the growth rate is not constant and is considered to be affected by the current state and other exogenous factors like global economic conditions (which are measured through the variable, t), then economic growth is described by

$$\frac{x(t+1)-x(t)}{x(t)}=g(x(t),t).$$

In general, it is not easy to explicitly solve the above function. There are different established methods of solving different types of difference equations. This book introduces concepts, theorems, and methods in difference equations theory which are widely used in contemporary economic analysis and provides many traditional as well as contemporary applications of the theory to different fields in economics.

1.1 Difference equations and economic analysis

This book is a unique blend of the theory of difference equations and their exciting applications to economics. First, it provides a comprehensive introduction to most important concepts and theorems in difference equations theory in a way that can be understood by anyone who has basic knowledge of calculus and linear algebra. In addition to traditional applications of the theory to economic dynamics, it also contains many recent developments in different fields of economics. It is mainly concerned with how difference equations can be applied to solve and provide insights into economic dynamics. We emphasize "skills" for application. When applying the theory to economics, we outline the economic problem to be solved and then derive difference equation(s) for this problem. These equations are then analyzed and/or simulated. We use computer simulation to demonstrate motion of economic systems. A large fraction of examples in this book are simulated with Mathematica.[1] Today, more and more researchers and educators are using computer tools such as Mathematica to solve – once seemingly impossible to calculate even three decades ago – complicated and tedious problems.

This book provides not only a comprehensive introduction to applications of linear and linearized difference equations theory to economic analysis, but also studies nonlinear dynamical systems which have been widely applied to economic

[1] Enns and McGuire (2001), Shone (2002), and Abell and Braselton (2004).

analysis only in recent years. Linearity means that the rule that determines what a piece of a system is going to do next is not influenced by what it is doing now. The mathematics of linear systems exhibits a simple geometry. The simplicity allows us to capture the essence of the problem. Nonlinear dynamics is concerned with the study of systems whose time evolution equations are nonlinear. If a parameter that describes a linear system is changed, the qualitative nature of the behavior remains the same. But for nonlinear systems, a small change in a parameter can lead to sudden and dramatic changes in both the quantitative and qualitative behavior of the system.

Nonlinear dynamical theory reveals how nonlinear interactions can bring about qualitatively new structures and how the whole is related to and different from its individual components. The study of nonlinear dynamical theory has been enhanced with developments in computer technology. A modern computer can explore a far wider class of phenomena than it could have been imagined even a few decades ago. The essential ideas about complexity have found wide applications among a wide range of scientific disciplines, including physics, biology, ecology, psychology, cognitive science, economics and sociology. Many complex systems constructed in those scientific areas have been found to share many common properties. The great variety of applied fields manifests a possibly unifying methodological factor in the sciences. Nonlinear theory is bringing scientists closer as they explore common structures of different systems. It offers scientists a new tool for exploring and modeling the complexity of nature and society. The new techniques and concepts provide powerful methods for modeling and simulating trajectories of sudden and irreversible change in social and natural systems.

Modern nonlinear theory begins with Poincaré who revolutionized the study of nonlinear differential equations by introducing the qualitative techniques of geometry and topology rather than strict analytic methods to discuss the global properties of solutions of these systems. He considered it more important to have a global understanding of the gross behavior of all solutions of the system than the local behavior of particular, analytically precise solutions. The study of the dynamic systems was furthered in the Soviet Union, by mathematicians such as Liapunov, Pontryagin, Andronov, and others. Around 1960, the study by Smale in the United States, Peixoto in Brazil and Kolmogorov, Arnold and Sinai in the Soviet gave a significant influence on the development of nonlinear theory. Around 1975, many scientists around the world were suddenly aware that there is a new kind of motion – now called chaos – in dynamic systems. The new motion is erratic, but not simply "quasiperiodic" with a large number of periods.[2] What is surprising is that chaos can occur even in a very simple system. Scientists were interested in complicated motion of dynamic systems. But only with the advent of computers,

[2] In the solar system, the motion traveled around the earth in month, the earth around the sun in about a year, and Jupiter around the sun in about 11.867 years. Such systems with multiple incommensurable periods are known as *quasiperiodic*.

with screens capable of displaying graphics, have scientists been able to see that many nonlinear dynamic systems have chaotic solutions. As demonstrated in this book, nonlinear dynamical theory has found wide applications in different fields of economics.[3] The range of applications includes many topics, such as catastrophes, bifurcations, trade cycles, economic chaos, urban pattern formation, sexual division of labor and economic development, economic growth, values and family structure, the role of stochastic noise upon socio-economic structures, fast and slow socio-economic processes, and relationship between microscopic and macroscopic structures. All these topics cannot be effectively examined by traditional analytical methods which are concerned with linearity, stability and static equilibrium points. Nonlinear dynamical theory has changed economists' views about evolution. For instance, the traditional view of the relations between laws and consequences - between cause and effect - holds that simple rules imply simple behavior, therefore complicated behavior must arise from complicated rules. This vision had been held by professional economists for a long time. But it has been recently challenged due to the development of nonlinear theory. Nonlinear theory shows how complicated behavior may arise from simple rules. To illustrate this idea, we consider the following model

$$\frac{x(t+1)-x(t)}{x(t)} = a - 1 - ax(t), \quad a > 0.$$

The growth rate is a linear function of the state variable $x(t)$. We may rewrite the above equation as follows

$$x(t+1) = ax(t)(1 - x(t)), \quad a > 0.$$

This is the well-studied *logistical map*. This seemingly simple map exhibits very complicated behavior as we will analyze later on. For instance, figure 1.1.1 depicts chaotic behavior of the difference equation with a given parameter value and initial condition.

The existence of chaos implies that no one can precisely know what will happen in society in the future, except that it will be changing in some bounded area. To illustrate why no one can precisely foresee the consequences of the intervention policy, let us try to find out what happen to the chaotic system when it starts from two different but very near states. In figure 1.1.2, we simulate the case of $a = 5.75$. Let us consider two cases of $x_0 = 0.400$ and

[3] For applications of nonlinear theory to economics, see Dendrinos and Sonis (1990), Rosser (1991), Zhang (1991, 2005a), Lorenz (1993), Azariadis (1993), Puu (1989), Ferguson and Lim (1998), Flaschel *et al* (1997), Chiarella and Flaschel (2000), and Shone (2002).

$x_0 = 0.405$ over 100 years. It can be seen that the two behaviors are varied over time.

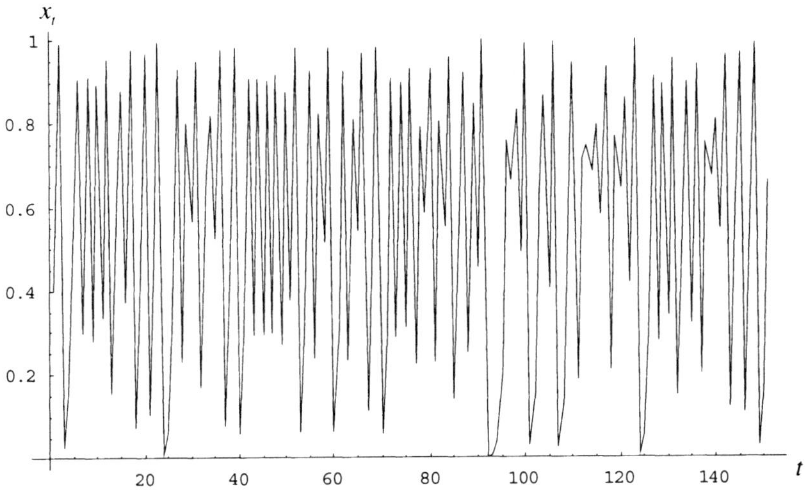

Figure 1.1.1: Chaos when $a = 5.75$ and $x_0 = 0.4$

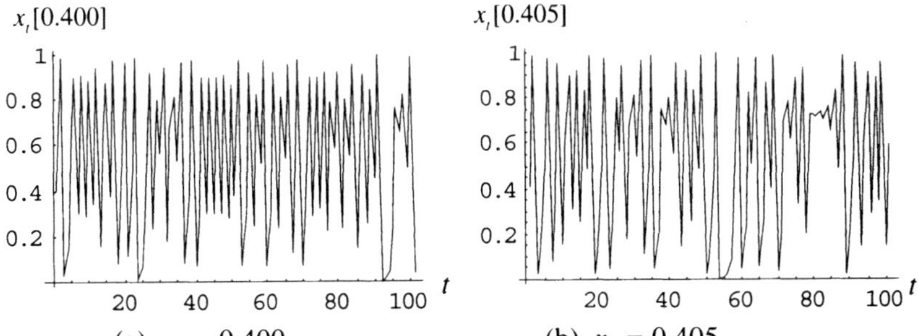

(a) $x_0 = 0.400$ (b) $x_0 = 0.405$

Figure 1.1.2: The dynamics with different initial conditions, $a = 5.75$

We calculate the difference $x[t, 0.400] - x[t, 0.405]$ between the path started at $x_0 = 0.400$ and the one at $x_0 = 0.405$ over 100 years as in figure 1.1.3.

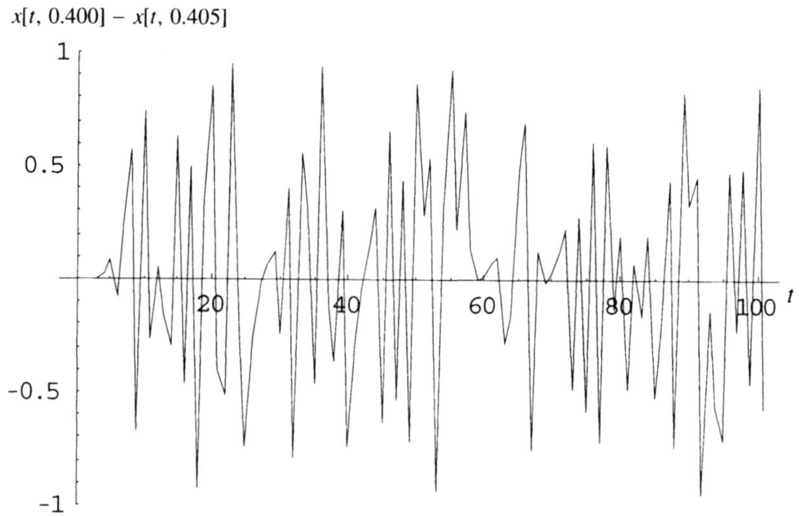

Figure 1.1.3: Small differences at the beginning signify much

It can be shown when the parameter a is small, the difference equation has a unique equilibrium point. As the parameter a exceeds a certain value, the steady state ceases being approached monotonically, and an oscillatory approach occurs. If a is increased further, the steady state becomes unstable and repels nearby points. As a increases, one can find a value of a where the system possesses a cycle of period k for arbitrary k (see figure 1.1.4). Also, there exists an uncountable number of initial conditions from which emanate trajectories that fluctuate in a bounded and aperiodic fashion and are indistinguishable from a realization of some stochastic (chaotic) process.

Nonlinear dynamical systems are sufficient to determine the behavior in the sense that solutions of the equations do exist, it is frequently difficult to figure out what behavior would be. It is often impossible to explicitly write down solutions in algebraic expressions. Nonlinear economics based on nonlinear dynamical theory attempts to provide a new vision of economic dynamics: a vision toward the multiple, the temporal, the unpredictable, and the complex. There is a tendency to replace simplicity with complexity and specialism with generality in economic research. The concepts such as totality, nonlinearity, self-organization, structural changes, order and chaos have found broad and new meanings by the development of this new science. According to this new science, economic dynamics are considered to resemble a turbulent movement of liquid in which varied and relatively stable forms of current and whirlpools constantly change one another. These changes consist of dynamic processes of self-organization along with the spontaneous formation of increasingly subtle and complicated structures. The accidental nature and the presence of structural changes like catastrophes and bifurcations, which are characteristic of nonlinear systems and whose further trajectory is determined by chance, make dynamics irreversible.

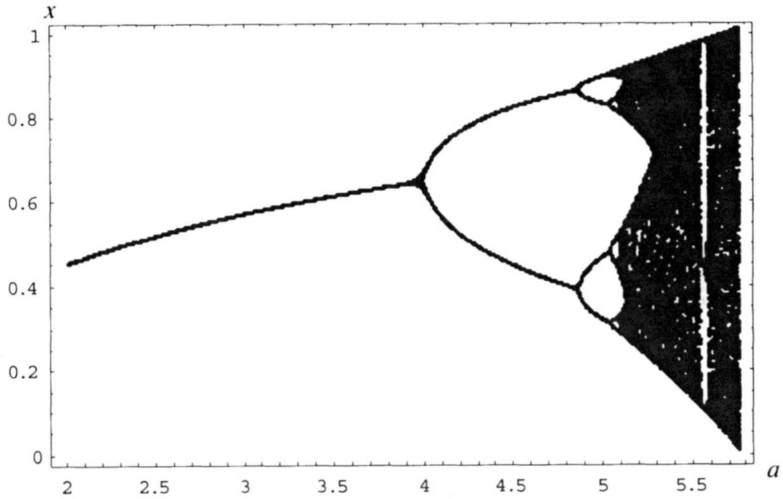

Figure 1.1.4: The map of bifurcations for $a \in [2, 5.75]$

1.2 Overview

This book presents the mathematical theory in linear and nonlinear difference equations and its applications to many fields of economics. The book is for economists and scientists of other disciplines who are concerned to model and understand the time evolution of economic systems. It is of potential interest to advanced undergraduates and graduate students in economics, economic professionals, applied mathematicians who are interested in social sciences, as well as researchers in social sciences. The book is organized as follows.

Chapter 2 is organized as follows. Section 2.1 deals with linear first-order difference equations. We also examine dynamics of two economic models, a price dynamic model with adaptation and a model of amortization. In section 2.2, we introduce some basic concepts which will be used not only for one dimensional problems but also for higher dimensional ones. Section 2.3 introduces concepts of stability and some basic theorems about stabilities of one-dimensional difference equations. We apply these results to the cobweb model and a dynamic model with inventory. Section 2.4 is concerned with conditions for stabilities of nonhyperbolic equilibrium points. Section 2.5 deals with dissipative maps. Section 2.6 introduces difference equations of higher order and provides general solutions to the system. Section 2.7 examines difference equations of higher orders with constant coefficients. In section 2.8, we are concerned with limiting behavior of linear difference equations. We examine limiting behavior of the Samuelson multiplier-accelerator interaction model.

Chapter 3 applies the concepts and theorems of the previous chapters to analyze different models in economics. Although the economic relations in these models are complicated, we show that the dynamics of all these models are determined by one-dimensional difference equations. Section 3.1 examines a traditional model of interactions between inflation and unemployment. The model is built on the expectations-augmented version of the Phillips relation and the adaptive expectations hypothesis. We solve the model and show that the characteristic equation may have either: (1) distinct real roots; or (2) repeated real roots; or (3) complex roots. Section 3.2 introduces the one-sector growth model. The model is different from most of the growth models in the literature in that it treats saving as an endogenous variable through introducing wealth into utility function. We demonstrate that the OSG model has a unique stable equilibrium. Section 3.3 generalizes the OSG model proposed in section 3.2. Section 3.4 deals with the overlapping-generation (OLG) model – one of the most popular models among economists. The model is essential for the reader to approach some of the models in this book as well. Different from the OSG models in the previous two sections, in the OLG analytical framework, each person lives for only two periods. This is the main shortcoming of the model; nevertheless, its popularity is sustained partly because this framework often simplifies complicated analytical issues. Section 3.5 introduces a growth model to demonstrate persistence of inequality. In this model, the evolution of income within each dynasty in society is governed by a dynamical system that generates a poverty trap equilibrium point along with a high-income equilibrium. Poor dynasties, those with the income at the threshold level, converge to a low income steady state, whereas dynasties with income above the threshold converge to a high income steady. Section 3.6 studies a model to provide insights into evolutionary processes of Schumpeterian creative destruction. Section 3.7 is concerned with interactions among human capital accumulation, economic growth, and inequality. The model exhibits three possible equilibrium points: a low-growth trap, a pair of equilibrium points in the intermediate and advanced development phase. If these equilibrium points exist, it can be shown that the poverty trap is stable, while in development phase, the first equilibrium point will be unstable and the second one stable. Section 3.8 studies an urban dynamic model to highlight how the trade-off between optimal and equilibrium city sizes behaves when human capital externalities are introduced into urban dynamics. Section 3.9 introduces a growth model of monetary economy. The model addresses the Tobin effect and the existence of monetary economy. Section 3.10

introduces endogenous time distribution between leisure and work into the OSG model.

Chapter 4 examines periodic, aperiodic, chaotic solutions of scalar systems. Section 4.1 defines concepts such as periodic or aperiodic solutions (orbits). This section also introduces some techniques to find periodic solutions and provides conditions for judging stability of periodic solutions. Section 4.2 is concerned with period-doubling bifurcations. This section introduces concepts such as branch, bifurcation values, period-doubling bifurcation route to chaos, Myrberg's number and Fiegenbaum's number. Section 4.3 deals with aperiodic orbits. This section introduces the Li-Yorke theorem and the Sharkovsky theorem, which are important for proving existence of chaos in scalar systems. Section 4.4 studies some typical types of bifurcations. They include supercritical fold, subcritical fold, supercritical pitchfork, subcritical pitchfork, transcritical bifurcations. Section 4.5 introduces theory of Liapunov numbers. In this section, we also examine behavior of a model of labor market. In section 4.6, we study chaos. We simulate a demand and supply model to demonstrate chaotic behavior.

Chapter 5 applies concepts and theorems of the previous chapters to analyze different models in economics. The models in this chapter exhibits periodic, aperiodic, or chaotic behavior. Section 5.1 studies a model of endogenous business cycles in the presence of knowledge spillovers. Many economic indicators, such as GDP, exhibit asymmetry as if they repeatedly switch between different regimes. For instance, it has been found that (i) positive shocks are more persistent than negative shocks in the United States and France; (ii) negative shocks are more persistent than negative shocks in the United Kingdom and Canada; and (iii) there is almost no asymmetry in persistence in Italy, Japan, and (former) Germany. The model in this section provides some insights into well-observed asymmetric nature of business cycles. Section 5.2 studies a nonlinear cobweb model with normal demand and supply, naïve expectations and adaptive production adjustment. The model exhibits a horseshoe. Section 5.3 examines an inventory model with rational expectations. In section 5.4, we discuss an economic growth model with pollution. The model is an extension of the standard neoclassical growth model which has a unique stable equilibrium point. Chaos exists in the model because of the effects of pollution upon production. It is know that the neoclassical growth theory based on the Solow growth model focuses accumulation as an engine of growth, while the neo-Schumpeterian growth theory stresses innovation. Section 5.5 studies a model to capture these two mechanisms within the same framework. The model generates an unstable balanced growth path and the economy achieves sustainable growth cycles, moving back and forth between the two phases – one is characterized by higher output growth, higher investment, no innovation, and a competitive market structure; the other by lower output growth, lower investment, high innovation, and a more monopolistic structure. Section 5.6 identifies economic fluctuations in a monetary economy within the OLG framework. Section 5.7 shows chaos in a model of interaction of economic and population growth.

Chapter 6 is organized as follows. Section 6.1 studies phase space analysis of planar linear difference equations. This section depicts dynamic behavior of the system when the characteristic equation has two distinct eigenvalues, or repeated eigenvalues, or complex conjugate eigenvalues. Section 6.2 studies autonomous linear difference equations. This section provides a procedure of finding general solutions of the system. Section 6.3 studies nonautonomous linear difference equations. This section provides a procedure of finding general solutions of the system. We also examine a few models of economic dynamics. They include a dynamic input-output model with time lag in production, a cobweb model in two interrelated markets, a duopoly model, a model of oligopoly with 3 firms, and a model of international trade between two countries. This section also shows how the one-dimensional difference equation of higher order can be expressed in multi-dimensional equations of first order. Section 6.4 defines concepts of stabilities and relations among these concepts. This section also provides conditions for stability or instability of difference equations. Section 6.5 studies Liapunov's second method. The theory of Liapunov functions is a global approach toward determining asymptotic behavior of solutions. Section 6.6 studies the theory of linearization of difference equations. There are two possible ways to simplify dynamical systems: one is to transform one complex system to another one which is much easier to analyze; and the other is to reduce higher dimensional problems to lower ones. The center manifold theorem helps us to reduce dimensions of dynamical problems. Section 6.7 defines the concept of conjugacy and shows how to apply the center manifold theorem. Section 6.8 studies the Hénon map, demonstrating bifurcations and chaos of the planar difference equations. Section 6.9 studies the Neimark-Sacker (Hopf) bifurcation. This section identifies the Hopf bifurcation in the discrete Kaldor model. Section 6.10 introduces the Liapunov numbers and discusses chaos for planar dynamical systems.

Chapter 7 applies the concepts and theorems of the previous chapter to examining behavior of different economic systems. Section 7.1 studies Dornbush's exchange rate model. We show how a monetary expansion will result in an immediate depreciation of the currency and sustain the inflation as the price level gradually adjusts upward. Section 7.2 studies a two-sector OLG model with the Leontief production functions. The economy produces two, consumption and investment, goods; it has two, consumption and investment sectors. We provide conditions when the system is determinate or indeterminate. Section 7.3 introduces a one-sector real business cycle model with mild increasing returns-to-scale with government spending. Section 7.4 introduces endogenous fertility and old age support into the OLG model. Section 7.5 examines a model to capture the historical evolution of population, technology, and output. The economy evolves three regimes that have characterized economic development: from a Malthusian regime (where technological progress is slow and population growth prevents any sustained rise in income per capita) into a post-Malthusian regime (where technological progress rises and population growth absorbs only part of output growth) to a modern growth regime (where population growth is reduced and income growth is sustained). The model is defined within the OLG framework with a single good and it exhibits the structural patterns observed over history. Section 7.6 examines a model of unemployment and inflation. We demonstrate that the model which is built on the well-accepted assumptions may behave chaotically. Section 7.7 provides a model of long-run competitive two-periodic OLG model with money and capital. Section 7.8 introduces heterogeneous groups to the OSG model. Section 7.9 examines interdependence between economic growth and human capital accumulation in the OSG modeling framework.

As concluding remarks to this book, we address two important issues, which have been rarely studied in depth in economic dynamical analysis, changeable speeds and economic structures. The understanding of these two issues is essential not only for appreciating validity and limitations of different economic models in the literature, but also for developing general economic theories. We also include a mathematical appendix. A.1 introduces matrix theory. A.2 shows how to solve linear equations, based on matrix theory. A.3 introduces metric spaces and some basic concepts and theorems related to metric spaces. A.4 defines some basic concepts in the study of functions and states the implicit function theorem. A.5 gives a general expression of the Taylor Expansion. A.6 is concerned with convexity of sets and functions and concavity of functions. A.7 shows how to solve unconstrained maximization problems. In A.8, we introduce conditions for constrained maximization. A.9 introduces theory of dynamic optimization.

Chapter 2

Scalar linear difference equations

In discrete dynamics, time, denoted by t, is taken to be a discrete variable so that the variable t is allowed to take only integer values. For example, if a certain population has discrete generations, the size of the $(t+1)$st generation $x(t+1)$ is a function of the t th generation $x(t)$. Different from the continuous-time dynamics where the pattern of change of a variable x is embodied in its derivatives with respect to the change of time t which is infinitesimal in magnitude, in the discrete dynamics the pattern of change of variable x is described by "differences", rather than by derivatives of x. Hence, the system in discrete time is in the form of difference equations, rather than differential equations. As the value of variable $x(t)$ will change only when the variable t changes from one integer value to the next, such as from 2004 to 2005 during which nothing is supposed to happen to $x(t)$, in difference equation theory the variable t is referred to as *period*, in the analytical sense, not necessarily in the calendar sense. The time interval between two successive states is usually suggested by the real process itself. For example, $x(t+1)$ could be separated from $x(t)$ by one hour, one day, one week, one month, etc.

To describe the pattern of change in x as a function of t, we introduce the difference quotient $\Delta x / \Delta t$. As t has to take integer values, we choose $\Delta t = 1$. Hence, the difference quotient $\Delta x / \Delta t$ is simplified to the expression Δx; this is called the first difference of x, and is denoted by

$$\Delta x(t) \equiv x(t+1) - x(t),$$

where $x(t)$ is the value of x in the t th period and $x(t+1)$ is its value in the period immediately following the t th period. The change in x during two consecutive time periods may be affected by many factors. We may express the pattern of change of x by, for instance

$$\Delta x(t+1) = -\, a x^{\alpha}(t).$$

Equations of this type are called *difference equations*. By the way, we use difference equations and discrete dynamical systems exchangeable.[1] There are other forms of difference, which are equivalent to the above equation, for instance

$$x(t+1) - x(t) = -\, a x^{\alpha}(t),$$

or

$$x(t+1) = x(t) - a x^{\alpha}(t) \equiv f(x(t)).$$

The evolution of the system starting from x_0 is given by the sequence

$$\begin{aligned} &x_0 \ (= x(0)), \\ &x(1) = f(x_0), \\ &x(2) = f(x(1)) = f(f(x_0)), \\ &x(3) = f(x(2)) = f(ff((x_0))), \cdots. \end{aligned}$$

We usually write $f^2(x),\ f^3(x), \cdots$ in place of

$$f(f(x)),\ f(ff((x))),\ \cdots.$$

Hence, we have

$$x(t+1) = f(x(t)) = f^{t+1}(x_0).$$

[1] When mathematicians talk about difference equations, they usually refer to the analytical theory of the subject, and when they talk about discrete dynamical systems, they generally refer to its geometrical and topological aspects.

$f(x_0)$ is called the *first iterate* of x_0 under f; $f^t(x_0)$ is called the *t th iterate* of x_0 under f. The set of all (positive) iterates

$$\{f^t(x_0): t \geq 0\}$$

where $f^0(x_0) = x_0$ is called the (positive) *orbit* of x_0 and is denoted by $O(x_0)$.

If the equation f is replaced by a function g of two variables, that is, $g: R \times Z^+$, where Z^+ is the set of nonnegative integers and R is the set of real numbers, then we have

$$x(t+1) = g(x(t), t).$$

This equation is called *nonautonomous* or *time-variant*, whereas

$$x(t+1) = f(x(t)),$$

is called *autonomous* or *time-invariant*.

This chapter is organized as follows. Section 2.1 deals with linear first-order difference equations. We also examine dynamics of two economic models, a price dynamic model with adaptation and a model of amortization. In section 2.2, we introduce some basic concepts, which will be used not only for one-dimensional problems but also for higher dimensional ones. Section 2.3 introduces concepts of stability and some basic theorems about stabilities of one-dimensional difference equations. We also apply these results to the cobweb model and a dynamic model with inventory. Section 2.4 is concerned with conditions for stabilities of nonhyperbolic equilibrium points. Section 2.5 deals with dissipative maps. Section 2.6 introduces difference equations of higher order and provides general solutions to the system. Section 2.7 examines difference equations of higher orders with constant coefficients. In section 2.8, we study limiting behavior of linear difference equations. We examine limiting behavior of the Samuelson multiplier-accelerator interaction model.

2.1 Linear first-order difference equations

A typical linear homogenous first-order equation is given by

$$x(t+1) = a(t)x(t), \quad x(t_0) = x_0, \quad t \geq t_0 \geq 0, \tag{2.1.1}$$

where $a(t) \neq 0$. One may obtain the solution of equation (2.1.1) by a simple iteration

$$x(t_0 + 1) = a(t_0)x_0,$$
$$x(t_0 + 2) = a(t_0 + 1)x(t_0 + 1) = a(t_0 + 1)a(t_0)x_0,$$
$$x(t_0 + 3) = a(t_0 + 2)x(t_0 + 2) = a(t_0 + 2)a(t_0 + 1)a(t_0)x_0.$$

And inductively, it is easy to see that

$$x(t) = x(t_0 + t - t_0) = \left[\prod_{i=t_0}^{t-1} a(i)\right]x_0. \tag{2.1.2}$$

Example There are $2n$ people. Find the number of ways, denoted as $p(n)$, to group these people into pairs.

To group $2n$ people into pairs, we first select a person and find that person a partner. Since the partner can be taken to be any of the other $2n - 1$ persons in the original group, there are $2n - 1$ ways to form this first group. We are left with the problem of grouping the remaining $2n - 2$ persons into pairs, and the number of ways of doing this is $p(n - 1)$. We have

$$p(n) = (2n - 1)p(n - 1).$$

Since two people can be paired only one way, we have $p(1) = 1$. To apply formula (2.1.2), we rewrite the above formula as

$$p(n + 1) = (2n + 1)p(n).$$

According to formula (2.1.2), we have

$$p(n) = \left(\prod_{i=1}^{n-1}(2i + 1)\right)p(1) = (2n - 1)(2n - 3)\cdots 1 = \frac{(2n)!}{2^n n!}.$$

The *nonhomogeneous first-order linear equation* associated with equation (2.1.1) is given by

$$x(t + 1) = a(t)x(t) + g(t), \quad x(t_0) = x_0, \quad t \geq t_0 \geq 0. \tag{2.1.3}$$

The unique solution to equation (2.1.3) is found as follows

$$\begin{aligned} x(t_0+1) &= a(t_0)x_0 + g(t_0), \\ x(t_0+2) &= a(t_0+1)x(t_0+1) + g(t_0+1) \\ &= a(t_0+1)a(t_0)x_0 + a(t_0+1)g(t_0) + g(t_0+1). \end{aligned}$$

Inductively, it can be shown that the solution is

$$x(t) = \left[\prod_{i=t_0}^{t-1} a(i)\right] x_0 + \sum_{r=t_0}^{t-1}\left[\prod_{i=r+1}^{t-1} a(i)\right] g(r). \tag{2.1.4}$$

Example Solve the equation

$$x(t+1) = (t+1)x(t) + 2^t (t+1)!,\ \ x(0) = 1,\ \ t > 0.$$

By formula (2.1.4), we have

$$x(t) = \left[\prod_{i=0}^{t-1} (i+1)\right] x_0 + \sum_{k=0}^{t-1}\left[\prod_{i=k+1}^{t-1} (i+1)\right] 2^k (k+1)! = t! + t!\sum_{k=0}^{t-1} 2^k = 2^t t!$$

A special case of equation (2.1.3) is that $a(t)$ is independent of t

$$x(t+1) = ax(t) + g(t),\ \ x(t_0) = x_0,\ \ t \geq t_0 \geq 0. \tag{2.1.5}$$

According to equation (2.1.4), the solution to equation (2.1.5) is

$$x(t) = a^t x_0 + \sum_{r=0}^{t-1} a^{t-r-1} g(r),$$

where we set $t_0 = 0$.

Example Find a solution to the equation

$$x(t+1) = 2x(t) + 3^t,\ \ x(1) = 0.5.$$

The solution is given by

$$x(t) = \left(\frac{1}{2}\right)2^{t-1} + \sum_{k=1}^{t-1} 2^{t-k-1}3^k = 3^t - 5 \cdot 2^{t-1}.$$

Example (price dynamics with adaptive expectations) There are two financial assets available to investors; a riskless bank deposit yielding a constant rate r in perpetuity, and a common share, that is, an equity claim on some firm, which pays out a known stream of dividends per share, $\{d(s)\}_{s=t}^{\infty}$. Let $p(s)$ be the actual market price of a common share at the beginning of period s, before the dividend $d(s) > 0$ is paid. Suppose also that the future share prices are unknown but that all investors have the common belief at $t = s$ that the price is going to be $p^e(s+1)$ at the beginning of the following period.

We consider the following arbitrage condition

$$(1 + r)p(t) = d(t) + p^e(t+1).$$

This equation means that if a monetary sum of $p(t)$ dollars were invested in the stock market at time t, it should yield at $t + 1$ an amount whose expected value

$$d(t) + p^e(t+1)$$

equals the principal plus interest on an equal sun invested in bank deposits. The adaptive expectation hypothesis is described by

$$p^e(t+1) = ap(t) + (1 - a)p^e(t),$$

where the parameter, $a \in [0, 1]$, describes the speed of learning. Using the arbitrage condition to eliminate expected prices from the adaptive expectation equation yields

$$p(t+1) = \lambda p(t) + b(t),$$

where

$$\lambda \equiv (1 + r)\frac{1 - a}{1 + r - a}, \quad b(t) \equiv \frac{d(t + 1) - (1 - a)d(t)}{1 + r - a}.$$

We have $\lambda \in (0, 1)$ if $r > 0$ and $a \in (0, 1)$. The solution of

$$p(t + 1) = \lambda p(t) + b(t)$$

is given by

$$p(t) = \lambda^t p(0) + \sum_{i=0}^{t-1} \lambda^i d(t - i).$$

Another special case of equation (2.1.3) is that both $a(t)$ and $g(t)$ are constant. That is

$$x(t + 1) = ax(t) + b, \quad x(t_0) = x_0, \quad t \geq t_0 \geq 0. \tag{2.1.6}$$

Using formula (2.1.4), we solve equation (2.1.6) as

$$x(t) = \begin{cases} a^t x_0 + b\left(\dfrac{a^t - 1}{a - 1}\right) & \text{if } a \neq 1, \\ x_0 + b, & \text{if } a = 1. \end{cases} \tag{2.1.7}$$

Example *Amortization* is the process by which a loan is repaid by a sequence of periodic payments, each of which is part payment of interest and part payment to reduce the outstanding principal. Let $w(t)$ denote the outstanding principal after the t th payment $m(t)$. Suppose that interest charges compound at the rate r per payment period. The outstanding principal $w(t + 1)$ after the $(t + 1)$st payment is equal to the outstanding principal $w(t)$ after the t th payment plus the interest $rw(t)$ incurred during the $(t + 1)$th period minus the t th payment $m(t)$, that is

$$w(t + 1) = w(t) + rw(t) - m(t).$$

Let w_0 stand for the initial debt. Then, determination of $w(t)$ is to solve the following difference equation

$$w(t+1) = (1+r)w(t) - m(t), \quad w(0) = w_0. \tag{2.1.8}$$

Equation (2.1.8) belongs to the type of equation (2.15). Applying the solution of equation (2.1.5) to equation (2.1.8), we get

$$w(t) = (1+r)^t w_0 - \sum_{k=0}^{t-1} (1+r)^{t-k-1} m(k).$$

In particular, if the payment $m(k)$ is constant, say M, then the above solution becomes

$$w(t) = (1+r)^t w_0 - \left((1+r)^t - 1\right)\frac{M}{r}.$$

If the loan is to be paid off in t payments (that is, $w(t) = 0$), the monthly payment M is given by

$$M = \frac{(1+r)^t}{(1+r)^t - 1} r w_0.$$

Example The Lees are purchasing a new house costing $200,000 with a down payment of $25,000 and a 30 -year mortgage. Interest on the unpaid balance of the mortgage is to be compounded at the monthly rate of 1%, and monthly payments will be $1800. How much will the Lees owe after t months of payments?

To answer this question, let $x(t)$ denote the balance in dollars that will be owed on the mortgage after t months payment. Then, as the previous example shows, we have

$$x(t) = 1.01x(t-1) - 1800, \quad t \geq 1.$$

The amount owed initially is the purchase price minus the down payment, so $x(0) = 175{,}000$. We thus solve

$$x(t) = \left(1.01^t\right)175{,}000 - 180{,}000\left(1.01^t - 1\right) = 180{,}000 - 5000\left(1.01^t\right).$$

For example, the balance of the loan after 20 years (240 months) of payments is $x(240) \approx 125{,}537$.

Exercise 2.1

1 Find the solutions of the following difference equations:

(a) $x(t+1) - (t+1)x(t) = 0, \quad x(0) = x_0;$

(b) $x(t+1) - e^{2t}x(t) = 0, \quad x(0) = x_0;$

(c) $x(t+1) - \dfrac{t}{1+t}x(t) = 0, \quad x(0) = x_0;$

(d) $x(t+1) - 0.5x(t) = 2, \quad x(0) = x_0;$

(e) $x(t+1) - x(t) = e^t, \quad x(0) = x_0.$

2 A debt of $\$12{,}000$ is to be amortized by equal payments of $\$380$ at the end of each month, plus a final partial payment one month after the last $\$380$ is paid. If interest is at an annual rate of 12% compounded monthly, construct an amortization schedule to show the required payments.

2.2 Some Concepts

Let us consider the one-dimensional difference equation

$$x(t+1) = f(x(t)) = f^{t+1}(x_0), \quad t = 0, 1, \cdots, \tag{2.2.1}$$

where $f : R \to R$ is a given nonlinear function in $x(t)$. When studying the motion of difference equations, we attempt to determine equilibrium points and periodic points, to analyze their stability and asymptotic stability, and to determine aperiodic points and chaotic behavior. We refer to equation (2.2.1) as a scalar (or one-dimensional) *dynamical system*. The function f is called the *map* associated with equation (2.2.1). A solution of equation (2.2.1) is a sequence $\{\phi_t\}_{t=0}^{\infty}$ that satisfies the equation for all $t = 0, 1, \cdots$. If an initial condition $x(0) = x_0$ is given, the problem of solving equation (2.2.1) so that the solution satisfies the initial condition is called the *initial value problem*. The *general solution* to equation (2.2.1) is a sequence $\{\phi_t\}_{t=0}^{\infty}$ that satisfies equation (2.2.1) for all $t = 0, 1, \cdots$ and involves a constant C that can be determined once an initial value is prescribed. A particular solution is a sequence $\{\phi_t\}_{t=0}^{\infty}$ that satisfies equation (2.2.1) for all $t = 0, 1, \cdots$.

Definition 2.2.1. The sequence

$$\{x_0, x_1, \cdots, x_t, \cdots\}$$

is denoted by $O(x_0)$ and is called the *orbit* or *trajectory* of the system starting from x_0.

Definition 2.2.2. A point x^* is called a *stationary point* of equation (2.2.1) if

$$x^* = f(x^*). \tag{2.2.2}$$

Each x^* can be regarded either as a state of the dynamical system

$$x(t+1) = f(x(t)),$$

satisfying equation (2.2.2) or as a solution to the system of equation

$$x = f(x).$$

We also call x^* a *fixed* (or *stationary* or *equilibrium*) *point* of f.

Example Every stationary state of the system

$$x(t+1) = ax(t)(1 - x(t))$$

must satisfy the equation

$$x = ax(1 - x).$$

We see that $x^* = 0$ is a stationary state regardless of the value of a. Another stationary point is given by

$$x = \frac{a-1}{a}.$$

For every a we can visualize the fixed points of

$$f(x) = ax(1-x),$$

since they are given by the intersection of the graph f with the line $y = x$. Figure 2.2.1 depicts the case of $a = 3$.

By the way, we introduce how to visualize solutions to one-dimensional difference equations. A frequently used plot is the so-called *stair-step diagram* or *staircase diagram*, or *cobweb diagram*. The diagram is a plot in a rectangular coordinate system of: (1) the graph of the function $y = f(x)$; (2) the identity line $y = x$; and (3) a polygonal line that results from joining the points

$$(x_0, x_1),\ (x_1, x_1),\ (x_1, x_2),\ (x_2, x_2),\ (x_2, x_3),\ (x_3, x_3), \cdots.$$

Figure 2.2.2 shows that the line segments of the polygonal line create the impression of stairs.

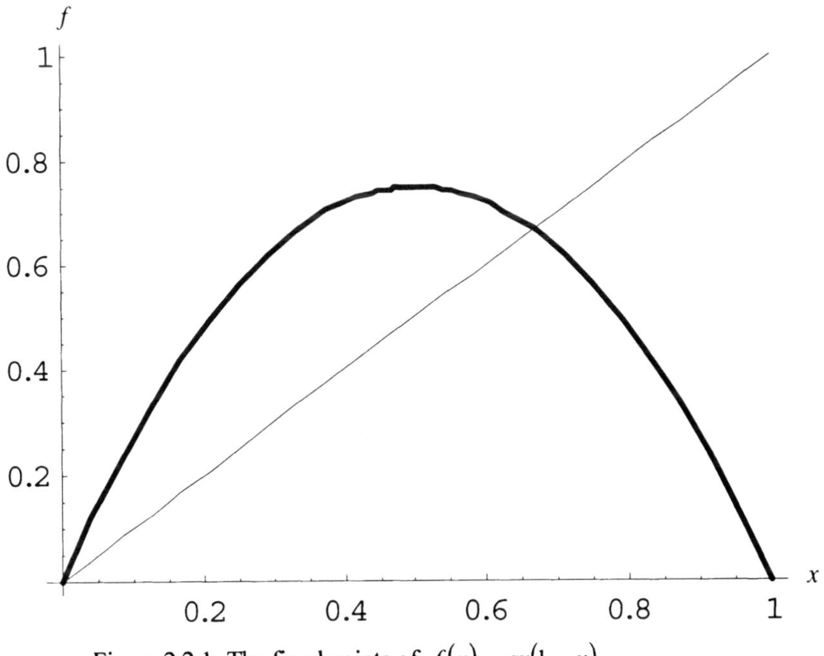

Figure 2.2.1: The fixed points of $f(x) = ax(1-x)$

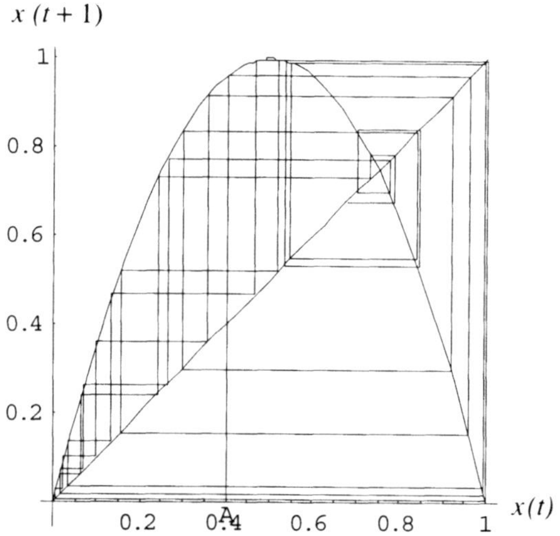

Figure 2.2.2: Staircase diagram of $O(0.4)$ for $f(x) = 4x(1 - x)$

Another plot used for visualizing solutions to one-dimensional difference equation is called *time series*. It consists of a representation of the variable $x(t)$ as a function of t. See figures 2.2.3 and 2.2.4.

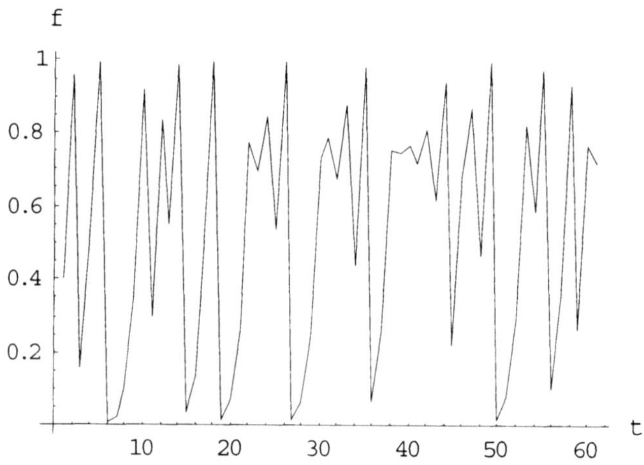

Figure 2.2.3: Time series plot of $O(0.4)$ for $f(x) = 4x(1 - x)$

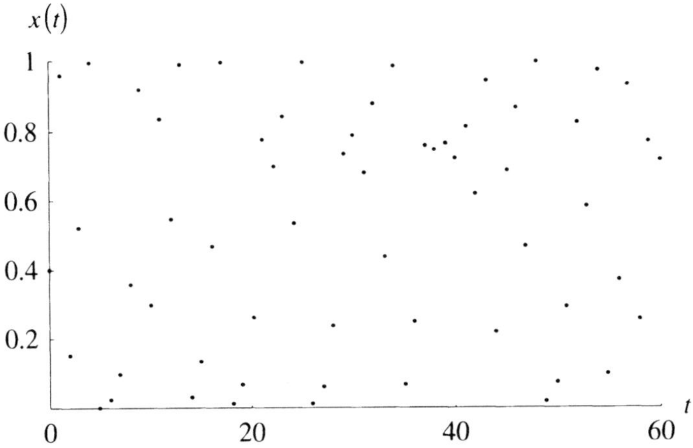

Figure 2.2.4: Plot of points of $O(0.4)$ for $f(x)=4x(1-x)$

Definition 2.2.3. A point $x^* \in R$ is said to be an *eventually equilibrium* (or *stationary*) *point* for equation (2.2.1) or an *eventually fixed point* for f if there exists a positive integer r and a fixed point $\bar{x}$ of f such that[2]

$$f^r(x^*)=\bar{x}, \quad f^{r-1}(x^*)\neq \bar{x}.$$

Example The logistical difference equation

$$f=4x(1-x)$$

has two fixed points, 0 and $3/4$. Finding eventually fixed points is to solve

$$f^r(x)=0, \frac{3}{4},$$

where r is a positive integer greater than 1. For instance, with $x=3/4$ and $r=2$, we obtain the algebraic equation

---

[2] A sequence $\{x(t)\}|_{t=0}^{\infty}$ is said to have eventually some property P, if there exists an integer $N \geq k$ such that every term of $\{x(t)\}|_{t=N}^{\infty}$ has this property.

$$16x(1-x)[1-4x(1-x)] = \frac{3}{4}.$$

It is easy to check that the equation has the following eventually fixed points

$$\frac{1}{4},\ \frac{1}{2}+\frac{\sqrt{3}}{4},\ \frac{1}{2},\ \frac{1}{2}+\frac{\sqrt{2}}{4},$$

because

$$f\left(\frac{1}{4}\right) = \frac{3}{4},$$

$$f\left(f\left(\frac{1}{2}+\frac{\sqrt{3}}{4}\right)\right) = \frac{3}{4},$$

$$f\left(f\left(\frac{1}{2}\right)\right) = 0,$$

$$f\left(f\left(f\left(\frac{1}{2}+\frac{\sqrt{2}}{4}\right)\right)\right) = 0.$$

Exercise 2.2

1 Show:

(a) $f(x) = \cos x$ has a unique fixed point for $x \in [0, 1]$;

(b) $f(x) = x^3 - 2x + 1$ has three fixed points.

2 A piecewise linear version of the logistic equation is the *tent equation*

$$x(t+1) = \begin{cases} 2x(t), & \text{if } x(t) \le \frac{1}{2}, \\ 2(1-x(t)), & \text{if } x(t) > \frac{1}{2}. \end{cases}$$

This map may be written in the form

$$f(x) = 1 - 2\left|x - \frac{1}{2}\right|.$$

Show: (a) there are two equilibrium points; and (b) the point $1/2$ is an eventually fixed point.

3 A population of birds is modeled by the difference equation

$$x(t+1) = \begin{cases} 3.2x(t), & \text{if } 0 \le x(t) \le 1, \\ 0.5x(t) + 2.7, & \text{if } x(t) > 1, \end{cases}$$

where $x(t)$ is the number of birds in year t. Find the equilibrium points and then determine their stabilities.

2.3 Stabilities

A dynamical system might have unpredictability property, which means that orbits starting at points very close to each other can be quite far apart at some later time, even though the orbits remain confined in a bounded region. Since, on experimental grounds, the initial state of an orbit is never known accurately, we cannot "predict" where the system will be at some later time. To explain unpredictability, we introduce the definition of stable and unstable orbits. The presence of unstable orbits plays an important role in dynamical systems.

Let us consider the one-dimensional difference equation

$$x(t+1) = f(x(t)), \quad t = 0, 1, \cdots. \tag{2.3.1}$$

Definition 2.3.1. A fixed point x^* of equation (2.3.1) is *locally stable* if for every $\varepsilon > 0$ there exists $\delta > 0$ such that

$$\|x_0 - x^*\| \le \delta \text{ implies that } \|x_t - x^*\| \le \varepsilon \text{ for all } t \ge 1.$$ [3]

[3] In this book, the norm $\|x\|$ denotes the *Euclidean norm* of x, defined by

$$\|x\| = \left(\sum_{i=1}^{n} x_i^2\right)^{1/2}.$$

A fixed point that is not stable is said to be *unstable*. An unstable equilibrium point is called a *source*, or a *repeller*.

Stability means that once we have chosen how close we want to remain x^* in the future, we can find how close we must start at the beginning. Figures 2.3.1 and 2.3.2 illustrate the two concepts.

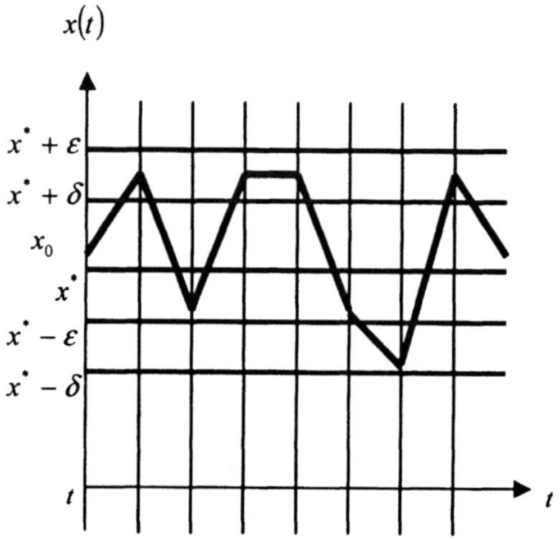

Figure 2.3.1: Stable x_0

Example Consider

$$x(t+1) = 1 - x(t).$$

The point $x^* = 0.5$ is the only fixed point of f. For every other initial state x_0 we have

$$x_1 = 1 - x_0, \quad x_2 = x_0.$$

We thus have

$$\|x_t - 0.5\| = \|x_0 - 0.5\|,$$

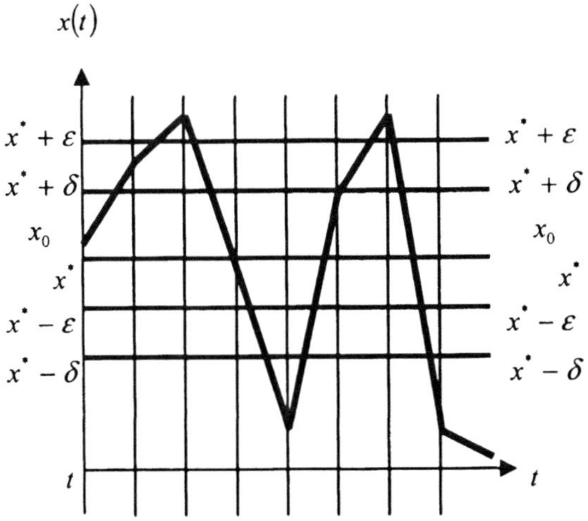

Figure 2.3.2: Unstable x_0

for all $t \geq 1$. Hence, the fixed point is stable (by selecting $\delta = \varepsilon$ in the definition).

Example Let $f : R \to R$ be defined as

$$f(x) = \begin{cases} 0.5x, & for \quad x \leq 0, \\ 2x, & for \quad x > 0. \end{cases}$$

The system has a unique equilibrium $x^* = 0$. Every orbit starting to the left of the origin will converge to x^*, while every orbit starting to the right of the origin will go to infinity. Thus, x^* is an unstable fixed point.

Definition 2.3.2. The point x^* is said to be *attracting* if there exists $\eta > 0$ such that

$$|x(0) - x^*| < \eta \text{ implies } \lim_{t \to \infty} x(t) = x^*.$$

If $\eta = \infty$, x^* is called a *global attractor* or *globally attracting*. The point x^* is an *asymptotically stable point* or a *sink* if it is stable and attracting. If $\eta = \infty$, x^* is said to be *globally asymptotically stable*.

Let f be a function defined on some interval Ω of R with values in Ω. It is known that under certain conditions the stability type of an equilibrium point x^* of difference equation (2.3.1) is the same as the stability type of the equilibrium point of the corresponding linearized equation

$$y(t+1) = f'(x^*)y(t). \tag{2.3.2}$$

A simple but not rigorous argument is that for x close to the value of an equilibrium point x^* we have

$$f(x) = f(x^*) + f'(x^*)(x - x^*) + HOT,$$

where HOT denotes the higher-order terms in $x - x^*$. Taking into account that x^* is an equilibrium point, neglecting HOT, and replacing $x = u_t$, we obtain the approximate equation

$$u(t+1) = x^* + f(x^*)(u(t) - x^*).$$

Finally, setting

$$y(t) = u(t) - x^*,$$

we obtain equation (2.3.2).

Definition 2.3.3. Assume f is continuously differentiable in Ω. Let x^* be a steady state $\in \Omega$. If

$$|f'(x^*)| = 1,$$

then x^* is *non-hyperbolic*. If

$$\left|f'(x^*)\right| \neq 1,$$

x^* is *hyperbolic*.

If x^* is non-hyperbolic, its stability type cannot be determined by its first derivative, as discussed later on.

Theorem 2.3.1. (linearized stability) Let x^* be a hyperbolic steady state $\in \Omega$. Then
(i) If $\left|f'(x^*)\right| < 1$, then x^* is asymptotically stable.
(ii) If $\left|f'(x^*)\right| > 1$, then x^* is unstable.

Proof. In the case of $\left|f'(x^*)\right| < 1$, consider β such that

$$\left|f'(x^*)\right| < \beta < 1.$$

By continuity of $f'(x)$, we have $\left|f'(x)\right| < \beta$ on some interval $(x^* - \varepsilon, x^* + \varepsilon)$ with $\varepsilon > 0$. For any x in this interval, applying the mean value theorem for derivatives yields

$$f(x) - f(x^*) = (x - x^*)f'(x^* + \theta(x - x^*)),$$

where $0 < \theta < 1$. Since $x^* + \theta(x - x^*)$ belongs $(x^* - \varepsilon, x^* + \varepsilon)$, we have

$$\left|f(x) - f(x^*)\right| < \beta\left|x - x^*\right|.$$

For any $x_0 \in (x^* - \varepsilon, x^* + \varepsilon)$, the sequence

$$x(t+1) = f(x(t))$$

starting at x_0 satisfies

$$\left|x(t+1)-x^*\right|=\left|f(x(t))-f(x^*)\right|<\beta\left|x_t-x^*\right|<\beta^{t}\left|x_0-x^*\right|.$$

We conclude that the sequence converges to x^* for all $x_0 \in (x^* - \varepsilon, x^* + \varepsilon)$, and x^* is locally stable. This also implies that the whole sequence remains in the neighborhood of x^*. The same proof applies to a corner steady state with $x_0 \in (0, \varepsilon)$.

The equilibrium point is asymptotically stable when $\left|f'(x^*)\right| < 1$. As $t \to \infty$, we have

$$\left|x(t)-x^*\right| \to 0.$$

We conclude asymptotical stability.

In the case of $\left|f'(x^*)\right| > 1$, consider β such that

$$\left|f'(x^*)\right| > \beta > 1.$$

By continuity of $f'(x)$, we have $\left|f'(x)\right| > \beta$ on some interval $(x^* - \varepsilon, x^* + \varepsilon)$ with $\varepsilon > 0$. For any x in this interval, applying the mean value theorem for derivatives yields

$$f(x)-f(x^*)=(x-x^*)f'(x^*+\theta(x-x^*)), \quad \text{with } 0<\theta<1.$$

Since $x^* + \theta(x - x^*)$ belongs $(x^* - \varepsilon, x^* + \varepsilon)$, we have

$$\left|f(x)-f(x^*)\right| > \beta\left|x-x^*\right|.$$

For any $x_0 \subset (x^* - \varepsilon, x^* + \varepsilon)$, the sequence

$$x(t+1)=f(x_t)$$

starting at x_0 satisfies

$$\left|x(t+1)-x^{*}\right|=\left|f(x(t))-f\left(x^{*}\right)\right|>\beta\left|x(t)-x^{*}\right|>\beta^{t}\left|x_{0}-x^{*}\right|,$$

which holds for the terms

$$x_{t}\in\left(x^{*}-\varepsilon, x^{*}+\varepsilon\right).$$

We conclude that the sequence does not converge to x^{*} for all

$$x_{0}\in\left(x^{*}-\varepsilon, x^{*}+\varepsilon\right),$$

and x^{*} is locally unstable.

From theorem 2.3.1, if the equilibrium point x^{*} of equation (2.3.1) is hyperbolic, then it must be either asymptotically stable or unstable, and the stability type is determined from the size of $f'\left(x^{*}\right)$. Figure 2.3.3 depicts different stability types.

Example (a cobweb model of demand and supply[4]) Consider the market for a single commodity. Assume that the output decision in period t is based on the then prevailing price $P(t)$ and that the output planned in period t will not be available for the sale, $Q^{s}(t+1)$, until period $t+1$. We thus have a lagged supply function

$$Q^{s}(t+1)=S(P(t)).$$

Equivalently

$$Q^{s}(t)=S(P(t-1)).$$

When such a supply function interacts with a demand function of the form

$$Q^{d}(t)=D(P(t)),$$

the price dynamics will be determined by the balance condition

[4] This is from chapter 16 in Chiang (1984).

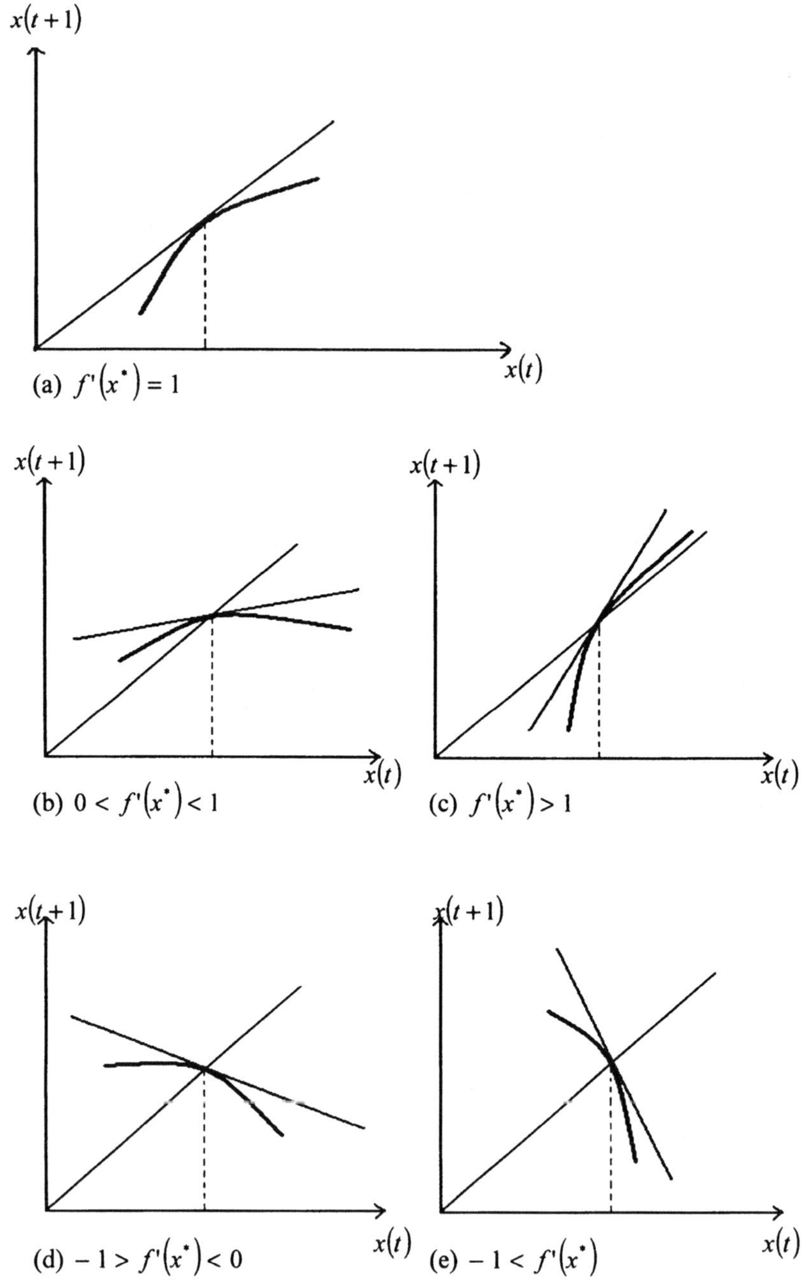

Figure 2.3.3: Stability types of scalar difference equations

$$Q^d(t) = Q^s(t).$$

Specify the demand and supply functions as follows

$$Q^d(t) = a_d - b_d P(t),$$
$$Q^s(t) = -a_s + b_s P(t-1), \quad a_d, b_d, a_s, b_s > 0.$$

The price dynamics is given by

$$P(t+1) + \frac{b_s}{b_d} P(t) - \frac{a_s + a_d}{b_d} = 0.$$

The solution is given by

$$P(t) = \left(P(0) - \frac{a_s + a_d}{b_s + b_d} \right) \left(-\frac{b_s}{b_d} \right)^t + \frac{a_s + a_d}{b_s + b_d}.$$

Introducing

$$P^* = \frac{a_s + a_d}{b_s + b_d},$$

where P^* is the fixed point, we may write the solution as

$$P(t) = \left(P(0) - P^* \right) \left(-\frac{b_s}{b_d} \right)^t + P^*.$$

Because b_s and b_d are positive, the time path will be oscillatory. We see

$$\left| f'(P^*) \right| = \left| \frac{b_s}{b_d} \right|.$$

According to the values of b_s and b_d, we have three possibilities, (i) explosive if $b_s > b_d$; (ii) uniform if $b_s = b_d$; and (iii) damped if $b_s < b_d$. We depict (i) and (iii) as in figure 2.3.4.

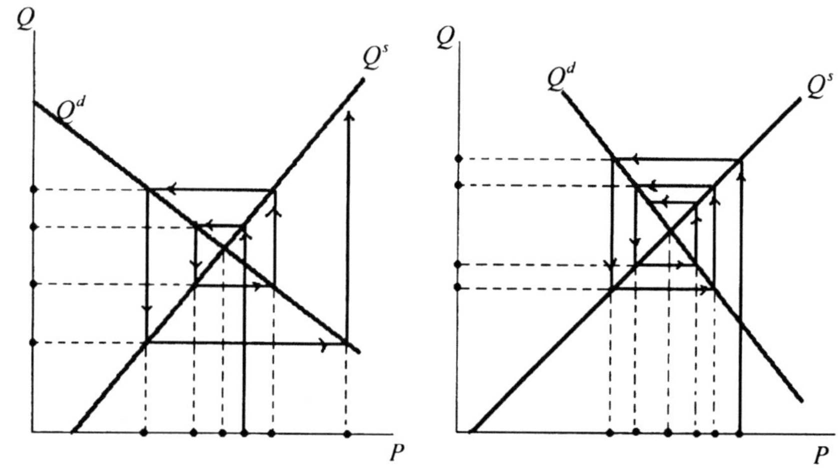

a) $b_s > b_d$ (the supply curve is steeper) b) $b_s < b_d$ (the supply curve is flatter)

Figure 2.3.4: The cobweb model

Example (a cobweb model with the normal-price expectation[5]) Consider the market for a single commodity. Assume that the output decision in period t is based on the then expected price $P^e(t)$. The supply function is specified as

$$Q^s(t) = -a_s + b_s P^e(t), \quad a_s, b_s > 0.$$

The demand function is

$$Q^d(t) = a_d - b_d P(t), \quad a_d, b_d > 0.$$

To form price expectations, we introduce a concept of "normal price", denoted as $\overline{P}$, as that price which producers would think sooner or later to obtain in the market. If the current price is different from the normal price, they think the former will modify, moving toward the latter. A simple way to formalize this is to specify

$$P^e(t) = P(t-1) + \alpha(\overline{P} - P(t-1)), \quad 0 < \alpha < 1.$$

The balance condition of

[5] This and the following example are from Gandolfo (1996: 38-43).

$$Q^d(t) = Q^s(t),$$

is now expressed as

$$-a_s + b_s P(t-1) + \alpha b_s \overline{P} - \alpha b_s P(t-1) = a_d - b_d P(t).$$

The solution to this equation is

$$P(t) = A\left(\frac{-b_s(1-\alpha)}{b_d}\right)^t + \overline{P}.$$

We see that the solution is oscillatory and the equilibrium point $P(t) = \overline{P}$ is stable if

$$b_s(1-\alpha) < b_d.$$

Example (a cobweb model with adaptive expectations) Like in the previous example, the supply and demand functions are specified as

$$Q^d(t) = a_d - b_d P(t), \quad a_d, b_d > 0,$$
$$Q^s(t) = -a_s + b_s P^e(t), \quad a_s, b_s > 0,$$

where $P^e(t)$ is the expected price. We assume that expectations are adapted in each period on the basis of the discrepancy between the observed value and the previous expected value, that is[6]

$$P^e(t) - P^e(t-1) = \alpha\left(P(t-1) - P^e(t-1)\right), \quad 0 < \alpha < 1.$$

We may also write the price expectation in the form of

$$P^e(t) - (1-\alpha)P^e(t-1) = \alpha P(t-1).$$

Substituting

[6] The adaptive expectation was initially proposed by Nerlove (1958).

$$P^e(t) = \frac{a_s + Q^s(t)}{b_s},$$

into the above equation yields

$$a_s + Q^s(t) - (1 - \alpha)(a_s + Q^s(t-1)) = b_s \alpha P(t-1).$$

By $Q^d(t) = Q^s(t)$, we get

$$Q^s(t) = a_d - b_d P(t).$$

Substituting

$$Q^s(t) = a_d - b_d P(t)$$

into the previous equation yields

$$P(t) + \left(\frac{b_s \alpha}{b_d} - 1 + \alpha\right) P(t-1) = \alpha\left(\frac{a_s + a_d}{b_d}\right).$$

The equilibrium point is given by

$$P^* = \frac{a_s + a_d}{b_s + b_d}.$$

The general solution of the problem is thus given by

$$P(t) = A\left(1 - \left(\frac{b_s}{b_d} + 1\right)\alpha\right)^t + P^*.$$

The stability condition is

$$\left|1 - \left(\frac{b_s}{b_d} + 1\right)\alpha\right| < 1.$$

Example (a market model with inventory) We now extend the cobweb model to a case that sellers can keep an inventory of the commodity. Assume that the quantity demanded, $Q^d(t)$, and the quantity currently produced, $Q^s(t)$, are unlagged linear functions of price $P(t)$. The adjustment of price is effected not through market clearance in every period, but through a process of price-setting by the sellers: at the beginning of each period, sellers set a price for that period after taking into consideration the inventory situation. If inventory accumulated as a result of the preceding-period price, the current-period price is set a lower level than before, and vice versa. Moreover, the price adjustment made from period to period is inversely proportional to the observed change in the inventory. With these conditions, we can write the following equations

$$Q^d(t) = a_d - b_d P(t),$$
$$Q^s(t) = -a_s + b_s P(t), \quad a_d, b_d, a_s, b_s > 0,$$
$$P(t+1) = P(t) - \sigma\left(Q^s(t) - Q^d(t)\right),$$

where σ denotes the stock-induced-price-adjustment coefficient. Substituting the first two equations into the last one yields

$$P(t+1) - [1 - \sigma(b_s + b_d)]P(t) = \sigma(a_s + a_d).$$

Let P^* be the fixed point. Then, we may write the above solution

$$P(t) = \left(P(0) - P^*\right)\left(1 - \sigma(b_s + b_d)\right)^t + P^*,$$

where

$$P^* \equiv \frac{a_s + a_d}{b_s + b_d}.$$

We see that the stability is determined by the term $1 - \sigma(b_s + b_d)$.

Example (the Newton-Raphson method) The *Newton-Raphson method* is a well-known numerical method for finding the roots of the equation

$$g(x) = 0.$$

Newton's algorithm for finding a zero x^* of $g(x)$ is given by the difference equation

$$x(t+1) = x(t) - \frac{g(x(t))}{g'(x(t))} \equiv f(x(t)),$$

where x_0 is the initial guest of the root x^*. To determine whether Newton's algorithm provides a sequence $\{x(t)\}$ that converges to x^* we calculate

$$|f'(x^*)| = \left|1 - \frac{[g'(x^*)]^2 - g(x^*)g''(x^*)}{[g'(x^*)]^2}\right| = 0,$$

where we use $g(x^*) = 0$. We conclude that

$$\lim_{t\to\infty} x(t) = x^*,$$

if x_0 is close enough to x^* and $g'(x^*) \neq 0$.

Example Consider

$$x(t+1) = 1 - \lambda x^2(t),$$

where

$$x \in [-1, 1], \quad \lambda \in (0, 2].$$

There are two equilibrium points

$$x_{1,2}^* = \frac{-1 \pm \sqrt{1 + 4\lambda}}{2\lambda}.$$

It is straightforward to check that x_1^* is unstable for all $\lambda \in (0, 2]$; x_2^* is asymptotically stable if $0 < \lambda < 3/4$ and unstable if $\lambda > 3/4$.

Definition 2.3.4. Let x^* be an asymptotically stable fixed point of a map f. The *basin of attraction* $\aleph(x^*)$ of x^* is defined as the maximal set J that contains x^* and is such that

$$f^t(x) \to x^* \quad \text{as} \quad t \to \infty \quad \text{for every} \quad x \in J.$$

The basin of attraction of an attracting periodic point of period p is defined in analogous fashion, with the map f replaced by the p-th iterate.

Example The map

$$f(x) = 2x(1 - x),$$

has an attracting fixed point $x^* = 1/2$ with a basin of attraction $\aleph(1/2) = (0, 1)$.

We will see later that basins of attraction may have complicated structures even for simple looking maps.

Definition 2.3.5. A set M is said to be *invariant* under a map f if $f(M) \subset M$, that is, if for every $x \in M$, the elements of $O(x)$ belong to M.

Theorem 2.3.2.[7] Let x^* be an attracting fixed point of a map f. Then the basin of attraction $\aleph(x^*)$ is an invariant open interval.

Exercise 2.3

1 Given demand and supply functions for the Cobweb model as follows:

(i)
$$Q^d(t) = 18 - 3P(t),$$
$$Q^s(t) = -3 + 4P(t-1);$$

(ii)
$$Q^d(t) = 19 - 6P(t),$$
$$Q^s(t) = -5 + 6P(t-1).$$

---

[7] See Elaydi (2000).

Find the intertemporal equilibrium prices, and determine whether or not the equilibrium points are stable.

2 Let

$$f(x) = x^2.$$

Show that $x^* = 0$ is stable and $x^* = 1$ is unstable.

3 Find the equilibrium points and determine their stability for the equation

$$x(t+1) = 5 - \frac{6}{x(t)}.$$

4 *Pielou's logistic equation* is defined as

$$x(t+1) = \frac{\alpha x(t)}{1 + \beta x(t)}, \quad \alpha > 1, \quad \beta > 0.$$

(a) Find the positive equilibrium point; and (b) demonstrate, using the stair step diagram, that the positive equilibrium point is stable, taking $\alpha = 2$ and $\beta = 1$.

2.4 Stabilities of nonhyperbolic equilibrium points

From the information about the sign of the first derivative value of the map f at an equilibrium point, theorem 2.3.1 determines the stability properties of the equilibrium point when it is hyperbolic. This section addresses the nonhyperbolic case where $|f'(x^*)| = 1$. Further information about derivatives is needed to determine stability properties. We discuss $f'(x^*) = 1$ and $f'(x^*) = -1$ separately. This section assumes that f has a continuous second derivative.

Definition 2.4.1. An equilibrium point x^* of

$$x(t+1) = f(x(t)), \tag{2.4.1}$$

is *semistable* (from the right) if given $\varepsilon > 0$ there exists $\delta > 0$ such that if $x(0) > x^*$, $x(0) - x^* < \delta$, then $x(t) - x^* < \varepsilon$. Semistability from the left is defined similarly. If in addition

$$\lim_{t \to \infty} x(t) = x^*,$$

whenever $x(0) - x^* < \eta$, then x^* is said to be *semiasymptotically stable* from the right. We can similarly define *semiasymptotical stability* from the left.

The following two theorems are referred to Elaydi.[8]

Theorem 2.4.1. Suppose that for an equilibrium point x^* of equation (2.4.1), $f'(x^*) = 1$. The following statements then follow:

(i) If $f''(x^*) \neq 0$, then x^* is unstable.
(ii) If $f''(x^*) = 0$ and $f'''(x^*) > 0$, then x^* is unstable.
(iii) If $f''(x^*) = 0$ and $f'''(x^*) < 0$, then x^* is asymptotically stable.

Theorem 2.4.1 is illustrated as in figure 2.4.1. If

$$f''(x^*) \neq 0,$$

then the curve is either concave upward if $f''(x^*) > 0$ or concave downward if $f''(x^*) < 0$ as shown in figures 2.4.1 a-d. If $f''(x^*) > 0$, then

$$f'(x) \geq M > 1,$$

for all x in a small interval $I = (x^*, x^* + \varepsilon)$. Then, we have

$$|x(1) - x^*| = |f(x(0)) - f(x^*)| = f'(\xi)|x(0) - x^*| > M|x(0) - x^*|,$$

where $\xi \in I$. By induction we conclude

[8] Section 1.4 in Elaydi (1999).

$$|x(t) - x^*| > M'|x(0) - x^*|.$$

We conclude that x^* is unstable. We can discuss other cases similarly. It should be noted that as proved in Sandefur,[9] if $f''(x^*) > 0$ ($f''(x^*) < 0$), then the equilibrium point x^* is semistable from the left (right).

Example Consider

$$x(t+1) = x^2(t) + x(t), \quad t = 0, 1, \cdots.$$

The difference equation has a unique equilibrium point $x^* = 0$. As

$$f'(0) = 1, \quad f''(0) = 2 > 0,$$

by theorem 2.4.1 we conclude that x^* is unstable. To examine its semistability, first assume $x_0 > 0$. Then

$$x(1) = x^2(0) + x(0) > x(0).$$

By induction we get $x(t) > x(t-1)$. The sequence $\{x(t)\}$ either converges to a fixed point or diverges to infinity. Since the only fixed point is zero, $\{x(t)\}$ diverges to infinity. On the other hand, if $x_0 \in (-1/2, 0)$

$$x(1) = x^2(0) + x(0) > x(0),$$

and $x(1) \in (-1/2, 0)$. By induction, we conclude

$$x(t) > x(t-1).$$

The sequence $\{x(t)\}$ converges to 0.

Example Consider

---

[9] Sandefur (1990).

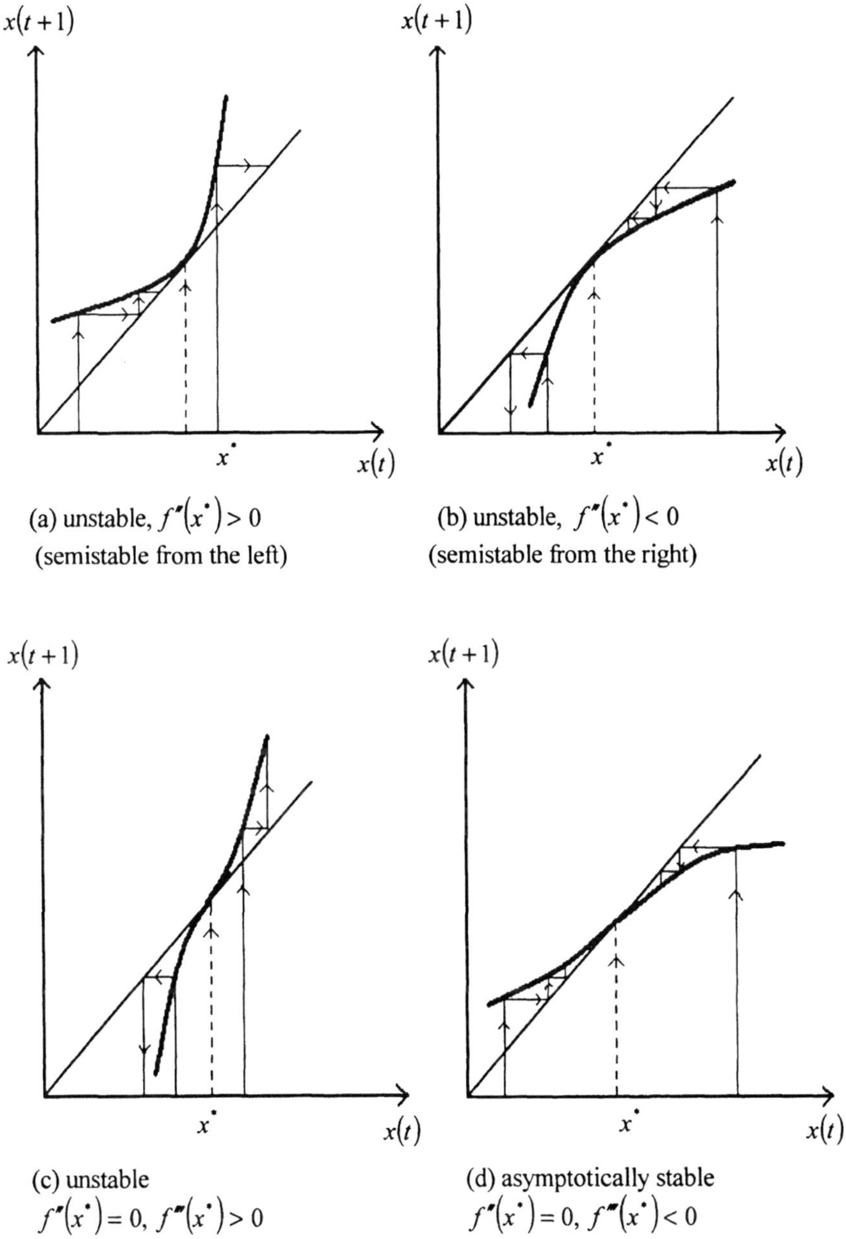

(a) unstable, $f''(x^*) > 0$
(semistable from the left)

(b) unstable, $f''(x^*) < 0$
(semistable from the right)

(c) unstable
$f''(x^*) = 0$, $f'''(x^*) > 0$

(d) asymptotically stable
$f''(x^*) = 0$, $f'''(x^*) < 0$

Figure 2.4.1: Stability types in the case of $f'(x^*) = 1$

$$x(t+1) = x^3(t) + x(t), \quad t = 0, 1, \cdots.$$

The difference equation has a unique equilibrium point $x^* = 0$. As

$$f'(0) = 1, \quad f''(0) = 0, \quad f'''(0) = 6 > 0,$$

by theorem 2.4.1 we conclude that x^* is unstable. We can determine this by induction. Assume that $x_0 > 0$. Then

$$x(1) = x^3(0) + x(0) > x(0).$$

By induction we get $x(t) > x(t-1)$. The sequence $\{x(t)\}$ either converges to a fixed point or diverges to infinity. Since the only fixed point is zero, $\{x(t)\}$ diverges to infinity. On the other hand, if $x_0 < 0$, we have

$$x(1) = x^3(0) + x(0) < x(0).$$

By induction $x(t) < x(t-1)$. The sequence $\{x(t)\}$ diverges to $-\infty$. Hence, the equilibrium point is a source.

Example Consider

$$x(t+1) = -x^3(t) + x(t), \quad t = 0, 1, \cdots.$$

The difference equation has a unique equilibrium point $x^* = 0$. As

$$f'(0) = 1, \quad f''(0) = 0, \quad f'''(0) = -6 < 0,$$

we conclude that x^* is asymptotically stable.

In the case of $f'(x^*) = -1$, the map f is not monotone but rather oscillatory, and it flips from a point close to x^* to the other side of x^*. If the equilibrium point x^* becomes unstable, and orbit cannot approach x^*. But if the iterates remain bounded, it is possible that the odd iterates converge to a limit point p, and the even iterates converge to a limit point $f(p)$. If this happens, then

$$f(f(p)) = p,$$

with p different from $f(p)$. Then p is a periodic point of period two. This change in global behavior of solutions of equation (2.4.1) is called *period-doubling* or *flip bifurcation.*

Example Consider

$$x(t+1) = -x(t), \quad t = 0, 1, \cdots.$$

The unique equilibrium point is zero. Every solution of this equation, except $x^* = 0$, is periodic with period two.

Example Consider

$$x(t+1) = -x(t) + x^2(t), \quad t = 0, 1, \cdots.$$

The equilibrium points of this equation are $x_1^* = 0$ and $x_2^* = 2$. To examine behavior of the sequence $\{x(2k)\}$, we note

$$x(2k+2) = -x(2k+1) + x^2(2k+1) = x(2k) - x^3(2k) + x^4(2k).$$

Setting

$$y(k) = x(2k),$$

we rewrite the above equation as

$$y(k+1) = y(k) - y^3(k) + y^4(k) \equiv g(y(k)). \tag{2.4.2}$$

Since

$$g'(0) = 1, \quad g''(0) = 0, \quad g'''(0) = -8,$$

by theorem 2.4.1 we conclude that $x_1^* = 0$ is a sink. Likewise, we conclude that $x_2^* = 2$ is a source.

Similarly, we conclude that the sequence $\{x(2k+1)\}$ satisfies equation (2.4.2). Hence it is convergent to the zero equilibrium. Consequently, both even-indexed terms and odd-indexed terms are convergent to zero, hence $x_1^* = 0$ is a sink, while $x_2^* = 2$ is a source. Also applying the above method, check that the origin is a source of the following difference equation

$$x(t+1) = -x(t) - x^3(t).$$

Before stating stability conditions for the case of $f'(x^*) = -1$, we introduce the notion of the *Schwarzian derivative* of a function f

$$Sf(x) = \frac{f'''(x)}{f'(x)} - \frac{3}{2}\left[\frac{f''(x)}{f'(x)}\right]^2.$$

For $f'(x^*) = -1$, then

$$Sf(x^*) = -f'''(x^*) - \frac{3}{2}[f''(x^*)]^2.$$

Theorem 2.4.2. Suppose that for an equilibrium point x^* of equation (2.4.1), $f'(x^*) = -1$. The following statements then follow:
(i) If $Sf(x^*) < 0$, then x^* is asymptotically stable.
(ii) If $Sf(x^*) > 0$, then x^* is unstable.

Example Consider

$$x(t+1) = x^2(t) + 3x(t).$$

The equilibrium points are 0 and -2. We have

$$f' = 2x + 3.$$

Since $f'(0) = 3$, the equilibrium point 0 is unstable. As $f'(-2) = -1$, theorem 2.4.2 applies. As

$$Sf(-2) = -12 < 0,$$

the equilibrium point -2 is asymptotically stable.

Exercise 2.4

1 Find the equilibrium points and determine their stability of the following difference equation

$$x(t+1) = ax^2(t) + bx(t) + c, \quad a \neq 0.$$

2 Suppose that if $f'(x^*) = 1$, then $f''(x^*) \neq 0$. Prove that x^* is (i) semiasymptotically stable from the right if $f''(x^*) < 0$; and (ii) semiasymptotically stable from the left if $f''(x^*) > 0$. Applying this result, determine whether the equilibrium point $x^* = 0$ of the following equations is semiasymptotically stable from the left or from the right

(a) $x(t+1) = x^3(t) + x^2(t) + x(t)$;
(b) $x(t+1) = x^3(t) - x^2(t) + x(t)$.

3 Find the equilibrium points and discuss the stability of the following difference equations

(a) $x(t+1) = -x^2(t) + x(t), \quad t = 0, 1, \cdots;$
(b) $x(t+1) = 1 - \frac{3}{4}x^2(t), , \quad t = 0, 1, \cdots.$

2.5 On dissipative maps

Consider the one-dimensional difference equation

$$x(t+1) = f(x(t)), \quad t = 0, 1, \cdots. \tag{2.5.1}$$

An interval $[a, b]$ is called an *absorbing interval*, or an *attracting interval* if every orbit of equation (2.5.1) eventually enters the interval. An interval is called an *invariant interval* if every orbit that enters the interval stays in it forever. A difference equation (map) with an absorbing interval is called *dissipative difference equation* (*dissipative map*).

Theorem 2.5.1.[10] Let $[a, b]$ be an interval of real numbers and assume that f is a continuous, nondecreasing function such that

$$f([a, b]) \subset [a, b],$$

and equation (2.5.1) has a unique equilibrium x^*. Then every solution of equation (2.5.1) with initial point in $[a, b]$ converges to x^*. If equation (2.5.1) is dissipative then every solution of equation (2.5.1) converges to x^*. □

Example Consider

$$x(t+1) = \frac{x(t)}{1 + x(t)}, \quad t = 0, 1, \cdots .$$

The map satisfies all conditions of theorem 2.5.1 on the interval $[0, 1]$. Thus, every solution of this equation that starts in the invariant $[0, 1]$. In addition, every solution that starts in $[1, \infty)$ in one step enters the invariant interval and converges to the zero equilibrium. It should be noted that the situation for negative initial conditions is more complicated as we face the problem of the existence of the solution.

Theorem 2.5.2. Let $[a, b]$ be an interval of real numbers and let f be a continuous, nonincreasing function such that $f([a, b]) \subset [a, b]$ and equation (2.5.1) has no prime period-2 solution. Then every solution of equation (2.5.1) that enters eventually $[a, b]$ converges to x^*. In addition, the subsequences $\{x(2t)\}$ and $\{x(2t+1)\}$ of every solution of equation (2.5.1) converge monotonically to x^* by oscillating about the equilibrium such that

$$\left(x(t+1) - x^*\right)\left(x(t) - x^*\right) < 0, \quad \text{for } n = 0, 1, \cdots . \qquad \square$$

Example The difference equation

---

[10] This theorem and the next one are referred to Kulenović and Ladas (2001).

$$x(t+1) = e^{-x(t)}, \quad t = 0, 1, \cdots$$

has an invariant interval. The corresponding map is decreasing everywhere and it does not possess a prime period-2 solution. Indeed, if such a solution would exist it would satisfy

$$e^{-e^{-x}} = x,$$

for some $p \neq x^*$. In fact, it is straightforward to show that the function $e^{-e^{-x}} = x$ has a unique solution on the interval $[0, 1]$. Thus, all conditions of theorem 2.5.2 on the interval $[0, 1]$ are satisfied and every solution of the equation that starts in the invariant interval $[0, 1]$ converges to the unique equilibrium point. In addition, every solution that starts in R enters the invariant interval in at most two steps and so converges to the equilibrium point.

Example Consider the difference equation

$$x(t+1) = 1 + \frac{A}{x^p(t)}, \quad t = 0, 1, \cdots \tag{2.5.2}$$

where A, p, and $x(0)$ are positive numbers. This equation has an invariant interval $[1, 1+A]$ and the corresponding map is decreasing for all positive values. The equation has a unique positive equilibrium, x^*, where x^* is the unique positive root of the equation

$$x^{p+1} - x^p - A = 0.$$

If $p \leq 1$, x^* is locally asymptotically stable. Similarly, if $p > 1$, x^* is locally asymptotically stable provided that

$$x^* < \frac{p}{p-1}.$$

One can show that this condition is equivalent to the condition

$$A < \frac{p^p}{(p-1)^{p+1}}. \tag{2.5.3}$$

When condition (2.5.3) holds, the Schwarzian derivative can be used to show that x^* is locally asymptotically stable. Thus, in both cases, (a) $p \le 1$, and (b) $p > 1$ and condition (2.5.3), the unique equilibrium point x^* is locally asymptotically stable. To show that equation (2.5.2) has no prime period- 2 solution we consider the second iterate

$$f^2(x) = f(f(x)) = 1 + \frac{A}{\left(1 + \frac{A}{x^p}\right)^p}.$$

Introduce

$$H(x) \equiv f^2(x) - x.$$

Clearly, the equilibrium x^* is a solution of the equation

$$H(x) = 0.$$

We now show that x^* is the only solution by checking that $H(x)$ is strictly decreasing. As

$$H'(x) = \frac{(pA)^2}{H_0^{p+1}(x)} - 1,$$

where

$$H_0(x) = x + \frac{A}{x^{p-1}}.$$

Since the function $H_0(x)$ has a minimum at

$$x = \tilde{x} = ((p-1)A)^{1/p},$$

we have

$$H_0(x) \geq H_0(\tilde{x}) = \frac{p}{p-1}((p-1)A)^{1/p}.$$

Thus we have

$$H'(x) \leq \frac{(pA)^2}{H_0^{p+1}(\tilde{x})} - 1 \leq \left(\frac{A(p-1)^{p+1}}{p^p}\right)^{(p-1)/p} - 1 \leq 0,$$

if condition (2.5.3) holds, where equality holds only if $x = \tilde{x}$. Hence, $H(x)$ has a unique zero and this is the unique positive equilibrium point. Hence, equation (2.5.2) does not possess a prime period-2 solution and by theorem 2.5.2 every solution of this equation that starts in the invariant interval $[1, 1 + A]$ converges to the unique equilibrium point, x^*. In addition, every solution that starts in $[0, \infty)$ in at most two steps enters the invariant interval, hence converges to the equilibrium. Thus the invariant interval $[1, 1 + A]$ is an absorbing interval, and the difference equation is dissipative.

The remaining case of

$$p > 1, \quad A < \frac{p^p}{(p-1)^{p+1}},$$

is qualitatively different. It can be shown that equation (2.5.2) has two periodic solutions, each if period 2, which are asymptotically stable and global attractors with a basin of attraction $(0, x^*)$ and (x^*, ∞) respectively.[11]

Exercise 2.5

1 Show that for the difference equation

$$x(t+1) = \frac{x^2(t)}{1 + x^2(t)}, \quad t = 0, 1, \cdots,$$

the interval $[0, 1]$ is an absorbing interval and zero is a global attractor.

---

[11] See DeVault *et al.* (1992).

2.6 Linear difference equations of higher order

The normal form of a k th-order *nonhomogeneous linear* difference equation is given by

$$x(t+k)+p_1(t)x(t+k-1)+\cdots+p_k(t)x(t)=g(t), \qquad (2.6.1)$$

where $p_i(t)$ and $g(t)$ are real-valued functions defined for $t \geq t_0$ and

$$p_k(t) \neq 0$$

for all $t \geq t_0$. Here, the sequence $g(t)$ is called the *forcing term*, the *external force*, the *control*, or the *input* of the system. If $g(t)$ is identically zero, then equation (2.6.1) is said to be a *homogeneous* equation. We can evaluate $x(k)$ with

$$x(k)=g(0)-p_1(0)x(k-1)-\cdots-p_k(0)x(0).$$

Similarly, we can evaluate $x(k+1)$, and so on. A sequence $\{x(t)\}_{t_0}^{\infty}$ is said to be a *solution* of equation (2.6.1) if it satisfies the equation. If

$$x(t_0), \cdots, x(t_0+k-1)$$

are specified, we have the *initial value problem.*

Example Consider

$$x(t+3)-\frac{t}{t+1}x(t+2)+tx(t+1)-3x(t)=t,$$

where the initial conditions are

$$x(1)=0, \quad x(2)=-1, \quad x(3)=1.$$

It is straightforward to evaluate

$$x(4)=\frac{5}{2}, \quad x(5)=-\frac{4}{3}, \quad x(6)=-\frac{3}{2}, \quad x(7)=20.9.$$

Consider the homogenous difference equation

$$x(t+k)+p_1(t)x(t+k-1)+\cdots+p_k(t)x(t)=0. \tag{2.6.2}$$

Definition 2.6.1. The functions $f_1(t), f_2(t), \cdots, f_r(t)$ are said to be *linearly dependent* for $t \geq t_0$ if there are constants $a_1, a_2, \cdots, a_r$, not all zero, such that

$$\sum_{i=1}^{r} a_i f_i(t) = 0, \quad t \geq t_0.$$

Definition 2.6.2. A set of k linearly independent solutions of equation (2.6.2) is called a *fundamental set of solutions*.

It is often difficult to check the linear independence of a set of functions using the definition. We now introduce a simple method to check linear independence of solutions using the so-called Casoratian.

Definition 2.6.3. The *Casoratian* $W(t)$ of the solutions

$$x_1(t), x_2(t), \cdots, x_r(t)$$

is given by[12]

$$W(t)=\begin{vmatrix} x_1(t) & x_2(t) & \cdots & x_r(t) \\ x_1(t+1) & x_2(t+1) & \cdots & x_r(t+1) \\ \vdots & & & \\ x_1(t+r-1) & x_2(t+r-1) & \cdots & x_r(t+r-1) \end{vmatrix}.$$

Example It is straightforward to check that the sequences

$$1, \ (-3)^t, \ 2^t,$$

are solutions of the difference equation

$$x(t+3)-7x(t+1)+6x(t)=0.$$

[12] This is the discrete analogue of the Wronskian in differential equations.

The following calculation verifies the linear independence of the solutions

$$W(t)=\begin{vmatrix}1 & (-3)^t & 2^t \\ 1 & (-3)^{t+1} & 2^{t+1} \\ 1 & (-3)^{t+2} & 2^{t+2}\end{vmatrix}=-20(2)^t(-3)^t.$$

The following Abel formula gives an effective method for verifying linear independence of solutions.

Lemma 2.6.1. (Abel's lemma)[13] Let

$$x_1(t), x_2(t), \cdots, x_k(t),$$

be solutions of equation (2.6.2) and let $W(t)$ be their Casoratian. Then for $t \geq t_0$

$$W(t)=(-1)^{k(t-t_0)}\left(\prod_{i=t_0}^{t-1} p_k(i)\right)W(t_0). \qquad \square$$

We see that if $p_k(t) \neq 0$ for all $t \geq t_0$, then the Casoratian $W(t) \neq 0$ if and only if $W(t_0) \neq 0$.

Theorem 2.6.1. The set of solutions

$$x_1(t), x_2(t), \cdots, x_k(t)$$

of equation (2.6.2) is a fundamental set if and only if for some $t_0 \in Z^+$, the Casoratian $W(t_0) \neq 0$. $\square$

Example Solve the difference equation

$$x(t+3)+3x(t+2)-4x(t+1)-12x(t)=0.$$

We verify that the solutions

---

[13] The proof is referred to Elaydi (1999).

$$2^t, \quad (-2)^t, \quad (-3)^t,$$

form a fundamental set of solutions of the equation. This is proved by calculating

$$W(t)=\begin{vmatrix} 2^t & (-2)^t & (-3)^t \\ 2^{t+1} & (-2)^{t+1} & (-3)^{t+1} \\ 2^{t+2} & (-2)^{t+2} & (-3)^{t+2} \end{vmatrix} \Rightarrow W(0)=\begin{vmatrix} 1 & 1 & 1 \\ 2 & -2 & 3 \\ 4 & 4 & 9 \end{vmatrix} = -20.$$

Theorem 2.6.2. (the fundamental theorem) If $p_k(t) \neq 0$ for all $t \geq t_0$, then equation (2.6.2) has a fundamental set of solutions for $t \geq t_0$. □

Superposition Principle. If $x_1(t), x_2(t), \cdots, \ x_k(t)$ are solutions of equation (2.6.2), then

$$x(t)=\sum_{i=1}^{k} a_i x_i(t)$$

is also a solution of equation (2.6.2), where a_i are constant. □

Let $x_1(t), x_2(t), \cdots, \ x_k(t)$ be a fundamental set of solutions of equation (2.6.2). Then the *general solution* of equation (2.6.2) is given by

$$x(t)=\sum_{i=1}^{k} a_i x_i(t).$$

for arbitrary constants a_i. Any solution of equation (2.6.2) may be obtained from the general solution by a suitable choice of the constants a_i.

We now examine nonhomogeneous equation (2.6.1). It is customary to refer to the general solution of the homogeneous equation (2.6.2) as the *complementary solution* to nonhomogeneous equation (2.6.1). Denote the complementary solution by $x_c(t)$. Then we have

$$x_c(t)=\sum_{i=1}^{k} a_i x_i(t),$$

where $x_1(t), x_2(t), \cdots, x_k(t)$ is a fundamental set of solutions of equation (2.6.2) and a_i are arbitrary constants. A solution of nonhomogeneous equation (2.6.1) is called a *particular solution* and will be denoted by $x_p(t)$.

Theorem 2.6.3. Any solution of equation (2.6.1) may be written as

$$x(t) = x_c(t) + x_p(t). \qquad \square$$

The general solution of equation (2.6.1) is given by

$$x(t) = x_c(t) + x_p(t).$$

As we studied how to find $x_c(t)$, we now learn how to find $x_p(t)$. We only introduce the *method of undetermined coefficients* to compute $x_p(t)$. In this method, we first guest the form of the particular solution and then substituting this function into the difference equation. This method is effective mainly when $g(x(t))$ is a linear combination or products of terms, each having one of the forms

$$a^t,\ \sin(bt),\ \cos(bt),\ t^k.$$

Example Solve the difference equation

$$x(t+2) + x(t+1) - 12x(t) = t2^t. \qquad (2.6.3)$$

The characteristic roots are $\rho_1 = 3$ and $\rho_2 = -4$. Hence,

$$x_c(t) = a_1 3^t + a_2(-4)^t.$$

We try

$$x_p(t) = c_1 2^t + c_2 t(2)^t.$$

Substituting this into equation (2.6.3) yields

$$\{c_1 2^{t+2} + c_2(t+2)(2)^{t+2}\} + \{c_1 2^{t+1} + c_2(t+1)(2)^{t+1}\}$$
$$- 12\{c_1 2^t + c_2 t(2)^t\} = t2^t,$$

or

$$(10a_2 - 6a_1)2^t - 6a_2 t2^t = t2^t.$$

As this is held for any t, we should have

$$10a_2 - 6a_1 = 0, \quad -6a_2 = 1.$$

Hence

$$a_1 = -\frac{5}{18}, \quad a_2 = -\frac{1}{6}.$$

The general solution is

$$x(t) = a_1 3^t + a_2(-4)^t - \frac{5}{18}2^t - \frac{1}{6}t(2)^t.$$

Exercise 2.6

1 Find the Casoratian of the following functions

(i) 5^t, 35^{t+1}, e^t;

(ii) 5^t, $t5^t$, $n^2 5^t$.

2 For the following equations and their accompanied solutions, determine whether these solutions are linearly independent and find, if possible, the general solution:

(i) $x(t+3) - 3x(t+2) + 3x(t+1) - x(t) = 0$; $1, t, t^2$;

(ii) $x(t+3) + x(t+2) - 8x(t+1) - 12x(t) = 0$; $3^t, (-2)^t, (-2)^{t+3}$.

3 Find the general solutions of the following difference equations

(i) $x(t+2)+4x(t)=8(2)^t\cos(t\pi/2)$, ≤ 1 to guarantee the right inequality. We thus

(ii) $x(t+2)+8x(t+1)+7x(t)=t(2)^t$.

2.7 Equations with constant coefficients

We consider equation (2.6.2) when p_i are constant

$$x(t+k)+p_1x(t+k-1)+\cdots+p_kx(t)=0, \tag{2.7.1}$$

where $p_k \neq 0$. Suppose that solutions of equation (2.7.1) are in the form ρ^t. Substituting ρ^t into equation (2.7.1) yields

$$\rho^k+p_1\rho^{k-1}+\cdots+p_k=0. \tag{2.7.2}$$

This is called the *characteristic equation* of equation (2.1.7) and its roots ρ are called the *characteristic roots.*

Case a: If the characteristics roots $\{\rho_1, \rho_2, \cdots, \rho_k\}$ are distinct, then it can be shown that the set

$$\{\rho_1^t, \rho_2^t, \cdots, \rho_k^t\}$$

is a fundamental set of solutions. Hence, the general solution of equation (2.7.1) is

$$x(t)=\sum_{i=1}^{k} a_i\rho_i^t.$$

Example (the Fibonacci sequence and the rabbit problem) The problem appeared in 1202 in a book about the abacus, written by the famous Italian mathematician Leonard di Pisa, known as Fibonacci. The Fibonacci problem is as follows: how many pairs of rabbits will there be after one year if starting with one pair of mature rabbits, if each pair of rabbits gives birth to a new pair each month starting when it reaches its maturity age of two months? If $x(t)$ is the number of pairs of rabbits at the end of t months, then the recurrence relation that represents this model is given by the second-order linear difference equation

$$x(t+2) = x(t+1) + x(t), \;\; x(0) = 1, \;\; x(2) = 2, \;\; 0 \le t \le 10.$$

This equation is a special case of the *Fibonacci sequence*

$$x(t+2) = x(t+1) + x(t), \;\; x(0) = 1, \;\; x(2) = 2, \;\; t \ge 0.$$

The characteristic equation is

$$\rho^2 - \rho - 1 = 0.$$

The characteristic roots are

$$\rho_{1,2} = \frac{1}{2} \pm \frac{\sqrt{5}}{2}.$$

The general solution is

$$x(t) = a_1\left(\frac{1}{2} + \frac{\sqrt{5}}{2}\right)^t + a_2\left(\frac{1}{2} - \frac{\sqrt{5}}{2}\right)^t.$$

Substituting the initial conditions into the general solution leads to $a_{1,2} = \pm\sqrt{5}$.

Case b: Suppose that the distinct characteristic roots are $\rho_1, \rho_2, \cdots, \rho_r$ with multiplicitics $m_1, m_2, \cdots, m_r$. Then, the general solution of equation (2.7.1) is given by

$$x(t) = \sum_{i=1}^{r} \rho_i^t \left(a_{i0} + a_{i1}t + a_{i2}t^2 + \cdots + a_{im_i-1}t^{m_i-1}\right).$$

Example Solve

$$x(t+3) - 3x(t+2) + 16x(t+1) - 12x(t) = 0,$$
$$x(0) = 0, \;\; x(1) = 1, \;\; x(2) = 1.$$

It is straightforward to check that the characteristic roots are

$$\rho_1 = \rho_2 = 2, \quad \rho_3 = 3.$$

The general solution is

$$x(t) = a_0 2^t + a_1 t 2^t + a_3 3^t.$$

Substituting the initial conditions into the general solution, we solve

$$a_0 = 3, \quad a_1 = 2, \quad a_3 = -3.$$

The solution of the initial value problem is

$$x(t) = 3(2)^t + t2^{t+1} - 3^{t+1}.$$

Example Kobayashi and Anders have agreed to bet one dollar on each flip of a fair coin and to continue playing until one of them wins all of the other's money. What is the probability that Kobayashi will win all of Anders' money if Kobayashi starts with a dollars and Anders with b dollars?

To analyze this game, we denote $x(t)$ the probability that Kobayashi will win all of Anders' money if Kobayashi currently has t dollars. Let $A = a + b$ be the total amount of money available to the players. Note that $x(0) = 0$ because Kobayashi has no money left, and $x(A) = 1$ because Kobayashi has all the money. Moreover, if

$$1 \le t \le A - 1,$$

then Kobayashi has a 0.5 probability of winning one dollar on the next flip (raising the amount of money he has to $t + 1$ dollars) and 0.5 probability of losing one dollar on the next flip (reducing the amount of money he has to $t - 1$ dollars. Hence

$$x(t) = 0.5x(t + 1) + 0.5x(t - 1), \quad 1 \le t \le A - 1.$$

Rearrange the above equation

$$x(t + 1) - 2x(t) + x(t - 1) = 0, \quad 1 \le t \le A - 1.$$

The characteristic equation

$$\rho^2 - 2\rho + 1 = 0,$$

has a double root, $\rho_{1,2} = 1$. Hence, the general solution is

$$x(t) = (a_1 + a_2 t)1^t.$$

From $x(0) = 0$ and $x(A) = 1$, we determine $a_1 = 0$ and $a_1 = 1/A$. We thus obtain

$$x(t) = \frac{t}{a + b}.$$

When Kobayashi has a dollars, the probability of Kobayashi's winning the game is

$$x(a) = \frac{a}{a + b},$$

and the probability of Anders' winning the game is

$$1 - \frac{a}{a + b} = \frac{b}{a + b}.$$

Case c: Suppose that the characteristic equation has complex roots. For simplicity, we are concerned with

$$x(t + 2) + p_1 x(t + 1) + p_2 x(t) = 0.$$

The complex roots are given by

$$\rho_{1,2} = \alpha \pm i\beta.$$

It can be shown that the general solution is given by

$$x(t) = Ar^t \cos(\theta t - \omega),$$

where A and ω are arbitrary constants and

$$r = \sqrt{\alpha^2 + \beta^2}, \quad \theta = \tan^{-1}\left(\frac{\beta}{\alpha}\right).$$

We have been concerned with finding solutions of linear difference equations. Sometimes nonlinear difference equations can be transformed to linear ones. We discuss some types of nonlinear equations transformable to linear equations.

Example (equations of Riccati type) Consider

$$x(t+1)x(t) + p(t)x(t+1) + q(t)x(t) = 0. \tag{2.7.3}$$

Introduce $z(t) = 1/x(t)$. Equation (2.7.3) becomes

$$q(t)x(t+1) + p(t)z(t) + 1 = 0.$$

For the nonhomogeneous equation

$$x(t+1)x(t) + p(t)x(t+1) + q(t)x(t) = g(t),$$

we introduce

$$x(t) = \frac{z(t+1)}{z(t)} - p(t).$$

The above equation becomes

$$z(t+2) + (q(t) - p(t+1))z(t+1) - (g(t) + p(t)q(t))z(t) = 0.$$

Consider the *Pielou logistic equation*

$$x(t+1) = \frac{\alpha x(t)}{1 + \beta x(t)}.$$

Under transformation $x(t) = 1/z(t)$, this equation becomes

$$\alpha z(t+1) = z(t) + \beta.$$

Example (equations of general Riccati type) Consider

$$x(t+1) = \frac{a_1(x)x(t) + a_2(x)}{a_3(x)x(t) + a_4(x)}, \tag{2.7.4}$$

where

$$a_3(x) \neq 0, \ \ a_1(x)a_4(x) - a_2(x)a_3(x) \neq 0, \ \ t \geq 0.$$

To solve this equation, let

$$a_3(t)x(t) + a_4(t) = \frac{y(t+1)}{y(t)}.$$

Under this transformation, equation (2.7.4) becomes

$$y(t+1) + p_1(t)y(t+1) + p_2(t)y(t) = 0,$$

where

$$p_1(t) \equiv -\frac{a_3(t)a_4(t+1) + a_1(t)a_3(t+1)}{a_3(t)},$$

$$p_2(t) \equiv (a_1(t)a_4(t) - a_2(t)a_3(t))\frac{a_3(t+1)}{a_3(t)}.$$

Consider

$$x(t+1) = \frac{2x(t) + 3}{3x(t) + 2}.$$

Using the transformation

$$3x(t) + 2 = \frac{y(t+1)}{y(t)},$$

we see that the equation becomes

$$y(t+2) - 4y(t+1) - 5y(t) = 0.$$

Example Homogeneous difference equation of the type

$$f\left(\frac{x(t+1)}{x(t)}, t\right) = 0.$$

Use the transformation

$$z(t) = \frac{x(t+1)}{x(t)}$$

to convert such an equation to an equation in $z(t)$.

Exercise 2.7

1 Introduce the symbol $E^k x(t) = x(t+k)$. Find the general solutions of the following difference equations

(i) $x(t+2) - 16x(t) = 0$;
(ii) $(E-3)^2 (E^2+4)x(t) = 0$;
(iii) $(E^2+2)^2 x(t) = 0$.

2 Solve the following difference equations

(i) $x^2(t+1) - (2+t)x(t+1)x(t) - 2tx^2(t) = 0$;
(ii) $x(t+1)x(t) - \frac{2}{3}x(t+1) + \frac{1}{6}x(t) = \frac{5}{18}$;
(iii) $x(t+1) = x^2(t)$;

(iv) $x(t+2) = \dfrac{x^3(t+1)}{x(t)}$;

(v) $x(t+1) = \dfrac{1 - x^2(t)/2}{1 - x(t)}$.

2.8 Limiting behavior

For simplicity, we first examine limiting behavior of two-dimensional difference equations. We now examine limiting behavior of solutions of the following second-order linear difference equation

$$x(t+2) + p_1 x(t+1) + p_2 x(t) = 0. \tag{2.8.1}$$

Suppose that ρ_1 and ρ_2 are the two characteristic roots of the equation. Following the previous section, we have the following three cases.

Case a: If the characteristic roots, ρ_1 and ρ_2, are distinct, then the general solution of equation (2.8.1) is

$$x(t) = a_1 \rho_1^t + a_2 \rho_2^t.$$

Without loss of generality, assume $|\rho_1| > |\rho_2|$. Then

$$x(t) = \left(a_1 + a_2 \left(\frac{\rho_2}{\rho_1} \right)^t \right) \rho_1^t.$$

Since $|\rho_2 / \rho_1| < 1$, it follows that

$$\left(\frac{\rho_2}{\rho_1} \right)^t \to 0 \text{ as } t \to \infty.$$

Consequently

$$\lim_{t \to \infty} x(t) = \lim_{t \to \infty} a_1 \rho_1^t.$$

There are six different situations that may arise depending on the value of ρ_1.

(1) If $\rho_1 > 1$, the sequence $\{a_1 \rho_1^t\}$ diverges to ∞ (unstable system).
(2) If $\rho_1 = 1$, the sequence $\{a_1 \rho_1^t\}$ is a constant sequence.
(3) If $0 < \rho_1 < 1$, the sequence $\{a_1 \rho_1^t\}$ is monotonically decreasing to zero (stable system).
(4) If $-1 < \rho_1 < 0$, the sequence $\{a_1 \rho_1^t\}$ is oscillating around zero (i.e., altering in sign) and converging to zero (stable system).
(5) If $\rho_1 = -1$, the sequence $\{a_1 \rho_1^t\}$ is oscillating between two values a_1 and a_2.
(6) If $\rho_1 < -1$, the sequence $\{a_1 \rho_1^t\}$ is oscillating but increasing in magnitude (unstable system).

Case b: $\rho = \rho_1 = \rho_2$. The general solution of equation (2.8.1) is given by

$$x(t) = (a_1 + a_2 t)\rho^t.$$

If $|\rho| \geq 1$, the solution $x(t)$ diverges either monotonically if $\rho \geq 1$ or by oscillating if $\rho \leq -1$. However, if $|\rho| < 1$, then solution converges to zero.

Case c: Suppose that the characteristic equation has complex roots, denoted by

$$\rho_{1,2} = \alpha \pm i\beta.$$

The general solution is given by

$$x(t) = Ar^t \cos(\theta t - \omega),$$

where A and ω are arbitrary constants and

$$r \equiv \sqrt{\alpha^2 + \beta^2}, \quad \theta \equiv \tan^{-1}\left(\frac{\beta}{\alpha}\right).$$

(1) If $r > 1$, the solution $x(t)$ is oscillating but increasing in magnitude (unstable system).
(2) If $r = 1$, the solution $x(t)$ is oscillating but constant in magnitude.
(3) If $r < 1$, the solution $x(t)$ is oscillating but decreasing in magnitude (stable system).

Example (gambler's ruin) A gambler plays a sequence games against an adversary in which the probability that the gambler wins \$1.00 in any given game is a known value q, and the probability of his losing \$1.00 is $1 - q$, where $0 \le q \le 1$. He quits gambling if he either loses all his money or reaches his goal of acquiring N dollars. If the gambler runs out of money first, we say that the gambler has been ruined. Let $p(t)$ denote the probability that the gambler will be ruined if he possesses t dollars. He may be ruined in two ways. First, winning the next game; the probability of this event is q; then his fortune will be $t + 1$, and the probability of being ruined will become $p(t + 1)$. Second, losing the next game, the probability of this event is $1 - q$, and the probability of being ruined is $p(t - 1)$. Applying the theorem of total probabilities, we have

$$p(t) = qp(t + 1) + (1 - q)p(t - 1).$$

We have

$$p(t + 2) - \frac{1}{q} p(t + 1) + \frac{1 - q}{q} p(t) = 0, \quad t = 0, 1, \cdots, N,$$

with $p(0) = 1$ and $p(N) = 0$. The characteristic equation is

$$\rho^2 - \frac{1}{q}\rho + \frac{1 - q}{q} = 0, \quad t = 0, 1, \cdots, N.$$

The two characteristic roots are

$$\rho_1 = \frac{1 - q}{q}, \quad \rho_2 = 1.$$

In the case of $q \ne 1/2$, the general solution is

$$p(t) = \frac{\tilde{q}^t - \tilde{q}^N}{1 - \tilde{q}^N}, \quad \tilde{q} \equiv \frac{1 - q}{q}.$$

In the case of $q = 1/2$, the two characteristic roots are equal. In this case,

$$p(x) = a_1 + a_2 t.$$

As $p(0) = 1$ and $p(N) = 0$, we have

$$p(n) = 1 - \frac{n}{N}.$$

For instance, if one starts with \$4 dollar and the probability that one wins a dollar is 0.3, and one will quit if one runs out of money or has a total of \$10. With

$$n = 4, \quad q = 0.3, \quad N = 10,$$

the probability of being ruined is $p(4) = 0.994$.

We now examine nonhomogeneous difference equations in which input is constant. We have

$$x(t + 2) + p_1 x(t + 1) + p_2 x(t) = b, \tag{2.8.2}$$

where b is a non-zero constant. A particular equation (equilibrium in this case) x^* is determined by

$$x^* + p_1 x^* + p_2 x^* = b.$$

That is

$$x^* = \frac{b}{1 + p_1 + p_2}.$$

Consequently, the general solution of equation (2.8.2) is given by

$$x(t) = x_c(t) + x^*, \tag{2.8.3}$$

where $x_c(t)$ is the general solution of the corresponding homogeneous equation (2.8.1). From formula (2.8.3), we conclude that $x(t) \to x^*$ if and only if $x_c(t) \to 0$ as $t \to \infty$. Furthermore, $x(t)$ oscillates around x^* if and only if $x_c(t)$ oscillates about zero.[14] Summarizing the above discussions, we obtain the following theorem.

Theorem 2.8.1. The following statements are held.
(1) All solutions of equation (2.8.2) oscillate about x^* if only if none of the characteristic roots of equation (2.8.1) is a positive real number.
(2) All solutions of equation (2.8.2) converge to x^* as $t \to \infty$ if and only if

$$Max\{|\rho_1|, |\rho_2|\} < 1,$$

where ρ_1, ρ_2 are the characteristic roots of homogeneous equation (2.8.1).

The following lemma provides criteria for stability based on the values of the coefficients p_1 and p_2.[15]

Lemma 2.8.1. The conditions

$$1 + p_1 + p_2 > 0, \ 1 - p_1 + p_2 > 0, \ 1 - p_2 > 0, \tag{2.8.4}$$

are necessary and sufficient conditions for the solution of equations (2.8.1) and (2.8.2) to be asymptotically stable (i.e., all solutions converge to x^*).

Example (the Samuelson multiplier-accelerator interaction model) Consider the national income model. The national income $Y(t)$ at year t is equal to the sum of consumer expenditure $C(t)$, private investment, $I(t)$, and constant government expenditure, G. That is

$$Y(t) = C(t) + I(t) + G.$$

---

[14] We say $x(t)$ oscillates about x^* if $x(t) - x^*$ alternates sign, i.e., if $x(t) > x^*$, then $x(t+1) < x^*$.

[15] The proof is referred to Elaydi (1999: 82).

Assume that consumption expenditure is proportional to the national income $Y(t-1)$ in the preceding year, that is

$$C(t) = \alpha Y(t-1),$$

where α ($0 < \alpha < 1$) is the marginal propensity to consume. Private investment is proportional to the increase in consumption, that is

$$I(t) = \beta[C(t) - C(t-1)], \quad \beta > 0.$$

Substituting this equation and

$$C(t) = \alpha Y(t-1)$$

into

$$Y(t) = C(t) + I(t) + G$$

yields

$$Y(t+2) - \alpha(1+\beta)Y(t+1) + \alpha\beta Y(t) = 1.$$

The equation has a unique equilibrium state Y^* given by

$$Y^* = \frac{1}{1-\alpha}.$$

Applying lemma 2.8.1 to the problem, we conclude that if $\alpha < 1$ and $\alpha\beta < 1$, the equilibrium point is asymptotically stable. The national income $Y(t)$ fluctuates around the equilibrium state Y^* if and only if

$$\alpha < \frac{4\beta}{(1+\beta)^2}.$$

As we can explicitly solve the problem, we can examine its stability properties. Figure 2.8.1 shows the stability properties of the Samuelson multiplier-accelerator interaction model.

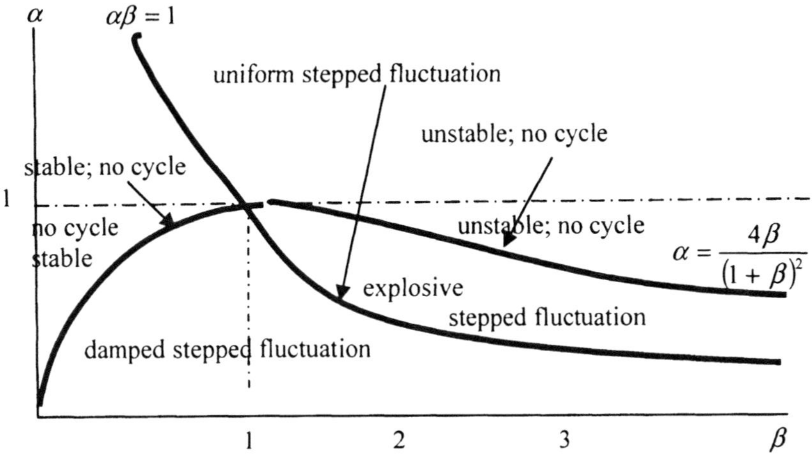

Figure 2.8.1: Stability of the multiplier-accelerator interaction model

We now consider the k th-order scalar equation

$$x(t+k)+p_1x(t+k-1)+p_2x(t+k-2)+\cdots+p_kx(t)=0, \qquad (2.8.5)$$

where p_i 's are real numbers and $p_k \neq 0$. The characteristic equation of equation (2.8.5) is

$$p(\rho)=\rho^k+p_1\rho^{k-1}+p_2\rho^{k-2}+\cdots+p_k=0. \qquad (2.8.6)$$

As discussed in section 2.7, if the distinct characteristic roots are $\rho_1, \rho_2, \cdots, \rho_r$ with multiplicities $m_1, m_2, \cdots, m_r$, then, the general solution of equation (2.7.1) is given by

$$x(t)=\sum_{i=1}^{r}\rho_i^t\left(a_{i0}+a_{i1}t+a_{i2}t^2+\cdots+a_{im_i-1}t^{m_i-1}\right).$$

It can be shown that the zero solution of equation (2.8.5) is asymptotically stable if and only if $|\rho|<1$ for every characteristic root ρ. Furthermore, the zero solution of equation (2.8.5) is stable if and only if $|\rho|\leq 1$ and those characteristic roots ρ with $|\rho|=1$ are simple (i.e., not repeated, $m_i=1$). On the other hand, if there is a multiple characteristic root with $|\rho| = 1$, then the zero solution of equation (2.8.5) is unstable. We see that it is important to judge whether the characteristic roots fall inside the unit cycle. The Schur-Cohn criterion gives an answer.

Before introducing the criterion, we define the inners of a matrix $A = \lfloor a_{ij} \rfloor$. The *inners* of a matrix are the matrix itself and all the matrices obtained by omitting successively the first and last rows and the first and last columns. For instance, the inners of $\lfloor a_{ij} \rfloor_{3\times3}$ consists of $\lfloor a_{ij} \rfloor_{3\times3}$ and the scalar a_{22}. For $\lfloor a_{ij} \rfloor_{5\times5}$, they are

$$\begin{bmatrix} a_{11} & a_{12} & a_{13} & a_{14} & a_{15} \\ a_{21} & a_{22} & a_{23} & a_{24} & a_{25} \\ a_{31} & a_{32} & a_{33} & a_{1134} & a_{35} \\ a_{41} & a_{42} & a_{43} & a_{44} & a_{45} \\ a_{51} & a_{52} & a_{53} & a_{54} & a_{55} \end{bmatrix}, \quad \begin{bmatrix} a_{22} & a_{23} & a_{24} \\ a_{32} & a_{33} & a_{34} \\ a_{42} & a_{43} & a_{44} \end{bmatrix}, \quad a_{33}.$$

A matrix A is said to be *positive innerwise* if the determinants of all of its inners are positive.

Theorem 2.8.2. (the Schur-Cohn criterion[16]) The zeros of characteristic equation (2.8.6) lie inside the unit cycle if and only if we have: (i) $p(1) > 0$; (ii) $(-1)^k p(-1) > 0$; and (iii) the $(k-1)\times(k-1)$ matrices

$$A_{k-1}^{\pm} = \begin{bmatrix} 1 & 0 & \cdots & & 0 \\ p_1 & 1 & & & 0 \\ \vdots & & & & \vdots \\ p_{k-3} & & & & \\ p_{k-2} & p_{k-3} & \cdots & p_1 & 1 \end{bmatrix} \pm \begin{bmatrix} 0 & 0 & \cdots & 0 & p_k \\ 0 & 0 & \cdots & p_k & p_{k-1} \\ \vdots & \vdots & & & \vdots \\ 0 & p_k & & & p_3 \\ p_k & p_{k-1} & \cdots & p_3 & p_2 \end{bmatrix},$$

are positive innerwise.

Example For the equation

$$x(t+2) + p_1 x(t+1) + p_2 x(t) = 0,$$

[16] See Jury (1964).

the Schur-Cohn criterion is given by

$$p(1) = 1 + p_1 + p_2 > 0,$$
$$(-1)^2 p(-1) = 1 - p_1 + p_2 > 0,$$
$$A_1^{\pm} = 1 \pm p_2 > 0.$$

We thus conclude that the zero solution is asymptotically stable if and only if

$$|p_1| < 1 + p_2 < 2.$$

Okuguchi and Irie show that the necessary and sufficient conditions for a third order difference equation

$$x(t+3) + p_1 x(t+2) + p_2 x(t+1) + p_3 x(t) = 0,$$

are given by[17]

$$\begin{aligned} &1 + p_1 + p_2 + p_3 > 0, \\ &1 - p_1 + p_2 - p_3 > 0, \\ &1 - p_2 + p_1 p_3 - p_3^2 > 0. \end{aligned} \tag{2.8.7}$$

Example (inventory cycles) The economy has a single commodity which can be used either for consumption and investment. The basic equation of Metzler equation is[18]

$$Y(t) = U(t) + \left(\hat{Q}(t) - Q(t-1)\right) + I_0,$$

where $Y(t)$ is the output in period t, $U(t)$ is output to be currently sold according to producers' expectations on sales, $\hat{Q}(t)$ is the desired level of inventories, $Q(t-1)$ is the inventory level of the previous period, and I_0 is exogenous output of investment goods. Assume that producers wish to maintain a constant ratio between inventories and sales (the inventory accelerator). Producers apply this ratio in forming the desired level of inventories as follows

---

[17] See Okuguchi and Irie (1990).

[18] Metzler (1941) proposed one of seminal models for explaining inventory cycles. The model here is referred to Gandolfo (1996: 95-98).

$$\hat{Q}(t) = kU(t-1),$$

where k is a constant. In any period t, the actual inventory level $Q(t)$ is equal to the level that producers had planned for that period $\hat{Q}(t)$, minus the unintended variation in inventories occurring because of the difference between realized and expected sales, $C(t) - U(t)$, i.e.

$$Q(t) = \hat{Q}(t) - (C(t) - U(t)).$$

Producers form their expectations on sales as follows

$$U(t) = C(t-1) + \alpha(C(t-1) - C(t-2)),$$

where $C(t)$ is the current consumption level. No lag is assumed to exist between current consumption and current income. The consumption function is thus given by

$$C(t) = bY(t), \quad 0 < b < 1,$$

where b is constant. We have established the model. After simple substitutions, the motion of the system is given by

$$Y(t) - b[(1+k)(1+\alpha)+1]Y(t-1) + b(1+k)(1+2\alpha)Y(t-2)$$
$$- \alpha b(1+k)Y(t-3) = I_0.$$

A particular solution is given by

$$Y^* = \frac{I_0}{1-b}.$$

Applying conditions (3.8.7) to this problem, we see that the stability conditions are given by

$$3 - b(2k+3) > 0,$$
$$(1+k)(2+k)\alpha b^2 - (1+k)(1+2\alpha)b + 1 > 0.$$

Exercise 2.8

1 Determine the equilibrium points, their stability, and oscillatory behavior of the solutions of the following equations

(a) $x(t+2) - 2x(t+1) + 2x(t) = 1$;
(b) $x(t+2) + x(t+1) + 0.5x(t) = -5$.

2 Determine the limiting behavior of solutions of the following equation

$$x(t+2) - \alpha\beta x(t+1) + \alpha\beta x(t) = \alpha\theta, \quad \alpha,\ \beta,\ \theta > 0,$$

if (i) $\alpha\beta = 1$; (ii) $\alpha\beta = 2$; and (i) $\alpha\beta = 0.5$.

3 Show that the zero solution of

$$x(t+3) + p_1 x(t+2) + p_2 x(t+1) + p_3 x(t) = 0,$$

is asymptotically stable if and only if

$$|p_1 + p_3| < 1 + p_2, \quad |p_2 - p_1 p_3| < 1 - p_3^2.$$

Chapter 3

One-dimensional dynamical economic systems

This chapter applies the concepts and theorems of the previous chapters to analyze different models in economics. Although the economic relations in these models are complicated, we show that the dynamics of all these models are determined by one-dimensional difference equations. Section 3.1 examines a traditional model of interactions between inflation and unemployment. The model is built on the expectations-augmented version of the Phillips relation and the adaptive expectations hypothesis. We solve the model and show that the characteristic equation may have either: (1) distinct real roots; or (2) repeated real roots; or (3) complex roots. Section 3.2 introduces the one-sector growth model. The model is different from most of the growth models in the literature in that it treats saving as an endogenous variable through introducing wealth into utility function. We demonstrate that the OSG model has a unique stable equilibrium. Section 3.3 generalizes the OSG model proposed in section 3.2. Section 3.4 deals with the overlapping-generation (OLG) model – one of the most popular dynamic models among economists. The model is essential for the reader to approach some of the models in this book as well. Different from the OSG models in the previous two sections, in the OLG analytical framework, each person lives for only two periods. This is the main shortcoming of the model; nevertheless, its popularity is sustained partly because this framework often simplifies complicated analytical issues. Section 3.5 introduces a growth model to demonstrate persistence of inequality. In this model, the evolution of income within each dynasty in society is governed by a dynamical system that generates a poverty trap equilibrium point along with a high-income equilibrium point. Poor dynasties, those with the income at the threshold level, converge to a low income steady state, whereas dynasties with income above the threshold converge to a high income level. Section 3.6 studies a model to provide insights into processes of Schumpeterian creative destruction. Section 3.7 is concerned with interactions between human capital accumulation, economic growth, and inequality. The model exhibits three possible equilibrium points: a low-growth trap, a pair of equilibrium points in the intermediate and advanced development phase. If these equilibrium points exist, it can be shown that the poverty trap is stable, while in development phase, the first equilibrium point will be unstable and the second one stable. Section 3.8 studies an urban dynamic model to highlight how the trade-off between optimal and equilibrium city sizes behaves when human capital externalities are introduced into urban dynamics. Section 3.9 introduces a growth model of monetary economy. The model addresses the Tobin effect and the existence of monetary economy. Section 3.10 introduces endogenous time distribution between leisure and work into the OSG model.

3.1 A model of inflation and unemployment

We now consider an interaction of inflation and unemployment.[1] Denote the rate of inflation $p(t)$, which is defined by

$$p(t) \equiv \frac{P(t+1) - P(t)}{P(t)},$$

where $P(t)$ is the price. The *expectations-augmented* version of the *Phillips relation* assumes the following relationship between the rate of inflation, the unemployment rate, $U(t)$, and the expected rate of inflation, $\pi(t)$

$$p(t) = a - bU(t) + h\pi(t), \quad (0 < h \le 1), \tag{3.1.1}$$

where a, b, and h are positive parameters. The *adaptive expectations hypothesis* establishes a rule of the expected rate of inflation as follows

$$\pi(t+1) - \pi(t) = j(p(t) - \pi(t)), \quad 0 < j \le 1, \tag{3.1.2}$$

which j is a parameter. The equation states that if the actual rate of inflation exceeds the expected rate of inflation, then the expected rate of inflation tends to rise.

Denote the nominal money balance by $M(t)$ and its rate of growth by

[1] The model is based on Chiang (1984: 591-596).

$$m(t) \equiv \frac{M(t+1) - M(t)}{M(t)}.$$

Assume that m is constant over time. The model contains a feedback from inflation to unemployment

$$U(t + 1) - U(t) = -k(m - p(t + 1)), \quad k > 0. \tag{3.1.3}$$

The model consists of three equations, (3.1.1), (3.1.2), and (3.1.3), with three variables, p, U, and π. We now show that the dynamics can be described by a second-order linear difference equation.

From equation (3.1.1), we get

$$\begin{aligned} &p(t + 1) - p(t) = -b(U(t + 1) - U(t)) + h(\pi(t + 1) - \pi(t)) \\ &= kb(m - p(t + 1)) + hj(p(t) - \pi(t)), \end{aligned} \tag{3.1.4}$$

where we use equations (3.1.2) and (3.1.3). To eliminate $\pi(t)$ from equation (3.1.4), we observe from equation (3.1.1)

$$h\pi(t) = p(t) - a + bU(t).$$

Substituting this into equation (3.1.4) yields

$$(1 + kb)p(t + 1) - (1 - j(1 - h))p(t) + bjU(t) = kbm + aj. \tag{3.1.5}$$

From equation (3.1.5), we get

$$\begin{aligned} &bj(U(t + 1) - U(t)) = \\ &-(1 + kb)p(t + 2) + (2 - j(1 - h) + kb)p(t + 1) - (1 - j(1 - h))p(t). \end{aligned}$$

Inserting equation (3.1.3) into this equation finally yields the equation involving only p

$$p(t + 2) + a_1 p(t + 1) + a_2 p(t) = c, \tag{3.1.6}$$

where

$$a_1 \equiv -\left(\frac{2 - j(1-h) + (1-j)kb}{1+kb}\right),$$

$$a_2 \equiv \frac{1 - j(1-h)}{1+kb},$$

$$c \equiv \frac{kbjm}{1+kb}.$$

The fixed point of equation (3.1.6) is given by

$$p^* = \frac{c}{1 + a_1 + a_2} = m.$$

The equilibrium rate of inflation is exactly equal to the rate of monetary expansion. The characteristic equation is

$$\rho^2 + a_1 \rho + a_2 = 0.$$

We have

$$a_1^2 \geq (<)\ 4a_2$$

if and only if

$$(2 - j(1-h) + (1-j)kb)^2 \geq (<)\ 4(1 - j(1-h))(1+kb).$$

It is straightforward to check that there may arise either: (1) distinct real roots (e.g., $h = 0.2$, $j = 1/3$ and $bk = 5 \Rightarrow a_1 = 31/5$ and $a_2 = 5$); or (2) repeated real roots (e.g., $h = 1$, $j = 1/3$, and $kb = 3 \Rightarrow a_1^2 = 4a_2$); or (3) complex roots (e.g., $h = 1$, $j = 1 \Rightarrow a_1 = 2$ and $a_2 = 1 + bk > 1$). Figures 3.1.1 and 3.1.2 depict respectively case (1) and case (3) with the following specified values of the parameters

$$m = 0.06, \quad b = 0.3, \quad a = 0.05,$$

on the $p - U$-plane. The reader is encouraged to carry our stability analysis of the model in detail.

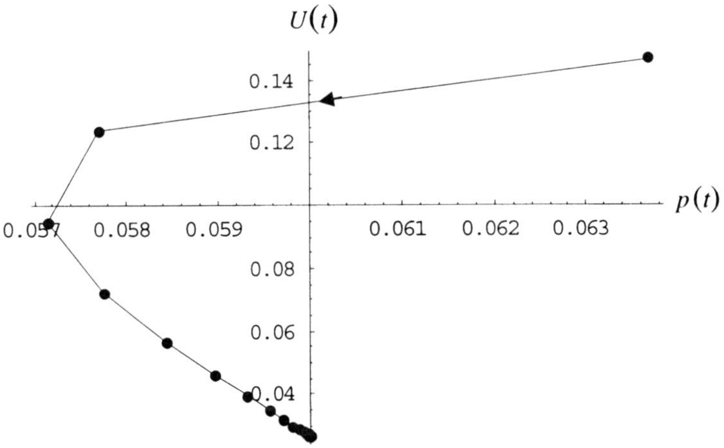

Figure 3.1.1: Different real eigenvalues

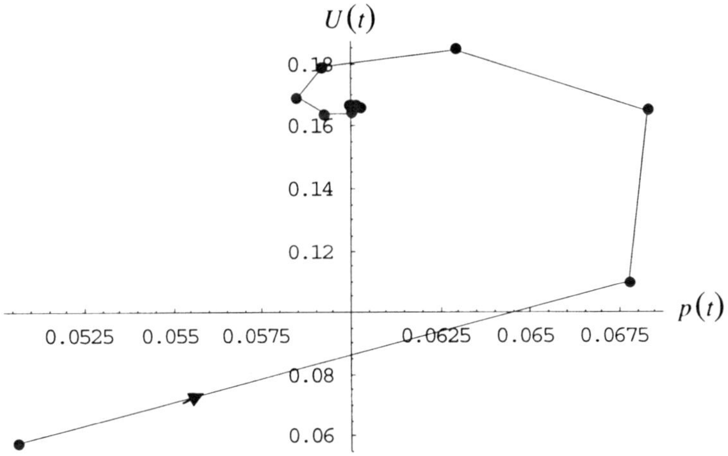

Figure 3.1.2: Complex eigenvalues

3.2 The one-sector growth (OSG) model

We now represent a one-sector growth (OSG) model in discrete time. The economy has an infinite future. We represent the passage of time in a sequence of periods, numbered from zero and indexed by $t = 0, 1, 2, \ldots$. Time 0, being referred to the beginning of period 0, represents the initial situation from which economy starts to grow. The end of period $t-1$ coincides with the beginning of period t; it can also be called time t. We assume that transactions are made in each period.

The model assumes that each individual lives forever. The population grows at rate n; thus

$$N(t) = (1+n)N(t-1).$$

Each individual supplies one unit of labor at each time t. Production in period t uses inputs amount $K(t)$ of capital and amount $N(t)$ of labor services. It supplies amount $Y(t)$ of goods. Here, production is assumed to be continuous during the period, but then use the same capital that existed at the beginning of the period.

The production function is

$$F(K(t), N(t)) = AK^{\alpha}(t)N^{\beta}(t), \quad \alpha + \beta = 1, \quad \alpha, \beta > 0.$$

The production function has constant returns to scale. Markets are competitive; thus labor and capital earn their marginal products, and firms earn zero profits. Let us assume that depreciation is proportional to capital and denote the rate of depreciation by δ_k. The total amount of depreciation is equal to $\delta_k K(t)$. The real interest rate and the wage of labor are given by

$$r(t) + \delta_k = \frac{\alpha F(t)}{K(t)},$$
$$w(t) = \frac{\beta F(t)}{N(t)}.$$

There is some initial capital stock K_0 that is owned equally by all individuals at the initial period. We write the marginal conditions in capital intensity

$$r(t) + \delta_k = \alpha A k^{-\beta}(t),$$
$$w(t) = \beta A k^{\alpha}(t), \tag{3.2.1}$$

where $k(t) \equiv K(t)/N(t)$.

We now model behavior of consumers. Consumers obtain income in period t from the interest payment, $r(t)K(t)$, and the wage payment, $w(t)N(t)$

$$Y(t) = r(t)K(t) + w(t)N(t).$$

We call $Y(t)$ the *current income*. The total value of wealth that consumers can sell to purchase goods and to save is equal to $K(t)$. The *gross disposable income* is equal to

$$\hat{Y}(t) = Y(t) + K(t).$$

The gross disposable income is used for saving and consumption. In period t, consumers would distribute the total available budget among savings, $S(t)$, and consumption of goods, $C(t)$. The budget constraint is given by

$$C(t) + S(t) = \hat{Y}(t).$$

We assume that utility level, $U(t)$, that the consumers obtain is dependent on the consumption level of commodity, $C(t)$, and the net saving, $S(t)$, in period t. We use the Cobb-Douglas utility function to describe consumers' preference

$$U(t) = C^{\xi}(t)S^{\lambda}(t), \quad \xi + \lambda = 1, \quad \xi, \lambda > 0,$$

in which ξ and λ are respectively the propensities to consume goods and to own wealth. Households maximize utility subject to the budget constraint. We solve the optimal choice of the consumers as

$$C(t) = \xi\hat{Y}(t), \quad S(t) = \lambda\hat{Y}(t).$$

Amount $K(t+1)$ in period t is equal to the savings made in period t, i.e.

$$K(t+1) = S(t).$$

Since the initial value K_0 and the labor force $N(t)$ are exogenously given, the above equation allow us to calculate recursively all the $K(t)$. Capital $K(1)$ is obtained from K_0 and N_0; $K(2)$ is obtained from $K(1)$ and $N(1)$, ..., etc. Then we directly calculate $r(t)$, $w(t)$, $S(t)$, and, $C(t)$ from the related equations.

We now rewrite the dynamics in per capita terms. With $S(t) = \lambda \hat{Y}(t)$ and

$$\hat{Y}(t) = AK^{\alpha}(t)N^{\beta}(t) + \delta K(t),$$

where $\delta \equiv 1 - \delta_k$, the capital accumulation,

$$K(t+1) = S(t),$$

is given by

$$K(t+1) = \lambda AK^{\alpha}(t)N^{\beta} + \lambda\delta K(t).$$

Dividing the above equation by N yields

$$\frac{K(t+1)}{N(t)} = \lambda Ak^{\alpha}(t) + \lambda\delta k(t).$$

Substituting

$$N(t) = \frac{N(t+1)}{1+n}$$

into the above equation, we have

$$(1+n)k(t+1) = \lambda Ak^{\alpha}(t) + \lambda\delta k(t). \tag{3.2.2}$$

This is a nonlinear difference equation in $k(t)$. We may rewrite this equation as

$$k(t+1) = \Psi(k(t)) \equiv \left(Ak^{\alpha}(t) + \delta k(t)\right)\frac{\lambda}{1+n}.$$

For this difference equation to be in steady state, we have

$$k(t+1) = k(t) = k^{*}.$$

Substituting this condition into equation (3.2.2) yields

$$(1 + n - \lambda\delta)k^* = \lambda A k^{*\alpha}.$$

The equation has a unique positive solution

$$k^* = \left(\frac{\lambda A}{1 + n - \lambda\delta}\right)^{1/\beta}.$$

It is straightforward to show

$$\left.\frac{d\Psi(k(t))}{dk(t)}\right|_{k(t)=k^*} = \alpha + \frac{\beta\delta\lambda}{1+n} < \alpha + \beta = 1.$$

We conclude that the unique equilibrium point is stable. Figure 3.2.1 shows the relation between $k(t+1)$ and $k(t)$, which we express by

$$k(t+1) = \Psi(k(t)).$$

The slope of $\Psi(k(t))$ is infinite at $k(t) = 0$ and diminishes toward a constant

$$\lambda\delta < 1 + n.$$

The function $\Psi(k(t))$ crosses the 45-degree line at the steady-state value, k^*. The capital stock monotonically approaches its unique equilibrium point as time passes. The equilibrium point is stable because the curve $\Psi(k(t))$ is always upward sloping, and it crosses the 45-degree line from above.

Summarizing the discussions in this section, we obtain the following theorem.

Theorem 3.2.1. Given the Cobb-Douglas production and utility functions, and a constant rate of population growth, the capital-labor ratio converges monotonically to a unique positive equilibrium point. The unique equilibrium point is stable. Moreover, the aggregate capital stock and real aggregate output converge to the balanced exponential growth paths proportional to the population growth. □

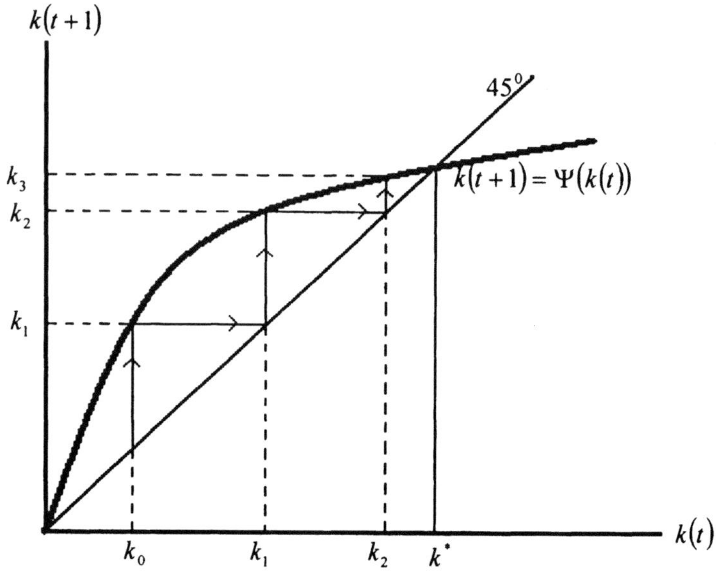

Figure 3.2.1: Dynamics in the OSG model

3.3 The general OSG model

We now generalize the OSG model in discrete time proposed in section 3.2 by replacing the Cobb-Douglas production and utility functions with more general production and utility functions.

As in section 3.2, population grows at rate n; thus

$$N(t) = (1 + n)N(t - 1).$$

Each individual supplies one unit of labor at each time t. Production in period t uses inputs amount $K(t)$ of capital and amount $N(t)$ of labor services. The production function is

$$F(t) = F(K(t), N(t)).$$

The marginal conditions can be expressed by

$$r(t) = f'(k(t)),$$
$$w(t) = f(k(t)) - k(t)f'(k_t), \qquad (3.3.1)$$

where

$$k(t) \equiv \frac{K(t)}{N(t)}, \quad f(t) \equiv \frac{F(t)}{N(t)} = F(k(t), 1).$$

The consumer is to choose his most preferred bundle $(c(t), s(t))$ of consumption and saving under his budget constraint. The utility maximizing problem at any time is defined by

$$\underset{c(t), s(t) \geq 0}{Max} \; U(c(t), s(t))$$
$$\text{s.t.: } c(t) + s(t) = \hat{y}(t) \equiv f(k(t)) + \delta k(t).$$

We require U a C^2 function, and satisfy $U_c > 0$ and $U_s > 0$ for any $(c(t), s(t)) > 0$. Construct the Lagrangian

$$L(c(t), s(t), \bar{\lambda}(t)) = U(c(t), s(t)) + \bar{\lambda}(t)(\hat{y}(t) - c(t) - s(t)).$$

The first-order condition for maximization is

$$U_c = U_s = \bar{\lambda}(t),$$
$$\hat{y}(t) - c(t) - s(t) = 0. \qquad (3.3.2)$$

The bordered Hessian for the problem is

$$|\bar{H}| = \begin{vmatrix} 0 & 1 & 1 \\ 1 & U_{cc} & U_{cs} \\ 1 & U_{sc} & U_{ss} \end{vmatrix} = 2U_{cs} - U_{cc} - U_{ss}.$$

The second-order condition tells that given a stationary value of the first-order condition, a positive $|\bar{H}|$ is sufficient to establish it as a relative maximum of U. Taking derivatives of equations (3.3.2) with respect to $\hat{y}_t$ yields

$$U_{cc}\frac{dc}{d\hat{y}} + U_{cs}\frac{ds}{d\hat{y}} = U_{sc}\frac{dc}{d\hat{y}} + U_{ss}\frac{ds}{d\hat{y}},$$

$$1 = \frac{dc}{d\hat{y}} + \frac{ds}{d\hat{y}}.$$

We solve these functions

$$\frac{ds}{d\hat{y}} = \frac{U_{sc} - U_{cc}}{2U_{cs} - U_{ss} - U_{cc}},$$

$$\frac{dc}{d\hat{y}} = \frac{U_{sc} - U_{ss}}{2U_{cs} - U_{ss} - U_{cc}}.$$

We see that

$$0 < \frac{ds}{d\hat{y}} < 1, \quad 0 < \frac{dc}{d\hat{y}} < 1,$$

in the case of $U_{sc} \geq 0$ under the second-order condition of maximization. We denote an optimal solution as function of the disposable income

$$(c(t), s(t)) = (c(\hat{y}(t)), s(\hat{y}(t))).$$

The capital accumulation equation is given by

$$K(t + 1) = s(\hat{y}(t))N(t).$$

Dividing the two sides of this equation by $N(t + 1)$ yields

$$k(t + 1) = \frac{s(\hat{y}(t))}{1 + n}.$$

This mapping controls the motion of the system. A stationary state for the growth progress is a capital-labor ratio, k^*, that satisfies

$$k = \frac{s(\hat{y})}{1 + n}. \tag{3.3.3}$$

Define

$$\Phi(k) \equiv \frac{s(\hat{y})}{(1+n)k} - 1, \quad k \ge 0. \tag{3.3.4}$$

When k is approaching zero, $\hat{y}$ $(= f(k) + \delta k)$ is also approaching zero, and hence $s(\hat{y})$ is coming near zero. As $0 < s'(\hat{y}) < 1$ and $f'(k) \to \infty$ as $k \to 0$

$$\lim_{k \to 0} \frac{s(\hat{y})}{(1+n)k} = \frac{s'(0)(f'(0) + \delta)}{(1+n)} > 1. \tag{3.3.5}$$

When k is approaching positive infinity, $\hat{y}$ is coming to positive infinity. As

$$0 < s'(\hat{y}) < 1, \quad f'(k) \to 0 \text{ as } k \to +\infty,$$

we have

$$\lim_{k \to +\infty} \frac{s(\hat{y})}{(1+n)k} = \frac{s'(+\infty)(f'(+\infty) + \delta)}{(1+n)} < 1. \tag{3.3.6}$$

Taking derivatives of equation (3.3.4) with respect to k yields

$$\frac{d\Phi}{dk} = \left[\frac{s'(\hat{y})k(f'(k) + \delta)}{s(\hat{y})} - 1\right]\frac{s(\hat{y})}{(1+n)k^2}.$$

We now show that $d\Phi/dk < 0$ for $k > 0$. To prove this, we use $ds/d\hat{y} < 1$ and the inequality

$$f'(k) < \frac{f(k)}{k},$$

which also guarantees $w > 0$. By equation (3.3.3) and the definition of $\hat{y}$, we have

$$\frac{s'(\hat{y})k(f'(k) + \delta)}{(1+n)s(\hat{y})} < \frac{s'(\hat{y})(f(k) + \delta k)}{(1+n)s(\hat{y})} = \frac{s'(\hat{y})\hat{y}}{(1+n)s(\hat{y})} < 1,$$

where we use $s'(\hat{y}) < 1$ and $\hat{y} / s(\hat{y}) \le 1$ to guarantee the right inequality. We thus conclude $d\Phi / dk < 0$ for $k > 0$. The equation,

$$\Phi(k) = 0,$$

for $k > 0$ has a unique solution because of equations (3.3.5), (3.3.6), and $d\Phi / dk < 0$. We now demonstrate that the unique stationary state is stable.

For the steady state to be stable, the following conditions must prevail

$$-1 < \frac{s'(\hat{y}^*)\left(f'(k^*) + \delta\right)}{1 + n} < 1.$$

As the steady-state values, k^* and $\hat{y}^*$ are positive, $s'(\hat{y}^*)$ and $f'(k^*)$ are positive. The left inequality always holds. To prove the right inequality, we use $ds / d\hat{y} < 1$ and the inequality

$$f'(k^*) < \frac{f(k^*)}{k^*}$$

(which also guarantees $w_t > 0$). By equation (3.3.3) and the definition of $\hat{y}$, we have

$$\frac{s'(\hat{y}^*)\left(f'(k^*) + \delta\right)}{1 + n} < \frac{s'(\hat{y}^*)\left(f(k^*)/k^* + \delta\right)}{1 + n} = \frac{s'(\hat{y}^*)\hat{y}^* / k^*}{1 + n} = \frac{s'(\hat{y}^*)\hat{y}^*}{s(\hat{y}^*)}.$$

Since $s'(\hat{y}) < 1$ and $\hat{y} / s(\hat{y}) \le 1$, we see that the right inequality of the inequalities is satisfied. We thus conclude that the conclusions for the Cobb-Douglas production and utility functions are also held for the general production and utility functions. Summarizing the discussions, we obtain the following theorem.

Theorem 3.3.1. Given a production function that is 'neoclassical' and a utility function that is a C^2 function, and satisfies $U_c > 0$ and $U_s > 0$ for any $(c_t, s_t) > 0$. Let the bordered Hessian be positive for any nonnegative. Then the capital-labor ratio converges monotonically to a unique positive steady state. The unique stationary state is stable. □

3.4 The overlapping-generations (OLG) model

This section introduces the model of finite-horizon households.[2] It is common to assume, in the OLG modeling framework, that each person lives for only two periods. People work in the first period and retire in the second. If one thinks of reality, one period perhaps last over 30 years. As we neglect any possibility of transfers from government or from members of other generations, people consume in both periods; they pay for consumption in the second period by saving in the first period. The cohort that is born at time t is referred to as generation t. Members of this generation are young in period t and old in period $t+1$. At each point in time, members of only two generations are alive. Each person maximizes lifetime utility, which depends on consumption in the two periods of life. It is assumed that people are born with no assets and don't care about events after their death; they are not altruistic toward their children, and therefore, do not provide bequests or other transfers to members of the next generations. Their lifetime utility is specified as

$$U(t) = \frac{c_1^{1-\theta}(t) - 1}{1-\theta} + \left(\frac{1}{1+\rho}\right)\frac{c_2^{1-\theta}(t+1) - 1}{1-\theta}, \tag{3.4.1}$$

where $\theta > 0$, $\rho > 0$, and $c_i(t)$ is consumption of generation i, $i = 1, 2$. Each individual supplies one unit of labor inelastically while young and receives the wage income $w(t)$; he does not work when old. Let s_t stand for the amount saved in period t. The budget constraint for period t is

$$c_1(t) + s(t) = w(t). \tag{3.4.2}$$

Let $r(t+1)$ denote the interest rate on one-period loans between periods t and $t+1$. In period $t+1$, the individual consumes the savings plus the accrued interest

$$c_2(t+1) = (1 + r(t+1))s(t). \tag{3.4.3}$$

---

[2] The model was initially examined by Samuelson (1958) and Diamond (1965). The model here is based on Blanchard (1985) and Barro and Sala-i-Martin (2004) in section 3.8. Some extensions of the model are referred to de la Croix and Michel (2002).

For each individual, $w(t)$ and $r(t+1)$ are exogenous; he chooses $c_1(t)$ and $s(t)$ (and c_{2t+1}) subject to equations (3.4.2) and (3.4.3). Substitute equations (3.4.2) and (3.4.3) into the utility to delete $c_1(t)$ and $c_2(t+1)$

$$U(t) = \frac{(w(t) - s(t))^{1-\theta} - 1}{1-\theta} + \left(\frac{1}{1+\rho}\right)\frac{(1 + r(t+1))^{1-\theta} s^{1-\theta}(t) - 1}{1-\theta}.$$

The first-order condition with respect to $s(t)$ yields

$$s^{-\theta}(t)(1 + r(t+1))^{1-\theta} = (1+\rho)(w(t) - s(t))^{-\theta}, \tag{3.4.4}$$

which also implies, under equations (3.4.2) and (3.4.3)

$$\frac{c_2(t+1)}{c_1(t)} = \left(\frac{1 + r(t+1)}{1+\rho}\right)^{1/\theta}. \tag{3.4.5}$$

Solve equation (3.4.4)

$$s(t) = \frac{w(t)}{\varphi(t+1)}, \tag{3.4.6}$$

where

$$\varphi(t+1) \equiv 1 + \frac{(1+\rho)^{1/\theta}}{(1 + r(t+1))^{(1-\theta)/\theta}} > 1.$$

We see

$$s_r \equiv \frac{\partial s(t)}{\partial r(t+1)} = \left(\frac{1-\theta}{\theta}\right)\left(\frac{1+\rho}{1 + r(t+1)}\right)\frac{s(t)}{\varphi(t+1)},$$

$$s_w \equiv \frac{\partial s(t)}{\partial w(t)} = \frac{1}{\varphi(t+1)}, \quad 0 < s_w < 1. \tag{3.4.7}$$

We have

$$s_r \begin{cases} > 0 & if \quad \theta < 1, \\ = 0 & if \quad \theta = 1, \\ < 0 & if \quad \theta > 1. \end{cases}$$

Firms' behavior is the same as in the OSG model in the previous section. They have the neoclassical production function

$$y(t) = f(k(t)),$$

where

$$y(t) \equiv \frac{Y(t)}{N(t)}, \quad k(t) \equiv \frac{K(t)}{N(t)},$$

where $N(t)$ is the number of young people, $Y(t)$ is the total output, $K(t)$ is the total capital. The marginal conditions are

$$\begin{aligned} r(t) &= f'(k(t)) - \delta_k, \\ w(t) &= f(k(t)) - k(t)f'(k(t)), \end{aligned} \tag{3.4.8}$$

where δ_k is the depreciation rate of capital.

In the closed economy, households' assets equal the capital stock. Aggregate net investment equals total income minus total consumption

$$K(t+1) - K(t) = w(t)N(t) + r(t)K(t) - c_1(t)N(t) - c_2(t)N(t-1). \tag{3.4.9}$$

Substituting equation (3.4.8) into equation (3.4.9), we get

$$K(t+1) - K(t) = F(K(t), N(t)) - C(t), \tag{3.4.10}$$

where

$$C(t) = c_1(t)N(t) + c_2(t)N(t-1).$$

From equations (3.4.9), (3.4.2), and (3.4.3), we obtain

$$K(t+1) = s(t)N(t) + (1 + r(t))(K(t) - s_{t-1}(t)N_{t-1}(t)).$$

Assume that the economy starts with the condition

$$c_2(1)N(0) = (1 + r(1))K(1),$$

which is equivalent to $K(2) = s(1)N(1)$ with equations (3.4.9) and (3.4.2). Hence, the above equation becomes

$$K(t+1) = s(t)N(t),$$

which means that the savings of the young equal the next period's capital stock. We can write this equation in per capita terms

$$k(t+1) = \frac{s(t)}{1+n}.$$

Substituting $s(t)$ in equation (3.4.6) into the above equation yields

$$k(t+1)\left[1 + \frac{(1+\rho)^{1/\theta}}{(1 + r(t+1))^{(1-\theta)/\theta}}\right] = \frac{w(t)}{(1+n)}, \tag{3.4.11}$$

in which we use the definition of $\varphi(t+1)$. In the case that the utility function is logarithmic ($\theta = 1$), it can be shown that equation (3.4.11) becomes

$$k(t+1) = \frac{1}{(1+n)(2+\rho)}\left[f(k(t)) - k(t)f'(k(t))\right]. \tag{3.4.12}$$

The analysis of equation (3.4.12) can be similarly conducted as for equation (3.2.2). The system has a unique equilibrium point and it is stable (check!).

Exercise 3.4.1

1 Discuss the existence of equilibrium point and find stability conditions when $\theta \neq 1$.

3.5 Persistence of inequality and development

This section introduces a growth model proposed by Moav to demonstrate persistence of inequality.[3] In this model, the evolution of income within each dynasty in society is governed by a dynamical system that generates a poverty trap equilibrium point along with a high-income equilibrium point. Poor dynasties, those with income at the threshold level, converge to a low income steady state, whereas dynasties with income above the threshold converge to a high income steady state.

Consider a small open overlapping-generations economy that operates in a one-good world. In each period the economy produces a single homogeneous product that can be used either for consumption or for investment. Production occurs within a period according to a concave, constant-returns-to-scale technology. The output produced at time t uses capital $K(t)$, and human capital efficiency units, $H(t)$, as follows

$$Y(t) = F(K(t), H(t)),$$

where investment in physical and human capital is made one period in advance. Assume that the world capital rate of return remains constant, denoted by R. Unrestricted international capital movement yields

$$F_K(K(t), H(t)) = R. \tag{3.5.1}$$

From the properties of the production function, we know that the wage per unit of human capital, w, is uniquely determined given the rate of return to capital, R, and is therefore, constant over time.[4] Like in the previous section, individuals live in two periods. Individuals, within as well as across generations, are identical in their preferences and their technology of human capital formation. They may differ in their initial wealth, inherited from their parents. Individuals cannot borrow in order to finance investment in human capital. Individual born in period t allocates her second life period income, $I_i(t+1)$, between

[3] This section is based on Moav (2002). See also other approaches by, for instance, Galor and Zeira (1993), Abe (1995), Benabou (1996), Durlauf (1996), Piketty (1997), to issues related to development and distribution. Aghion and Bolton (1997), Aghion and Howitt (1998) and Matsuyama (2000) also provided some interesting insights into evolution of inequality due to "institutional frictions".

[4] As $R = f'(k)$, where $k \equiv K/H$, $f \equiv F(k, 1)$, we have $k = g(R)$. Hence, k is constant. As $w = f - Rk$, w is constant.

household consumption, $c_i(t+1)$, and a bequest to the offspring, $b_i(t+1)$. The budget constraint is given by

$$I_i(t+1) = c_i(t+1) + b_i(t+1). \tag{3.5.2}$$

Preferences are defined by the utility function

$$U_i(t) = (1-\beta)\log c_i(t+1) + \beta\log\{\bar{\theta} + b_i(t+1)\},$$

where $\beta \in (0,1)$ and $\bar{\theta} > 0$. The optimal transfer of individual i born in period t is given by

$$b_i(t+1) = b(I_i(t+1)) = \begin{cases} 0 & \text{if } I_i(t+1) \le \theta, \\ \beta(I_i(t+1) - \theta) & \text{if } I_i(t+1) > \theta, \end{cases} \tag{3.5.3}$$

where $\theta \equiv \bar{\theta}(1-\beta)/\beta$.

We now discuss formation of human capital. In the first period of their lives in period t, individuals devote their entire time for the acquisition of human capital. Individuals acquire one efficiency unit of labor-basic skills. The level of human capital of an individual i, $h_i(t+1)$, is an increasing concave function of real resources invested in education, $e_i(t)$

$$h_i(t+1) = h(e_i(t)) = \begin{cases} 1 + \gamma e_i(t) & \text{if } e_i(t) < \bar{e}, \\ 1 + \gamma e_i & \text{if } e_i(t) \ge \bar{e}. \end{cases} \tag{3.5.4}$$

It is assumed that the marginal return to human capital, for $e_i(t) < \bar{e}$, is larger than the marginal return to physical capital, $w\gamma > R$. We also require that the income level below which individuals choose a zero bequest, θ, is larger than the wage rate, w, $\theta > w$. We also assume that the return to physical capital, R, is sufficiently low, i.e., $\beta R < 1$. In summary, we assume

$$\theta > w > \frac{R}{\gamma}, \quad \beta R < 1. \tag{3.5.5}$$

Under the above constraint, the second life period income, $I_i(t+1)$, is uniquely determined by first life period bequest $b_i(t)$

$$I_i(t+1) = I(b_i(t)) = \begin{cases} w(1+\gamma b_i(t)) & \text{if } b_i(t) \leq \bar{e}, \\ w(1+\gamma b_i(t)) + R(b_i(t) - \bar{e}) & \text{if } b_i(t) \geq \bar{e}. \end{cases} \tag{3.5.6}$$

From equations (3.5.3) and (3.5.6), the evolution of income within a dynasty is uniquely determined by

$$I_i(t+1) = \varphi(I_i(t)) = \begin{cases} w & \text{if } \beta(I_i(t) - \theta) < 0, \\ w(1+\gamma\beta(I_i(t) - \theta)) & \text{if } \beta(I_i(t) - \theta) \in [0, \bar{e}], \\ w(1+\gamma\bar{e}) + R(\beta(I_i(t) - \theta) - \bar{e}) & \text{if } \beta(I_i(t) - \theta) > \bar{e}, \end{cases} \tag{3.5.7}$$

where $I_i(0)$ is given. We have $I_i(t) \geq w$ for all t. Under equation (3.5.5), from equation (3.5.7) we conclude that there exists a low income, locally stable, poverty trap steady state $\bar{I} = w$ because

$$I_i(t+1) = \varphi(I_i(t)) = w, \quad \text{for all } I_i(t) \in [w, \theta].$$

It is assumed that the return to human capital, γw, and its potential magnitude, $\bar{e}$, are sufficiently large, such that an individual who receives a bequest $b_i(t) = \bar{e}$ will transfer her offspring a higher bequest

$$b_i(t+1) > b_i(t) = \bar{e}.$$

This assumption is expressed as

$$\beta[w(1+\gamma\bar{e}) - \theta] > \bar{e}. \tag{3.5.8}$$

This assumption assures a range of income, above the poverty trap range of income, in which

$$I_i(t+1) > I_i(t).$$

From equation (3.5.5), we see that there is a higher income steady state $\tilde{I}$ (higher than poverty trap) which is determined by

$$\tilde{I} = \frac{w(\gamma\bar{e} + 1) - R(\beta\theta + \bar{e})}{1 - \beta R}.$$

This steady state is locally stable.

In the dynamic economy, there exists an income threshold, $\hat{I}$, $\bar{I} < \hat{I} < \tilde{I}$. From equation (3.5.7), the threshold is given by

$$\hat{I} = \frac{\gamma\beta\theta - 1}{\gamma\beta w - 1} w. \tag{3.5.9}$$

This steady is locally unstable. It can be seen that dynasties with income below the threshold ($I_0(t) < \hat{I}$) converge to the poverty trap income level, and dynasties with income above the threshold ($I_0(t) > \hat{I}$) converge to the high income steady state.

In summary, the dynamical system, $I_i(t + 1) = \phi(I_i(t))$, depicted in figure 3.5.1, generates three steady states, a poverty trap, a high income steady state, and a threshold income. Dynasties with initial income below the threshold level converge to a low income steady state; dynasties with income above the threshold level converge to a high income steady state.

3.6 Growth with creative destruction

The year 1883 is special for the history of the world economy and the history of economic analysis. The year saw the death of Karl Marx (1918-1883), the birth of John Maynard Keynes (1883-1946), and the birth of Josephy Alois Schumpeter (1883-1950). In the *Theory of Economic Development* published in 1911, Schumpeter argued that development should be understood as only such changes in economic life as are not forced upon it from without but arise by its own initiative, from within. Schumpeter held that successful carrying out of new combinations of productive services is the essence of this process. Schumpeter's ideas about development and creative destruction have recently been modeled. This section represents such a model by Aghion and Howitt.[5]

[5] Aghion and Howitt (1992).

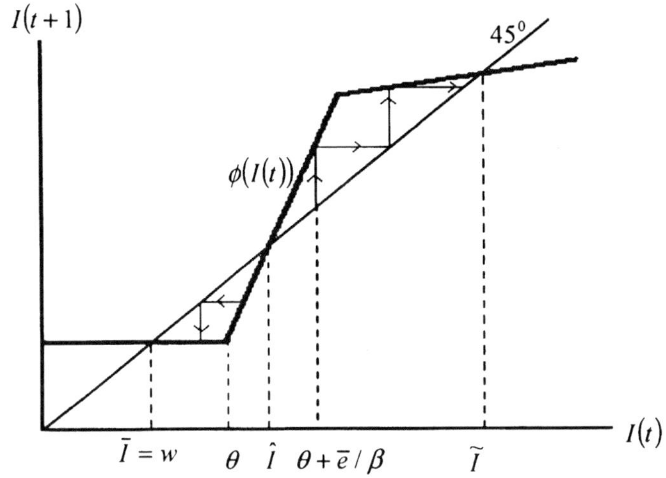

Figure 3.5.1: Multiple steady states and persistent income inequality

Each innovation consists of a new intermediate good that can be used to produce final output more efficiently than before. Research firms are motivated by the process by the prospect of monopoly rents that can be captured when a successful innovation is patented. But those rents in turn will be destroyed by the next innovation, which will render obsolete the existing intermediate good. Equilibrium is determined by a forward difference equation, according to which the amount of research in any period depends upon the expected amount of research next period. Here, we represent a simplified version of the model by Solow.[6] A period is the time between two successive innovations. The length of each period is random because of the stochastic nature of the innovation process, but the relationship between the amount of research in two successive periods can be modeled as deterministic. The amount of research this period depends negatively upon the expected amount next period, through two effects. The first is that of creative destruction. The payoff from research in this period is the prospect of monopoly rents in the next period. Those rents will last only until the next innovation occurs. The expectation of more research next period will discourage research this period. The second effect is that of a general equilibrium in the labor market. Workers can be used either in research or in manufacturing. To maintain labor market equilibrium, the expectation of more research next period must correspond to an expectation of higher demand for labor in research next period, which implies the expectation of a higher wage

[6] Solow (2000).

rate. Higher wages next period will reduce the monopoly rents that can be gained by exclusive knowledge of how to produce the best products. Thus the expectation of more research will discourage research this period.

The model assumes a constant employment. There are three classes of tradable objects: labor, a consumption good, and an intermediate good. There is a continuum of infinitely-lived individuals, with identical intertemporally additive preferences defined over lifetime consumption, and the constant rate of time preference. The marginal utility of consumption is equal to the rate of interest. There are three categories of labor: unskilled labor, which can be used only in producing the consumption good; (skilled) labor, which can be used either in research or in the intermediate sector; and specialized labor, which can be used only in research. Each individual is endowed with a unit flow of labor. Only one final good is produced by the fixed quantity of unskilled labor and skilled labor x. The production function is

$$F(t) = Af(x),$$

where we omit expressing unskilled labor and $f(x)$ is increasing $(f'(x) > 0)$ and concave $(f''(x) < 0)$. The variable A is a technological variable. Final good is used as *numeraire*.

Some skilled labor is devoted to R&D. When successful, the innovation is a new intermediate good that allows a higher value of A and thus makes the old intermediate good obsolete. Let t stand for the t th innovation (not time). For convenience of description, we consider that each successful innovation increases the final output producible with any x by a multiplicative factor γ, i.e.

$$\frac{A(t+1)}{A(t)} = \gamma.$$

Suppose n units of labor are devoted to R&D and innovations arrive according to a Poisson process with arrival rate λn. It should be noted that in the original model the specialized labor affects the arrival rate. Since the number of specialized labor is prefixed, we omit mentioning this type of labor. The probability of an innovation in a given short unit of time equals λn, and the probability of no innovation is equal to $1 - \lambda n$, and the probability of two or more innovations is equal to zero. The assumption of the Poisson process says that the probability of making an innovation of given size depends only on n, independent of past history of innovation. In fact, innovation is hard to model even in probability sense.

The innovating firm acquires a monopoly on the final production that is useful until the next innovation. Thus the t th innovation brings a negative externality through killing the rents of the firm that produced the $(t-1)$ st innovation and a positive externality through making the $(t+1)$ st innovation possible. A successful innovator is a monopoly of the intermediate good and is faced with a demand curve from the final-goods industry

$$Af'(x(t)) = P(t),$$

where $P(t)$ is the price of the intermediate good.

We introduce $V(t)$ and $\Pi(t)$ to respectively stand for the expected discounted rents associated with the t th successful innovation and the constant flow of rent expected by the t th innovator during the profitable life of the innovation. Let ρ stand for the discount rate of the rent expected by the t th innovator. Then the Fisher equation tells that the interest on the value of innovation equals the current income plus the expected capital gain, which equals

$$\lambda n(t)(-V(t)) + (1 - \lambda n(t)) \cdot 0.$$

That is

$$\rho V(t) = \Pi(t) - \lambda n(t) V(t).$$

The above equation is solved as

$$V(t) = \frac{\Pi(t)}{\rho + \lambda n(t)}. \tag{3.6.1}$$

The equation says that a large value of $n(t)$ reduces $V(t)$. In other words, research, like capital investment, is discouraged by the prospect of future R&D. Free entry and risk neutrality in R&D guarantees that entry will occur until the cost of conducting R&D is equal to the expected value of the innovation

$$w(t)n(t) = \lambda n(t)V(t+1) + (1 - \lambda n(t)) \cdot 0 \Rightarrow w(t) = \lambda V(t+1). \tag{3.6.2}$$

Labor market is cleared for every t

$$n(t) + x(t) = N, \tag{3.6.3}$$

where N is the constant volume of employment. Solow holds that equation (3.6.3) contains a major limitation of this model: "one of the true risks of R&D is that economic conditions should be cyclically weak during the effective life of an innovation, so that it turns out to be unprofitable because sales of final product are poor".

The intermediate good is produced using skilled labor alone. The production function is specified in such a way that the intermediate product is equal to the flow of skilled labor used in the intermediate sector. With the one-to-one technology for producing intermediate good, the monopolist maximizes

$$P(t)x(t) - w(t)x(t) = A(t)f'(x(t))x(t) - w(t)x(t). \tag{3.6.4}$$

Provided that marginal revenue falls with $x(t)$, the optimal $x(t)$ is a decreasing function of $w(t)/A(t)$, and the best achievable value of $\Pi(t)/A(t)$ falls as $w(t)/A(t)$ rises. Denote this function by

$$x(t) = \bar{\phi}\left(\frac{w(t)}{A(t)}\right).$$

From equation (3.6.3), $n(t)$ is an increasing function of $w(t)/A(t)$. We have

$$n(t) = N - x(t) = N - \bar{\phi}\left(\frac{w(t)}{A(t)}\right).$$

Solving the above equation with $w(t)/A(t)$ as the variable yields

$$\frac{w(t)}{A(t)} = \phi(n(t)), \tag{3.6.5}$$

where $\phi(n(t))$ rises in $n(t)$. Now equations (3.6.1) and (3.6.2) imply

$$w(t) = \lambda V(t+1) = \frac{\lambda \Pi(t+1)}{\rho + \lambda n(t+1)}.$$

Inserting equation (3.6.5) and $A(t+1)=\gamma A(t)$ into the above equation yields

$$\frac{w(t)}{A(t)}=\frac{\gamma((\Pi(t+1)/A(t+1))}{\rho+\lambda n(t+1)}. \tag{3.6.6}$$

As $\Pi(t+1)/A(t+1)$ is a decreasing function of $w(t+1)/A(t+1)$ and $w(t+1)/A(t+1)$ is an increasing function of $n(t+1)$, $\Pi(t+1)/A(t+1)$ falls as $n(t+1)$ rises. Consequently, the right-hand side of equation (3.6.6) is a decreasing function of $n(t+1)$, denoted by $\psi(n(t+1))$. Equations (3.6.5) and (3.6.6) imply

$$\phi(n(t))=\psi(n(t+1)).$$

This dynamic equation closes the model. As $\psi(n(t+1))$ falls in the variable, $n(t+1)$ and $\phi(n(t))$ rises in the variable $n(t)$, we rewrite the above map in the following form

$$n(t+1)=h(n(t)),\quad h'<0. \tag{3.6.7}$$

An equilibrium point is a solution of

$$n^*=h(n^*).$$

In general, as shown in figure 3.6.1, there will be a unique steady state. We know that $n(t)$ tends to n^* if $|h'(n)|<1$ for all n and will converge locally if the following inequality holds

$$|h'(n^*)|<1.$$

Once we determine the value of $n(t)$, we determine all the other variables in the system.

We may conduct usual comparative statics analysis with respect different parameters in the system.

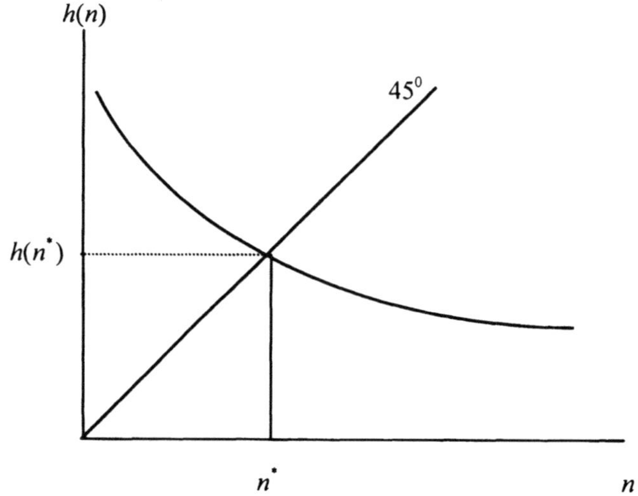

Figure3.6.1: A steady state in the Aghion-Howitt model

3.7 Economic evolution with human capital

This section is concerned with interactions between human capital accumulation, economic growth, and inequality.[7] In the economy, a competitive final goods sector produces one homogenous output using human capital and a variety of intermediate inputs with the following production function

$$Y(t) = \sum_{i=1}^{D(t)} n^{1-\delta}(i, t) H^{\delta}(t),$$

where δ ($0 < \delta < 1$) is a parameter, $D(t)$ is the number of different intermediate goods used in production, $n(i, t)$ is the quantity of the ith intermediate good employed, and H is the skilled adjusted stock of labor in the economy. The intermediate goods sector is competitive, and each intermediate good requires one unit of capital, k, to transform a new technology into a new intermediate good. The symmetric use of inputs implies

$$n(i, t) = n(t).$$

[7] The model of this section is based on Either and García-Peñalosa (2001).

Hence, the production function becomes

$$Y(t) = D(t)n^{1-\delta}(t)H^{\delta}(t).$$

There are two, skilled and unskilled, types of workers. The skilled-adjusted stock of labor is given by

$$H(t) = \left(\frac{D(t-1)}{D(t)}U^{-\alpha} + S_p^{-\alpha}(t)\right)^{1/\alpha}, \quad 0 > \alpha > -1, \tag{3.7.1}$$

where $S_p(t)$ denotes skilled labor employed in production and $U(t)$ unskilled labor. The elasticity of substitution, $1/(1+\alpha)$, falls in the interval $(1, \infty)$. Here, the rate of technological change is measured by $D(t)/D(t-1)$. The above information takes account of the hypothesis that the greater the speed of technological change, the relatively more productive skilled labor becomes, compared to unskilled labor. Let $w(t)$, $w_s(t)$, and $w_u(t)$ stand for, respectively, the relative wage of skilled to unskilled labor, the wage of the skilled and unskilled labor. Perfect competition in labor markets yields

$$w(t) \equiv \frac{w_s(t)}{w_u(t)} = \left(\frac{U(t)}{S_p(t)}\right)^{1+\alpha} \frac{D(t)}{D(t-1)}. \tag{3.7.2}$$

Agents live two periods; work when young and consume only when old. Let $S(t)$ be the total number of skilled labor. We normalize the total population of each generation

$$S(t) + U(t) = 1.$$

Skilled workers can be employed either in production, $S_p(t)$, or in research $S_r(t)$. When net technologies are introduced, agents must learn to work with these technologies to become skilled labor. Hence, at the start of their working lives, agents have to decide whether to invest in education or to remain unskilled. We assume that agents differ in their abilities to learn, μ, and that their abilities are distributed uniformly, $\mu \in [0, 1]$. Education is instantaneous and the cost of education is equal to $c(t)w_u(t)/\mu$. It is further assumed that the direct cost of education $c(t)w_u(t)$ is a decreasing function of the number of agents being educated, i.e.

$$c(t) = c(S(t)), \ c' < 0.$$

We specify

$$c(t) = \rho S^{-\sigma}(t).$$

Assume that an income tax τ is imposed only on skilled wages. The income of a skilled worker born at t is

$$Y_s(t) = (1 - \tau)w_s(t) - \frac{\rho S^{-\sigma} w_u(t)}{\mu}. \tag{3.7.3}$$

Agents choose to invest in skills if the income in equation (3.7.3) exceeds that of remaining unskilled

$$Y_u(t) = w_u(t).$$

Equality between these two expressions gives the level of ability of the agent that is indifferent between investing in education and working as unskilled

$$\tilde{\mu}(t) = \frac{\rho S^{-\sigma}(t)}{(1 - \tau)w(t) - 1}.$$

Substituting the uniform ability distribution, i.e.

$$S(t) = 1 - \tilde{\mu}(t)$$

into the above condition yields the following inverse labor supply equation

$$w(t) = \frac{1}{1 - \tau}\left(1 + \frac{\rho S^{-\sigma}(t)}{1 - S(t)}\right). \tag{3.7.4}$$

From equation (3.7.4), we conclude that externalities in education, σ, generate a U -shaped relationship between the relative wage and the supply of skilled labor. The greater the externality, the more prolonged the initial decline of the relative wage. Increases in the direct cost of education, ρ, or in the differential tax on skilled labor, τ, require higher wage for each level of skilled labor supply.

We take account of two ways of technological change, learning-by-doing and costly investment in research and development (R&D). Learning-by-doing takes place as skilled workers serendipitously discover new types of intermediate goods during production process. This source of technological change is given by

$$\frac{D(t)}{D(t-1)} = 1 + \gamma S(t-1).$$

Research which is undertaken only by the government is financed by a public entity that raises revenues through tax collection, τ. When $S_r(t)$ researchers are employed, the economy produces technological blueprints according to

$$\frac{D(t)}{D(t-1)} = (1 + \beta S(t-1))\beta S_r(t-1).$$

As skilled workers are fully employed, we have

$$S_p(t) + S_r(t) = S(t).$$

The budget constraint of the government is

$$\tau w_s(t) S(t) = w_s(t) S_r(t).$$

The above two conditions yield

$$S_r(t) = \tau S(t),$$
$$S_p(t) = (1 - \tau) S(t).$$

We assume that technological change takes place either via LBD or R&D, but not both in one period. It is not difficult to show that utilizing the equations above, we can express the inverse relative labor demands under the "LBD regime" and "R&D regime" as follows (check!)

$$w(t)\big|_{LBD} = \left(1 + \gamma S(t-1)\right)\left(\frac{1 - S(t)}{S(t)}\right)^{1+\alpha}, \tag{3.7.5}$$

$$w(t)\big|_{R\&D} = \tilde{\tau}\left(1 + \gamma S(t-1)\right)S(t-1)\left(\frac{1 - S(t)}{S(t)}\right)^{1+\alpha}, \tag{3.7.6}$$

where

$$\tilde{\tau} = \frac{\beta\tau}{(1-\tau)^{1+\alpha}}.$$

The general equilibrium in the goods and factor markets is attained by equating the labor supply, expressed in equation (3.7.4), to the labor demand conditions (3.7.5) and (3.7.6). That is

$$\Phi(S(t)) = \begin{cases} 1 + \gamma S(t-1), & \text{for LBD,} \\ \tilde{\tau}\left(1 + \gamma S(t-1)\right)S(t-1), & \text{for } R\&D, \end{cases} \tag{3.7.7}$$

where

$$\Phi(S(t)) \equiv \frac{1}{1-\tau}\left(1 + \frac{\rho S^{-\sigma}(t)}{1 - S(t)}\right)\left(\frac{S(t)}{1 - S(t)}\right)^{1+\alpha}.$$

The evolution of the economy is governed by equation (3.7.7). It is straightforward to see that for any given level of $S(t)$, all the other variables are uniquely determined. In steady state

$$S(t-1) = S(t) = S^*,$$

the steady growth rate is given by

$$\frac{Y(t) - Y(t-1)}{Y(t)} = \frac{D(t) - D(t-1)}{D(t)} = \begin{cases} \gamma S^*, & \text{for LBD,} \\ \beta\tau\left(1 + \gamma S^*\right)S^* - 1, & \text{for } R\&D. \end{cases}$$

We now analyze behavior of equation (3.7.7). To simplify the analysis, in the remainder of this section we impose

$$1 + \alpha > \sigma,$$

which means that the education externality is not strong given the elasticity of substitution between the two types of labor. Under this requirement, we have

$$\frac{d\Phi}{dS(t)} > 0$$

for $S(t) \in [0, 1]$. This also leads to $\partial S(t)/dS(t-1) > 0$. It can be shown that $\Phi(S(t))$ is first concave and then convex, $\Phi(0) = 0$ and $\Phi(1) = \infty$. Figure 3.7.1 depicts the functions of equation (3.7.7). We see that there are three possible equilibrium points: a low-growth trap in the LBD phase and a pair of equilibrium points in the intermediate and advanced development phase. If these equilibrium points exist, it can be shown that the LBD equilibrium will be stable, while in the R&D phase, the first equilibrium point will be unstable and the second one stable. Checking these properties is left to the reader as exercises. Economic interpretations are referred to the original article.

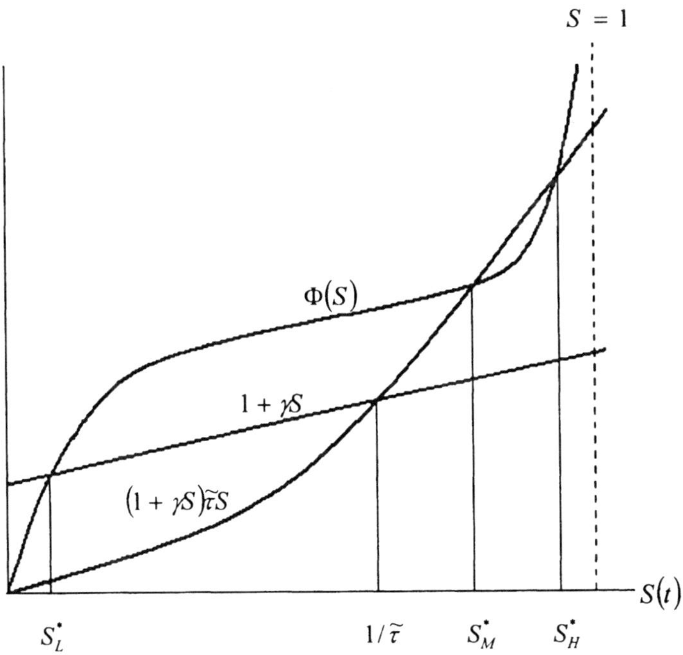

Figure 3.7.1: Three possible equilibrium points

3.8 Urbanization with human capital externalities

This section introduces an urban dynamic model to highlight how the trade-off between optimal and equilibrium city sizes behaves when human capital externalities are introduced into urban dynamics.[8] The economy consists of N *ex ante* identical workers. All workers are members of single-worker firms who live for one period, supply labor inelastically, and consume what they produce. In the beginning of the period workers make a joint location and human capital investment decision. In the middle of the period they work, and at the end of the period they consume. Workers may choose to reside either in urban areas or in rural areas. Workers in rural location have no incentive to invest in human capital. Workers in the rural areas receive the earning levels

$$I_r(t) = G_r A(t),$$

where $A(t)$ describes the economy's level of technology at time t and G_r is an exogenous productivity factor, representing, for instance, the quality of rural infrastructure.

Gross earnings received by a worker in urban location are

$$w(t) = GA(t)h^{\alpha}(t), \quad \alpha < 1,$$

where G is the quality of public infrastructure in urban areas and $h(t)$ is the amount of human capital possessed by the urban worker, which can be obtained through a per unit level of effort μ. Assume that the utility levels of workers in the urban location are negatively affected by the size of the local population, due to urban external diseconomies. The urban residents' earnings net of educational effort and congestion are

$$I(t) = GA(t)h^{\alpha}(t) - \mu h(t) - b(z(t)N)^{1/\delta}, \quad \delta > 1,$$

where z $(0 \le z \le 1)$ is the proportion of the population that locates in urban areas, and b is a constant. Urban workers maximize $I(t)$ by choosing $h(t)$. The optimal solution is

[8] This section is based on Bertinelli and Black (2004).

$$h^{*}(t)=\left(\frac{\alpha G}{\mu}\right)^{\beta}A^{\beta}(t),$$

where $\beta \equiv 1/(1-\alpha)$. With costless migration, the proportion of the population that chooses to locate in the urban areas, $\tilde{z}(t)$, adjusts until the point where $I(t)=I_r(t)$, i.e.

$$\tilde{z}(t)=\left[\frac{GA(t)h^{*}(t)-\mu h^{*}(t)-G_r A(t)}{bN^{1/\delta}}\right]^{\delta}, \tag{3.8.1}$$

or until $z(t)=1$ of full urbanization. It is straightforward to prove that the following city size $\delta_0\tilde{z}(t)N$, where

$$\delta_0 \equiv \left(\frac{\delta}{1+\delta}\right)^{\delta},$$

maximizes per capita net output $\tilde{I}(t)$.[9] When choosing to locate in the city, individual workers do not take account of the impact of their location on the costs of living for all other workers. The optimal city size is different from the city size given by equation (3.8.1). If the size of the city is restricted to $\delta_0\tilde{z}(t)N$, earning in the city will be larger than that in the rural area, making city dwellers better off.

Substituting

$$h^{*}(t)=\left(\frac{\alpha G}{\mu}\right)^{\beta}A^{\beta}(t)$$

into equation (3.8.1) gives the optimal level of net earnings for urban workers

$$I^{*}(t)=\alpha_0 A^{\beta}(t)-b(z(t)N)^{1/\delta},$$

---

[9] A planner chooses $z(t)$ to maximize $z(t)I(t)+(1-z(t))I_r(t)$. The solution of the optimal problem is given by $\delta_0\tilde{z}(t)N$.

where

$$\alpha_0 \equiv \left(\frac{\alpha^{\alpha} G}{\mu}\right)^{\beta}.$$

The motion of the overall level of technology $A(t)$ is specified as

$$A(t+1) = \max\left[A(t),\ (z(t)h(t))^{\theta}\right]. \tag{3.8.2}$$

There is no depreciation in A. Provided that

$$A(t) < (z(t)h(t))^{\theta},$$

we see

$$h^*(t) = \left(\frac{\alpha G}{\mu}\right)^{\beta} A^{\beta}(t),$$

can be rewritten as

$$h(t) = \left(\frac{\alpha G}{\mu}\right)^{\beta} (z(t-1)h(t-1))^{\theta\beta}. \tag{3.8.3}$$

Urbanization acts through two channels on technological change: directly via $z(t)$ in expression (3.8.2) and indirectly through the impact of $z(t-1)$ on $h(t)$ as expressed in equation (3.8.3).

For the dynamic path of technology $\{A(t)\}_{t=0}^{\infty}$, three cases may be considered, no-urbanization $z(t) = 0$, partial urbanization $0 < z(t) < 1$, and full urbanization

$$z(t) = 1.$$

When $z(t) = 0$ and $I(t) < I_r(t)$, there is no incentive for urbanization to occur. This will be the case whenever the initial level of technology A_0 satisfies

$$A_0 < \left(\frac{G_r}{\alpha}\right)^{1/\alpha\beta}.$$

The economy remains in a development trap, experiencing no growth. It can be seen that leaving aside the case of $z(t) = 0$, the dynamic path of $\{A(t)\}_{t=0}^{\infty}$ is given by

$$A(t+1) = G(A(t)) \equiv \begin{cases} G_1(A(t)) & \text{if } z(t) < 1,\ G_1(A(t)) > 0,\ (z(t)h(t))^{\theta} > A(t), \\ G_2(A(t)) & \text{if } z(t) = 1,\ G_2(A(t)) > 0,\ (z(t)h(t))^{\theta} > A(t), \\ A(t) & \text{otherwise,} \end{cases}$$

where

$$G_2(A(t)) \equiv b_0 A^{\theta\beta}(t),$$
$$b_0 \equiv \left(\frac{\alpha G}{\mu}\right)^{\theta\beta},$$
$$G_1(A(t)) \equiv b_0 \left(bN^{1/\delta}\right)^{-\theta\beta} A^{(\beta-\delta)\theta}(t)\left[\alpha_0 A^{\alpha\beta}(t) - G_r\right]^{\delta\theta}.$$

The function $G_1(A)$ governs the dynamic evolution of $A(t)$ in the case of partial urbanization. The full urbanization dynamics is characterized by $G_2(A)$. It is easy to prove that at the point

$$z(t) = 0, \quad G_1(A(t)) = G_2(A(t)),$$

the slope of $G_2(A)$ is strictly less than that of $G_1(A)$. We now illustrate different cases. The technical parts are not difficult and left the reader to check.

Case 1: The economy always remains at its initial levels of urbanization and aggregate output, for all initial levels of technology. This case occurs whenever the $G_1(A)$ function is concave and remains below the 45° line. This happens when A_0 is too low and/or when the incentives to invest in human capital are too low. The situation is illustrated in figure 3.8.1.[10]

[10] The plot is for concave $G_1(A)$. When G is convex, similar conclusions are held.

Case 2: This refers to the case when $G_1(A(t))$ intersects the 45^0 line. If the $G_1(A(t))$ is concave, there are two equilibrium points: a lower unstable one and an upper stable equilibrium. For all levels of technology above the lower intersection of the $G_1(A(t))$ function and the 45^0 line, the economy experiences growth of technology and urbanization. If $G_1(A(t))$ is concave, the upper equilibrium point is stable and partly urbanized, as illustrated in figure 3.8.2.

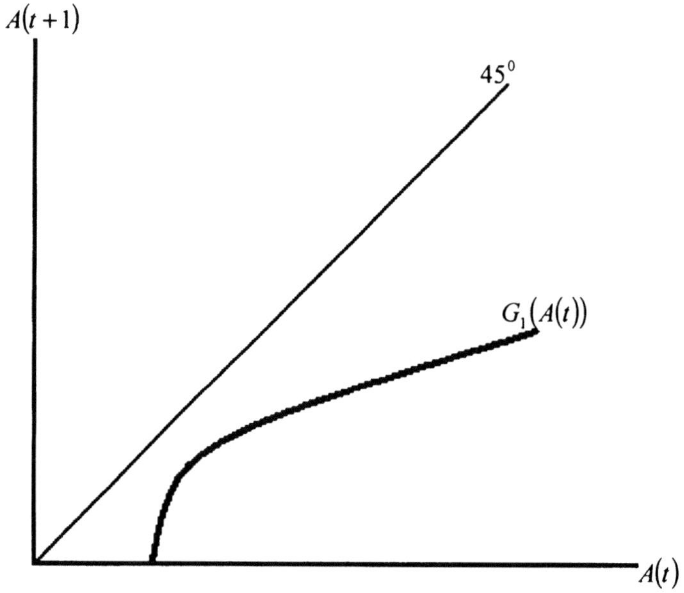

Figure 3.8.1: No urbanization

If $G_1(A(t))$ is convex, the economy experiences full urbanization. After full urbanization, the economy is characterized by $G_2(A(t))$. If $G_2(A(t))$ is concave, the economy will grow up to a steady state level of technology as depicted in figure 3.8.3a. If $G_2(A(t))$ is convex, the economy experiences unbounded growth with full urbanization as illustrated in figure 3.8.3b.

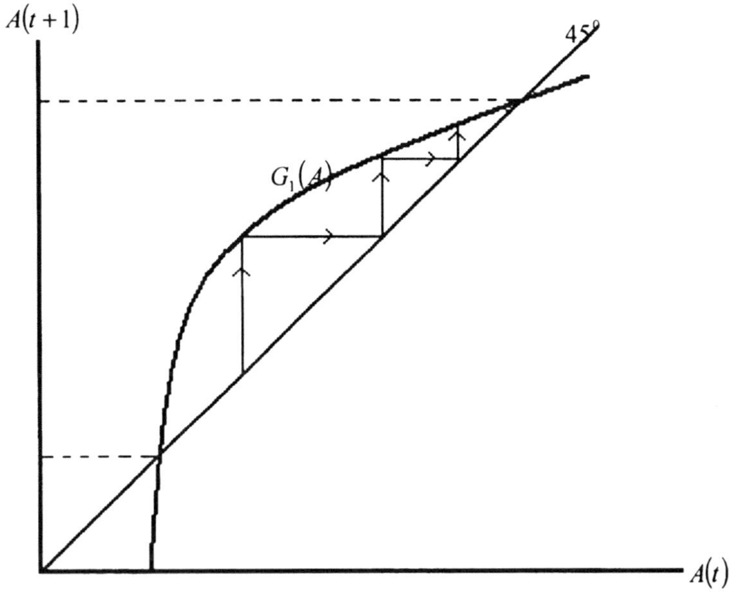

Figure3.8.2: Partial urbanization with two equilibrium points

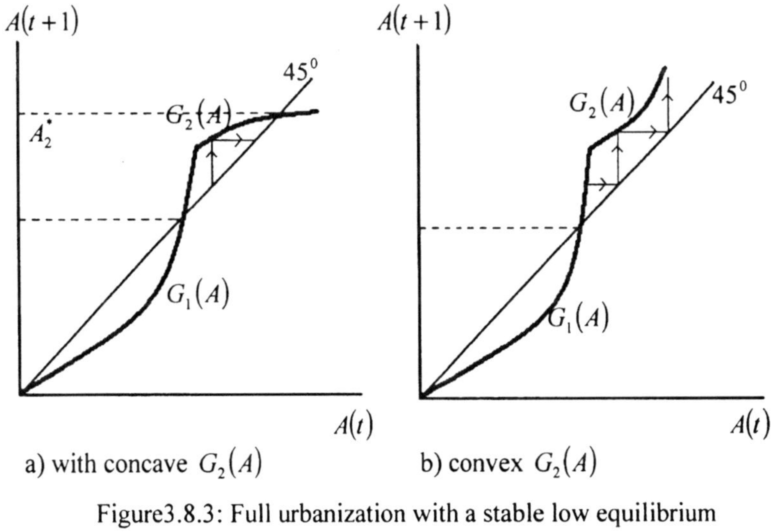

a) with concave $G_2(A)$ b) convex $G_2(A)$

Figure3.8.3: Full urbanization with a stable low equilibrium

3.9 The OSG model with money[11]

Modern analysis of the long-term interaction of inflation and capital formation begins with Tobin's seminal contribution in 1965. Tobin deals with an isolated economy in which "outside money" competes with real capital in the portfolios of agents within the framework of the Solow model.[12] Since then, many growth models of monetary economies are built within the OLG framework.[13] This section introduces money within the OSG framework proposed in section 3.2. We present the model in discrete time, numbered from zero and indexed by $t = 0, 1, 2, \cdots$. Time 0, being referred to the beginning of period 0, represents the initial situation from which economy starts to grow. The end of period $t-1$ coincides with the beginning of period t; it can also be called time t. We assume that transactions are made in each period. The model assumes that each individual lives forever. The population, N, is constant. Each individual supplies one unit of labor at each time t. Production in period t uses amount $K(t)$ of capital and amount N of labor services as inputs. It supplies amount $Y(t)$ of goods. Here, production is assumed to be continuous during the period, but then use the same capital that existed at the beginning of the period.

Production is made through a neoclassical constant return to scale technology. The real interest rate and the wage of labor are given as before by

$$r(t) + \delta_k = f'(k(t)), \quad w(t) = f(k(t)) - k(t)f'(k(t)). \tag{3.9.1}$$

For simplicity of expression, we normalize $N = 1$. There is some initial capital stock k_0 that is owned equally by all individuals at the initial period.

We assume that agents have perfect foresight with respect to all future events and capital markets operate frictionless. The government levies no taxes. Money is introduced by assuming that a central bank distributes at no cost to the population a per capita amount of fiat money, $M(t) > 0$. The scheme according to which the money stock evolves over time is deterministic and known to all agents. With μ being its constant net growth rate, the money stock, $M(t)$, evolves over time according to

$$M(t) = (1 + \mu)M(t-1), \quad \mu > 0.$$

---

[11] This model is referred to Zhang (2005b: chapter 3).

[12] See Tobin (1965). Outside money is the part of money stock which is issued by the government. See also Burmeister and Dobell (1970) and Zhang (1999).

[13] See Tobin (1969).

At the beginning of period t the government brings $M(t) - M(t-1)$ additional units of money per capita into circulation in order to finance all government expenditures via seigniorage. For the seigniorage mechanism to work, injections of the additional units of money take place before the other markets open. Let $m(t)$ stand for the real value of money per capita measured in units of the output good. Then, we can rewrite the above equation as

$$g(t) = \frac{M(t) - M(t-1)}{P(t)} = \frac{\mu}{1+\mu} m(t). \tag{3.9.2}$$

In this model, money acts as a pure store of value. The demand for money relies exclusively on the speculative conjecture that the future exchange value of money in terms of goods will be positive because people will express a positive demand for money in the future. In the absence of uncertainty and with money not being explicitly required for transaction purposes, this "bubbly view" implies that in a monetary equilibrium point the return on money needs to be equal to the return on competing assets such as claims on productive investment projects. In period t, each consumer receiving the per capita nominal money stock M believes that money will be exchanged at the expected future price $p^m(t+1) > 0$. The price of money $p^m(t)$ is in terms of goods or it expresses the amount of goods that can be purchased by one unity of money in each period t. According to the definitions, we have

$$P(t) = \frac{1}{p^m(t)}.$$

A necessary condition for a monetary economy to exist is that the price of money must be positive. If money has a positive value, people have an opportunity to save in money. According to the definition of the price of money, the deflation rate $P(t)/P(t+1)$ coincides with the real return on money. In a competitive economy the absence of arbitrage opportunities entails equality of return on assets

$$\frac{P(t)}{P(t+1)} = 1 + f'(k(t+1)) - \delta_k.$$

As

$$\frac{P(t)}{P(t+1)} = \frac{P(t)}{P(t+1)} \frac{M(t+1)}{(1+\mu)M(t)} = \frac{m(t+1)}{(1+\mu)m(t)},$$

the above condition becomes

$$1 + f'(k(t+1)) - \delta_k = \frac{m(t+1)}{(1+\mu)m(t)}. \tag{3.9.3}$$

The consumer obtains income in period t from the interest payment $r(t)k(t)$ and the wage payment $w(t)$

$$y(t) = r(t)k(t) + w(t), \tag{3.9.4}$$

where $y(t)$ is called the *current income*. We define the *disposable income* as the sum of $k(t) + m(t)$ and the current income, i.e.

$$\hat{y}(t) \equiv y(t) + k(t) + m(t). \tag{3.9.5}$$

The budget constraint is given by

$$c(t) + s(t) = \hat{y}(t).$$

For simplicity, we take the Cobb-Douglas utility function to describe consumers' preference

$$U(t) = c^{\xi}(t)s^{\lambda}(t), \quad \xi + \lambda = 1, \quad \xi, \ \lambda > 0,$$

in which ξ and λ are respectively the propensities to consume goods and to own wealth. A typical household maximizes the utility subject to the budget constraint. We solve the optimal choice of the consumers as

$$c(t) = \xi\hat{y}(t), \quad s(t) = \lambda\hat{y}(t). \tag{3.9.6}$$

According to the definitions, the per capita capital in period $t + 1$, $k(t+1)$, is equal to the saving made in period t minus the real value of money in period t, i.e.

$$k(t+1) = s(t) - m(t). \tag{3.9.7}$$

The economy starts to operate in $t = 0$. Each member in period 0 is endowed with $M(-1)$ units of money and owns k_0 $(= K(0)/N(0))$ units of physical capital. The labor force N is exogenously given. We will show that the above equations allow us to calculate recursively all the $k(t)$ and $m(t)$.

The government spending is given by equation (3.9.2). From equations (3.9.5) and (3.9.6), we obtain

$$s(t) = \lambda(y(t) + k(t) + m(t)).$$

Substituting equations (3.9.1) and (3.9.4) into the above equation yields

$$s(t) = \lambda(f(k(t)) + \delta k(t) + m(t)), \tag{3.9.8}$$

where $\delta \equiv 1 - \delta_k$. Substituting equation (3.9.8) into equation (3.9.7), we obtain

$$k(t+1) = \lambda(f(k(t)) + \delta k(t) + m(t)) - m(t). \tag{3.9.9}$$

From equation (3.9.3), we obtain

$$m(t+1) = (1+\mu)[\delta + f'(k(t+1))]m(t). \tag{3.9.10}$$

Definition 3.9.1. Given the initial capital stock k_0 and the initial money stock $M(-1)$, a *competitive equilibrium* is given by a sequence of quantities

$$\{m(t),\ c(t),\ s(t),\ g(t),\ k(t+1)\}$$

and a sequence of prices $\{r(t),\ w(t),\ P(t)\}$ such that for all periods $t = 0, 1, 2, \cdots$: (i) competition ensures that factors get paid their marginal products according to equations (3.9.1); (ii) given

$$M(t) = (1+\mu)M(t-1),$$

the budget constraint (3.9.2) of the government is satisfied; (iii) given the price consequence, agents solve optimally the decision problem; (iv) the evolution of money balances satisfies equation (3.9.10); (v) investments and savings are matched by equation (3.9.7); and (vi) all the markets clear with the equilibrium conditions being as following:

Labor market:

$$w(t) = f(k(t)) - k(t)f'(k(t));$$

Money market:

$$m(t) = M(t)/P(t);$$

Goods market:

$$f(k(t)) + k(t) = g(t) + \delta_k k(t) + c(t) + k(t+1).$$

Equilibrium can be further classified as *inside* and *outside money equilibrium*: an inside money equilibrium is associated with a situation in which outside balances are zero; an outside money equilibrium is associated with positive outside money balances.

We now examine dynamic properties of equations (3.9.9) and (3.9.10). At inside money equilibrium with $m(t) = 0$, we have

$$k(t+1) = \lambda(f(k(t)) + \delta k(t)). \tag{3.9.11}$$

Government spending is reduced to zero. As shown before, this system has a unique stable equilibrium point given by

$$\frac{f(\bar{k})}{\bar{k}} = \frac{\xi + \delta_k \lambda}{\lambda}. \tag{3.9.12}$$

We are mainly interested in monetary economy which is characterized by the three difference equations (3.9.2), (3.9.9) and (3.9.10). Equations (3.9.9) and (3.9.10) do not contain $g(t)$. In fact, once $m(t)$ is determined, we determine $g(t)$ by (3.9.2).[14] The dynamics of the monetary economy are described by equations (3.9.9) and (3.9.10). We may rewrite the system as follows

[14] Here, we consider that the conduct of monetary policy takes precedence over all fiscal matters. Gale (1983: chapter 2) provides a discussion of how to resolve the coordination problem between monetary and fiscal measures in a similar context.

$$k(t+1) = \lambda(f(k(t)) + \delta k(t)) - \xi m(t),$$
$$m(t+1) = (1+\mu)[\delta + f'(k(t+1))]m(t). \tag{3.9.13}$$

A steady state of the monetary economy needs to satisfy

$$k = \lambda(f(k) + \delta k) - \xi m,$$
$$1 = (1+\mu)(\delta + f'(k)). \tag{3.9.14}$$

From $1 = (1+\mu)(\delta + f'(k))$, we get

$$f'(k) = \delta_k - \frac{\mu}{1+\mu}. \tag{3.9.15}$$

From the properties of f, we see that if $\delta_k > \mu/(1+\mu)$, equation (3.9.15) has a unique solution, denoted by k^*. From the first equation in (3.9.14), we solve

$$m^* = \left(\frac{f(k^*)}{k^*} - \frac{\xi}{\lambda} - \delta_k\right)\frac{\lambda k^*}{\xi}. \tag{3.9.16}$$

For $m^* > 0$, we should require

$$\frac{f(k^*)}{k^*} > \frac{\xi}{\lambda} + \delta_k.$$

Comparing equations (3.9.12) and (3.9.16), we see that the monetary economy has a unique equilibrium point if

$$\frac{f(k^*)}{k^*} > \frac{f(\bar{k})}{\bar{k}}. \tag{3.9.17}$$

As

$$\frac{d(f/k)}{dk} = \frac{f' - f/k}{k} < 0,$$

because of strict concavity of f, f/k strictly decreases in k. Hence, equation (3.9.17) implies $\bar{k} > k^*$ if both $\bar{k}$ and k^* exist. As f' decreases in k, we see that if

$$f'(\bar{k}) < \delta_k - \frac{\mu}{1+\mu},$$

then $\bar{k} > k^*$. In summary, we have the following proposition.

Proposition 3.9.1. Suppose that the inside competitive equilibrium has a unique non-trivial steady state $\bar{k} > 0$. Then, the monetary economy has a unique steady state k^* if the following inequality holds:

$$f'(\bar{k}) < \delta_k - \frac{\mu}{1+\mu}.$$

The steady state is a saddle point.

Proof. We only need to examine the stability of the steady state. The Jacobian at steady state for equation (3.9.13) is given by

$$J = \begin{bmatrix} a_{11} & a_{22} \\ a_{21} & a_{22} \end{bmatrix},$$

in which

$$a_{11} = \lambda(f' + \delta),$$
$$a_{12} = -\xi,$$
$$a_{21} = \lambda m f'',$$
$$a_{22} = 1 - (1+\mu) m f'' \xi.$$

The characteristic equation is given by

$$H(\rho) = \rho^2 - Trace(J)\rho + Det(J),$$

where

$$Trace(J) = a_{11} + a_{22} = \lambda(f' + \delta) + 1 - (1 + \mu)mf''\xi > 0,$$
$$Det(J) = a_{11}a_{22} - a_{12}a_{21} = \lambda(f' + \delta) > 0.$$

We have

$$H(1) = 1 - Trace(J) + Det(J) = -1 + (1 + \mu)mf''\xi < 0.$$

We thus conclude that the two eigenvalues are real and satisfy

$$0 < \rho_1 < 1 < \rho_2.$$

The unique monetary equilibrium point is a saddle point.

From equation (3.9.15), we get

$$f''\frac{dk^{*}}{d\mu} = -\frac{1}{(1 + \mu)^2} < 0.$$

If the rate of monetary expansion μ is permanently raised, then the per capita capital stock in the new monetary steady state is higher than before. That is, the Tobin effect prevails.[15] This is illustrated as in figure 3.9.1. As μ is shifted, the accumulation equation

$$k = \lambda(f(k) + \delta k + m) - m$$

is not affected; the arbitrage relation

$$1 = (1 + \mu)(\delta + f'(k))$$

is shifted by lowering the return on real balances via increased inflation taxation. To rebalance the arbitrage relation, the composition of the portfolios of agents needs to readjust in favor of capital. In the new steady state, the capital stock will be higher than before.

[15] The substitution of capital for fiat money in reaction to an increase in anticipated inflation is called the Tobin effect, as described in Tobin (1965). See also Champ and Freeman (2001).

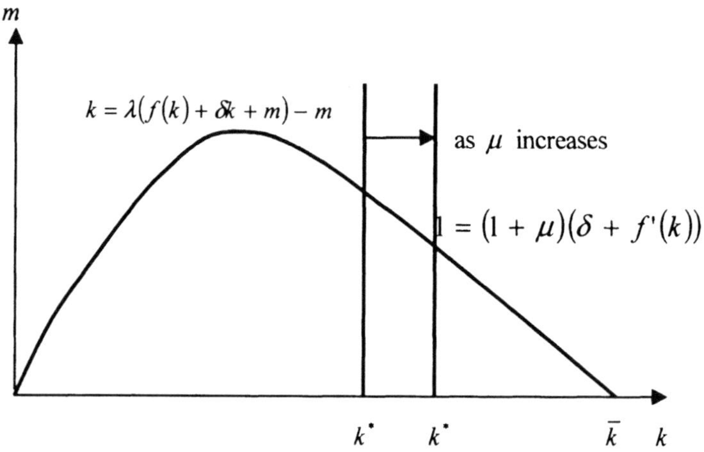

Figure 3.9.1: The Tobin effect

3.10 The OSG model with labor supply[16]

This section extends the OSG model to treat time distribution among various activities as endogenous variable. Most of the assumptions and variables are the same as in section 3.2. Let $N(t)$ stand for the flow of labor services used at time t for production. The total labor force $N(t)$ is given by $N(t) = T(t)N_0$, where $T(t)$ is the work time of a representative household and N_0 is the population. Let $k(t) \equiv K(t)/N(t)$. The marginal conditions are given by

$$r(t) + \delta_k = f'(k(t)), \quad w(t) \equiv f(k(t)) - k(t)f'(k(t)). \tag{3.10.1}$$

Let $k^*(t) (\equiv K(t)/N_0)$ stand for per capita wealth. According to the definition of $k(t)$ and $k^*(t)$, we have $k^*(t) = k(t)T(t)$. Per capita current income from the interest payment $r(t)k^*(t)$ and the wage payment $w(t)T(t)$ is given by

$$y(t) = r(t)k^*(t) + w(t)T(t).$$

---

[16] This section is based on Zhang (2005c).

The per capita *disposable income* is given by

$$\hat{y}(t) = y(t) + k^{*}(t).$$

The disposable income is used for saving and consumption. At each point of time, a consumer would distribute the total available budget among saving, $s(t)$, consumption of goods, $c(t)$. The budget constraint is given by

$$c(t) + s(t) = \hat{y}(t) = r(t)k^{*}(t) + w(t)T(t) + k^{*}(t). \tag{3.10.2}$$

Denote $T_h(t)$ the leisure time at time t and the (fixed) available time for work and leisure by T_0. The time constraint is expressed by

$$T(t) + T_h(t) = T_0.$$

Substituting this function into the budget constraint yields

$$w(t)T_h(t) + c(t) + s(t) = \bar{y}(t) \equiv r(t)k^{*}(t) + w(t)T_0 + k^{*}(t). \tag{3.10.3}$$

In our model, at each point of time, consumers have three variables to decide. We assume that utility level $U(t)$ that the consumers obtain is dependent on the leisure time, T_h, the consumption level of commodity, $c(t)$, and the saving, $s(t)$, as follows

$$U(t) = T_h^{\sigma}(t)c^{\xi}(t)s^{\lambda}(t), \quad \sigma, \xi, \lambda > 0, \quad \sigma + \xi + \lambda = 1,$$

where σ is called the propensity to use leisure time, ξ, the propensity to consume, and λ, the propensity to own wealth. For any individual, the wage rate, $w(t)$, and the rate of interest, $r(t)$, are given in markets and the level of wealth, $k(t)$, is predetermined before decision.

Maximizing $U(t)$ subject to the budget constraint (3.10.3) yields

$$w(t)T_h(t) = \sigma\bar{y}(t), \quad c(t) = \xi\bar{y}(t), \quad s(t) = \lambda\bar{y}(t). \tag{3.10.4}$$

According to the definitions, the household's wealth in period $t + 1$ is equal to the savings made in period t

$$k^*(t + 1) = s(t) = \lambda \bar{y}(t). \tag{3.10.5}$$

We have thus completed the model. We now analyze dynamic properties of the system.

We now examine the existence of equilibrium and stability. First, from the definition of $\bar{y}(t)$ and equations (3.10.1), we obtain

$$\bar{y} = (f'(k) - \delta_k)k^* + (f(k) - kf'(k))T_0 + k^*. \tag{3.10.6}$$

Substituting this equation into $w(t)T_h(t) = \sigma\bar{y}(t)$ yields

$$(\xi + \lambda)(f(k) - kf'(k))T_0 = (f'(k) + \delta)\sigma k^* + (f(k) - kf'(k))T,$$

where we use equation (3.10.1) and $T(t) + T_h(t) = T_0$. Inserting $k^*(t) = T(t)k(t)$ into the above equation, we solve

$$T(t) = \frac{(f(k) - kf'(k))(\xi + \lambda)T_0}{(f'(k) + \delta)\sigma k + (f(k) - kf'(k))}, \tag{3.10.7}$$

where

$$\delta \equiv 1 - \delta_k, \quad f - kf' > 0$$

for any positive k. We see that for any positive k

$$0 < T(t) < T_0.$$

This also guarantees

$$0 < T_h(t) \, (= T_0 - T(t)) < T_0.$$

We see that $T(t)$ is uniquely determined as a function of $k(t)$. From $k^*(t) = T(t)k(t)$ and equation (3.10.5), we obtain

$$T(t+1)k(t+1) = \lambda \bar{y}(t).$$

Inserting $k^*(t) = T(t)k(t)$ and equation (3.10.6) into this equation, we obtain

$$\Lambda_0((t+1))k(t+1) = \Lambda(k(t))(f(k(t)) - k(t)f'(k(t))), \tag{3.10.8}$$

where

$$\Lambda_0(k(t+1)) \equiv \frac{(f(k(t+1)) - k(t+1)f'(k(t+1)))(\xi + \lambda)}{(f'(k(t+1)) + \delta)\sigma k(t+1) + (f(k(t+1)) - k(t+1)f'(k(t+1)))},$$

$$\Lambda(k(t)) \equiv \left[\frac{(f'(k(t)) + \delta)k(t) + (f(k(t)) - k(t)f'(k(t)))}{(f'(k(t)) + \delta)\sigma k(t) + (f(k(t)) - k(t)f'(k(t)))}\right]\lambda > 0.$$

We calculate

$$\frac{d[k(t+1)\Lambda_0(k(t+1))]}{dk(t+1)} = \frac{(f(k) - kf'(k))\Lambda_0(k)}{(f'(k) + \delta)\sigma k + (f(k) - kf'(k))}$$
$$- \frac{\sigma k^2 f''\Lambda_0(k)((f'(k) + \delta)k + (f(k) - kf'(k)))}{(f(k) - kf'(k))[(f'(k) + \delta)\sigma k + (f(k) - kf'(k))]} > 0.$$

As $d(\Lambda_0 k)/dk > 0$ for any positive k, according to the implicit function theorem, from equation (3.10.8) we can express $k(t+1)$ as a function of $k(t)$ as follows

$$k(t+1) = \Lambda^*(k(t)), \tag{3.10.9}$$

where $\Lambda^*(k)$ has the same degree of smoothness of $f''(k)$. The difference equation involves a single variable $k(t)$. With a positive initial condition, the one-dimensional difference equation determines $k(t)$ in any period of time.

Lemma 3.10.1. For any positive solution $k(t)$ of equation (3.10.9), all the other variables are uniquely determined by the following procedure: $T(t)$ by equation (3.10.7) $\rightarrow$ $T_h(t) = T_0 - T(t) \rightarrow k^*(t) = T(t)k(t)$ $\rightarrow$ $r(t)$ and $w(t)$ by equations (3.10.1) $\rightarrow$ $\hat{y}(t)$ by equation (3.10.2) $\rightarrow$ $\bar{y}(t)$ by equation (3.10.3) $\rightarrow c(t)$ and $s(t)$ by equations (3.10.4) $\rightarrow$ $f(k(t))$ $\rightarrow$ $F(t) = N(t)f(k(t))$.

Lemma 3.10.1 guarantees that once we determine the dynamic properties of equation (3.10.9), we can determine the behavior of all the other variables in the system. Hence, it is sufficient for us to be concerned with equation (3.10.9).

By equation (3.10.8), an equilibrium point is given by

$$\Lambda_0(k)k = \Lambda(k)(f(k) - kf'(k)).$$

Inserting $\Lambda_0(k)$ and $\Lambda(k)$ into the above equation yields

$$\frac{f(k)}{k} - \left(\frac{\xi}{\lambda} + \delta_k\right) = 0. \tag{3.10.10}$$

The equation has a unique positive solution if

$$f'(0) > \frac{\xi}{\lambda} + \delta_k.$$

With lemma 3.10.1, we see that the system has a unique positive solution. We denote the equilibrium value of k by k^*. To check stability of the unique equilibrium, we calculate $d\Lambda_1 / dk$ at k^*. Taking derivatives of the two sides of equation (3.10.8), we have

$$\frac{d[\Lambda_0((t+1))k(t+1)]}{dk(t+1)} \frac{dk(t+1)}{dk(t)} = \frac{d[\Lambda(k(t))(f(k(t)) - k(t)f'(k(t)))]}{dk(t)}, \quad k = k^*. \tag{3.10.11}$$

Substituting

$$\frac{d(k\Lambda_0)}{dk} = \frac{(f - kf')\Lambda_0}{(f' + \delta)\sigma k + (f - kf')} - \frac{\sigma k^2 f''\Lambda_0((f' + \delta)k + (f - kf'))}{(f - kf')[(f' + \delta)\sigma k + (f - kf')]},$$

$$\frac{d\Lambda}{dk} = \left[\frac{f' + \delta}{(f' + \delta)k + (f - kf')} - \frac{(\sigma f' + \sigma\delta) - (\xi + \lambda)kf''}{(f' + \delta)\sigma k + (f - kf')}\right]\Lambda,$$

into equation (3.10.11), we have

$$F_1(k)\frac{d\Lambda_1}{dk} = F_2(k),$$

in which

$$F_1(k) \equiv f - kf' - \sigma k^2 f'' - \sigma k^3 f'' F^*,$$
$$F_2(k) \equiv \frac{(\xi + \lambda)(f' + \delta)k}{(kF^* + 1)} - (kF^* + 1)\sigma k^2 f'',$$
$$F^* \equiv \frac{f' + \delta}{f - kf'},$$

at k^*. Using $f' > 0$, $f'' < 0$ and $f - kf' > 0$, we see that $F_1 > 0$ and $F_2 > 0$.

It is straightforward to check $F_1 > F_2$. Hence, we conclude

$$\frac{d\Lambda_1}{dk} = \frac{F_2}{F_1} \in (0, 1).$$

This implies that the equilibrium point is stable.

Theorem 3.10.1. The dynamic system has a unique stable equilibrium.

We showed that the dynamic system has a unique stable equilibrium. We now examine impact of change in some parameters. First, we introduce technological change by specifying

$$f(k) = Ah(k),$$

where A describes the level of technology. Taking derivatives of equation (3.10.10) with respect to A yields

$$\frac{dk}{dA} = \frac{kf}{(f - kf')A} > 0,$$

where $f / k - f' > 0$. As technology is improved, the capital intensity, k, is increased. From equation (3.10.7), we obtain

$$\frac{dT}{dA} = -\frac{((f - kf')f' + kff'')(f + \delta k)kT\sigma}{A(f - kf')^2 [(f' + \delta)\sigma k + (f - kf')]}.$$

The sign of dT / dA is the same as that of

$$(f - kf')f' + kff''.$$

As $(f - kf')f' > 0$ and $kff'' < 0$, the impact is ambiguous. If f takes on the Cobb-Douglas form, i.e.

$$f = Ak^{\alpha},$$

then $dT / dA = 0$. If the production function takes on the CES form

$$f = A(ak^{\rho} + 1)^{1/\rho},$$

where $\rho < 1$, a and A are positive, we calculate

$$(f - kf')f' + kff'' = a\rho A^2 k^{\rho-1}(ak^{\rho} + 1)^{2/\rho-2}.$$

We see that if $\rho > 0$, then $dT / dA > 0$; if $\rho = 0$, then $dT / dA = 0$; and if $\rho < 0$, then $dT / dA < 0$. By $k^{*} = kT$, we have

$$\frac{dk^{*}}{dA} = k\frac{dT}{dA} + T\frac{dk}{dA}.$$

If the improvement in technology increases work time, then per capita wealth definitely increases; otherwise the impact is ambiguous. From equation (3.10.1), we obtain

$$\frac{dr}{dA} = \frac{f'}{A} + f''\frac{dk}{dA},$$
$$\frac{dw}{dA} = \frac{w}{A} - kf''\frac{dk}{dA} > 0.$$

The wage rate is increased due to technological improvement; but the impact on the rate of interest is ambiguous. The impact on the output level is given by

$$\frac{df}{dA} = \frac{f}{A} + f'\frac{dk}{dA} > 0.$$

We now examine the impact of change in preference. As

$$\sigma + \xi + \lambda = 1,$$

a change in the propensity to consume leisure has to be associated with some changes in other propensities. For simplicity, we specify the preference change as follows

$$d\sigma = -d\lambda, \quad d\xi = 0.$$

Taking derivatives of equation (3.10.10) with respect to σ yields

$$\frac{dk}{d\sigma} = -\frac{\xi k^2}{(f - kf')\lambda^2} < 0.$$

As the propensity to use leisure increases and the propensity to save declines, the capital intensity declines in the long term. It is important to note that if $d\sigma = -d\xi$ and $d\lambda = 0$, then we have $dk / d\sigma > 0$. That is, if the propensity to use leisure time increases and the propensity to consume declines, then the capital intensity increases. Another possible pattern of preference change is given by

$$d\sigma = -d(\xi + \lambda), \quad \frac{d\xi}{\xi} = \frac{d\lambda}{\lambda}.$$

In this case, we have $dk / d\sigma = 0$. As preference may change in different ways, its impact on k is dependent on the specified pattern. In what follows, we limit our discussion to the case of

$$d\sigma = - d\lambda, \quad d\xi = 0.$$

From equation (3.10.7), we obtain

$$\frac{1}{T}\frac{dT}{d\sigma} = \left[\frac{(f' + \delta)(f - kf')\sigma - (f + \delta k)\sigma k f''}{[(f' + \delta)\sigma k + (f - kf')](f - kf')}\right]\frac{dk}{d\sigma} - \frac{1}{(\xi + \lambda)}$$
$$- \frac{(f' + \delta)k}{(f' + \delta)\sigma k + (f - kf')} < 0.$$

We conclude that as the propensity to stay at home is increased, the time of staying at home increases. It should be remarked that this conclusion may not hold if we specify different patterns of preference change. By $k^* = kT$, we have

$$\frac{dk^*}{d\rho} = k\frac{dT}{d\rho} + T\frac{dk}{d\rho} < 0.$$

As the propensity to use leisure increases, per capita wealth declines in the long term. From equation (3.10.1), we obtain

$$\frac{dr}{d\sigma} = f''\frac{dk}{d\sigma} > 0, \quad \frac{dw}{d\sigma} = - kf''\frac{dk}{d\sigma} < 0.$$

The wage rate is increased and the rate of interest is increased. The impact on the output level is given by

$$\frac{df}{d\sigma} = f'\frac{dk}{d\sigma} > 0.$$

Chapter 4

Time-dependent solutions of scalar systems

This chapter examines periodic, aperiodic, chaotic solutions of scalar systems. Section 4.1 defines concepts such as periodic or aperiodic solutions (orbits). This section also introduces some techniques to find periodic solutions and provides conditions for judging stability of periodic solutions. Section 4.2 is concerned with period-doubling bifurcations. This section introduces concepts such as branch, bifurcation values, period-doubling bifurcation route to chaos, Myrberg's number and Fiegenbaum's number. Section 4.3 deals with aperiodic orbits. This section introduces the Li-Yorke theorem and the Sharkovsky theorem, which are important for proving existence of chaos in scalar systems. Section 4.4 studies some typical types of bifurcations. They include supercritical fold, subcritical fold, supercritical pitchfork, subcritical pitchfork, transcritical bifurcations. Section 4.5 introduces theory of Liapunov numbers. In this section, we also examine behavior of a model of chaotic labor market. In section 4.6, we study chaos. We simulate a demand and supply model to demonstrate economic chaos.

4.1 Periodic orbits

Definition 4.1.1. A sequence $\{x(t)\}$ is said to *oscillate about zero* or simply to *oscillate* if the terms $x(t)$ are neither eventually all positive nor eventually all negative. Otherwise the sequence is called *nonoscillatory*. A sequence $\{x(t)\}$ is called *strictly oscillatory* if for every $t_0 \geq 0$, there exist $t_1, t_2 \geq t_0$ such that

$$x(t_1) \cdot x(t_2) < 0.$$

A sequence $\{x(t)\}$ is said to oscillate about x^* if the sequence $x(t) - x^*$ oscillates. The sequence $\{x(t)\}$ is called strictly oscillatory about x^* if the sequence $x(t) - x^*$ is strictly oscillatory.

Theorem 4.1.1. (linearized oscillation theorem) Let $f : R \to R$ be a continuous function such that

$$uf(u) > 0,$$

for $u \neq 0$, and $f'(0) = 1$. If there exists $r > 0$ such that either

$$f(u) \leq u,$$

for $r \in [0, r]$ or $f(u) > u$ for $r \in [-r, 0]$, then every solution of

$$x(t+1) = x(t) - pf(x(t)),$$

oscillates if and only if every solution of the corresponding linearized equation

$$y(t+1) = y(t) - py(t),$$

oscillates, that is, if and only if $p \geq 1$.

Definition 4.1.2. A point $p \in R$ is called a *periodic point of period* k if $f^k(p) = p$. The point p is called a *periodic point of minimal period* k or *prime period* k *point* if

$$f^k(p) = p,$$

and k is the smallest positive number for which this holds. If p is a periodic point, then $O(p)$ is called the *periodic orbit*. Orbits that are not periodic are sad to be *aperiodic*.

The choice of "minimal" is motivated by the fact that the orbit cannot be decomposed into smaller loops. A fixed point can be regarded as a periodic point of period 1. Since

$$x_0 = f^p(x_0),$$

whenever p is the period of $O(x_0)$ we also have $f^m(x_0) \neq x_0$ for every $m < p$. Every point

$$x(t),\ t = 0, 1, \cdots, p - 1,$$

of the period orbit $O(x_0)$ of period p is periodic of the same period. Thus, $O(x_0)$ contains exactly p distinct periodic points of period p. Sometimes, we use

$$\{x(0), \cdots, x(p-1)\}$$

to denote a periodic orbit of period p.

The periodic orbits of period 2 of f are given by those intersections of the graph of $f(f(x))$ with the line

$$y = x,$$

which are not on the graph of $f(x)$. Plotting all three functions $f(x)$, $f(f(x))$, and x solves this problem. Suppose

$$f(x) = 3.2x(1 - x).$$

Figure 4.1.1 shows that f has two fixed points, which are also fixed for its second iterate. The two points where only the graph of the second iterate crosses the line $y = x$ are periodic points of period 2.

The function

$$f(x) = 3.82x(1 - x),$$

has a stable periodic orbit of period 3. Figure 4.1.2a describes the orbit with the initial condition $x_0 = 0.3$. The orbit appears chaotic, but this impression is incorrect. The orbit is asymptotically periodic of period 3, as shown in figure 4.1.2b.

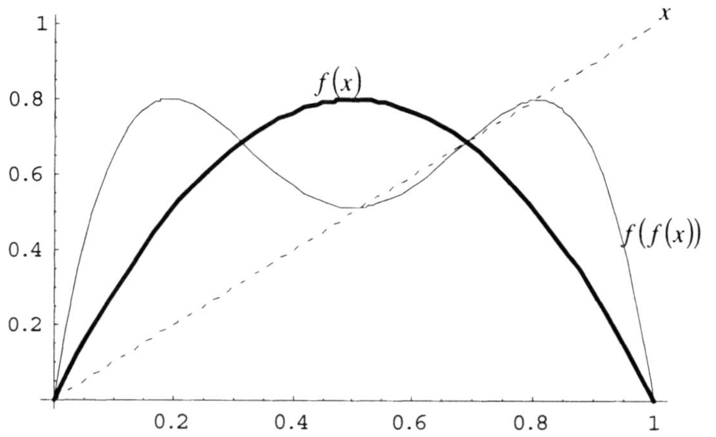

Figure 4.1.1: The solution of period 2 to $f(x) = 3.2x(1-x)$

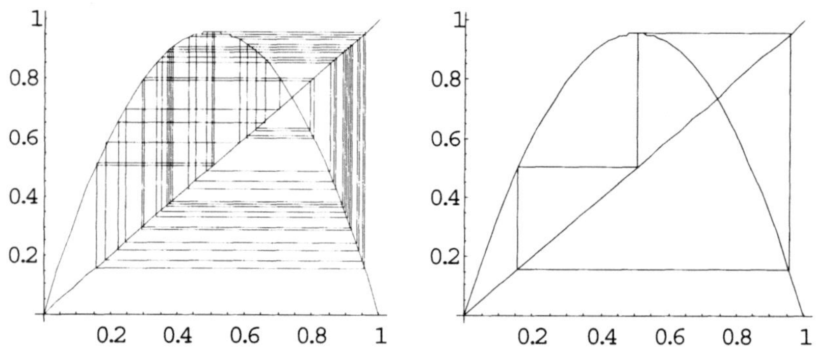

a) the orbit b) asymptotically periodic of period 3

Figure 4.1.2: The orbit of $f(x) = 3.82x(1-x)$ with $x_0 = 0.3$

Definition 4.1.3. A point $p^* \in R$ is called an *eventually periodic point* of minimal period k for difference equation (2.2.1) or an *eventually periodic point* for the map f if there exists a positive integer r and a periodic point p of minimal period k such that

$$f^r(p^*) = p, \quad f^{r-1}(p^*) \neq p.$$

Example The logistical difference equation

$$f = 4x(1 - x),$$

has two period-two solutions, $5/8 \pm \sqrt{5}/5$, which are the solutions, except the solutions of

$$x = 4x(1 - x),$$

of the following equation

$$x = f(f(x)) = 16x(1 - x)[1 - 4x(1 - x)].$$

Finding eventually periodic points leads to the equation $f^r(x) = p$, where r is a positive integer greater than 1. For $f = 4x(1 - x)$ with

$$p = \frac{5}{8} \pm \frac{\sqrt{5}}{5}, \quad r = 2,$$

we obtain the algebraic equation

$$16x(1 - x)[1 - 4x(1 - x)] = \frac{5}{8} \pm \frac{\sqrt{5}}{5}.$$

Example Consider the difference equation

$$x(t + 1) = \begin{cases} \frac{1}{2}x(t), & \text{if } x(t) \text{ is an even positive number} \\ 3x(t) + 1, & \text{if } x(t) \text{ is an odd positive number}, \quad t = 0, 1, \cdots, \end{cases}$$

where x_0 is a positive integer. The difference equation has no equilibrium points. The equation has a period-three solution $\{1, 4, 2\}$ and many eventually period-three solutions. For example, the solution that starts at 8 has the corresponding orbit

$$\{8, 4, 2, 1, 4, 2, 1, \cdots\}.$$

Calculate $O(7)$, etc. It is know that there are infinitely many eventually periodic solutions.[1] The *Collatz conjecture*, one of the most open difficult problems in the theory of difference equations, establishes that every solution of the above equation is eventually periodic to the period-three solution $\{1, 4, 2\}$.

Since a periodic point p of minimal period k is an equilibrium point of the map f^k, the notion of the stability of p follows from the definition of an equilibrium point, and the linearized stability result can be applied to f^k to determine the stability type of p.

Definition 4.1.4. A periodic point p of

$$x(t+1) = f(x(t))$$

of minimal period k is said to be *stable*, *asymptotically stable*, *unstable*, or a *global attractor* if p is, respectively, stable, asymptotically stable, an unstable equilibrium point, or a global attractor of f^k.

Applying the chain rule and theorem 2.3.1 to map f^2, we obtain that a two-periodic solution p is stable if

$$|f'(p)f'(f(p))| < 1,$$

and unstable if

$$|f'(p)f'(f(p))| > 1.$$

The number

$$\lambda - f'(p)f'(f(p))$$

is called a *multiplier of the orbit*. Likewise, the multiplier of a periodic orbit of any period k can be defined by

---

[1] See Bernstein and Lagarias (1996) and Lagarias (1985).

$$\lambda = f'(p)f'(f(p))\cdots f'(f^{k-1}(p)).$$

Example Consider

$$x(t+1) = \begin{cases} -x & \text{for } -1 \le x \le 1, \\ -2x+1 & \text{for } x > 1, \\ -2x-1 & \text{for } x < -1. \end{cases}$$

Notice that $f(1) = -1$ and $f(-1) = 1$. The period orbit $\{1, -1\}$ is unstable. In fact, for $x > 1$, we have

$$f(f(x)) = 4(x-1) + 1.$$

It is easy to show that the obit $O(x(0))$ with $x(0) > 1$ is going to $+\infty$. Thus, the state 1 is an unstable fixed point for the second iterate of f. Hence, the periodic orbit $\{1, -1\}$ is unstable.

Theorem 4.1.2. If

$$O(p) = \{p, f(p), \cdots, f^{k-1}(p)\}$$

is the orbit of the k-periodic point p, where f is a continuously differentiable function at p, then we have: (i) p is asymptotically stable if the multiplier of the orbit is less than unit, that is, $\lambda < 1$; and (ii) p is unstable if $\lambda > 1$.

Example (a genotype selection model) Consider a locust with two alleles: allele A with frequency p and allele a with frequency

$$q = 1 - p,$$

in a diploid population.[2] The change in gene frequency between one generation t and the next generation $(t+1)$ is given by

[2] See Section 1.9 in Elaydi (2000)

$$p(t+1) = \frac{p(t)g(p(t))}{p(t)g(p(t)) + 1 - p(t)} \equiv f(p(t)), \tag{4.1.1}$$

where $g(p(t))$ is the frequency-dependent fitness that the genotype AA has. The difference equation has three equilibrium points:

$$p_1^* = 0, \quad p_2^* = 1, \quad p_3^* \text{ determined by } g(p_3^*) = 1.$$

Now

$$f'(p) = \frac{g(p) + p(1-p)g'(p)}{[pg(p) + 1 - p]^2}.$$

We have $f'(0) = g(0)$ and

$$f'(1) = \frac{1}{g(1)} = g(0).$$

The two equilibrium points, $p_1^* = 0$ and $p_2^* = 1$, are unstable if $g(0) > 1$, which is true when

$$g(p) = e^{\beta(1-2p)}$$

with $\beta > 0$. We assume

$$g(p) = e^{\beta(1-2p)}$$

henceforth. Under this assumption f becomes

$$f = \frac{pe^{\beta(1-2p)}}{pe^{\beta(1-2p)} + 1 - p}. \tag{4.1.2}$$

Hence

$$p_3^* = \frac{1}{2}, \quad f'\left(\frac{1}{2}\right) = 1 - \frac{\beta}{2}.$$

Hence, for $0 < \beta < 4$, p_3^* is an asymptotically stable gene frequency.[3] If $\beta = 4$, then $f'(1/2) = -1$. A direct calculation shows that the Schwarzian derivative

$$Sf\left(\frac{1}{2}\right) < 0.$$

By theorem 2.4.2, the equilibrium point p_3^* is asymptotically stable when $\beta = 4$. At $\beta = 4$, p_3^* loses its stability and a new 2 -cycle is born. To simplify the analysis, introduce

$$x(t) = \frac{p(t)}{1 - p(t)}, \tag{4.1.3}$$

which transforms equation (4.1.1) under $g(p) = e^{\beta(1 - 2p)}$ into

$$x(t + 1) = x(t)\exp\left(\beta\frac{1 - x(t)}{1 + x(t)}\right) \equiv F(x(t)). \tag{4.1.4}$$

Let $\{x_1, x_2\}$ be a 2 -cycle of equation (4.1.4). Then

$$x_2 = x_1 \exp\left(\beta\frac{1 - x_1}{1 + x_1}\right),$$

$$x_1 = x_2 \exp\left(\beta\frac{1 - x_2}{1 + x_2}\right).$$

Multiplying the two equations yields

---

[3] It is proved that for $0 < \beta \neq 4$, the equilibrium point $p_3^* = 1/2$ is globally asymptotically stable on the interval $(0, 1)$. See Kocic and Ladas (1993).

$$x_1x_2 = x_1x_2 \exp\left(\beta\frac{1 - x_1}{1 + x_1}\right)\exp\left(\beta\frac{1 - x_2}{1 + x_2}\right).$$

Hence

$$\exp\left(2\beta\frac{1 - x_1x_2}{(1 + x_1)(1 + x_2)}\right) = 1.$$

We obtain

$$x_1x_2 = 1.$$

From equation (4.1.3), the corresponding 2 -cycle of equation (4.1.2) is $\{\overline{p}, 1 - \overline{p}\}$. To check the stability of this cycle, we first calculate

$$F'(x) = \left[1 - \frac{2x\beta}{(1 + x)^2}\right]\exp\left(\beta\frac{1 - x}{1 + x}\right).$$

We have

$$F'(x_1)F'(x_2) = F'(x_1)F'\left(\frac{1}{x_1}\right) = \left(1 - \frac{2\beta x_1}{(1 + x_1)^2}\right)^2 < 1.$$

By theorem 4.1.2, it follows that the cycle is asymptotically stable.

Example Consider the map

$$f(x) = x^2 - 0.85,$$

defined on the interval $[-2, 2]$. Find the 2 cycles and determine their stability. The 2 -periodic points are obtained by solving

$$f^2(x) = x,$$

i.e.

$$x^4 - 1.7x^2 - x - 0.1275 = 0. \tag{4.1.5}$$

This equation has four roots, two of which are fixed points of f. These two fixed points are the roots of the equation

$$x^2 - x - 0.85 = 0.$$

To eliminate these fixed points of f from equation (4.1.5), we divide the left-hand side of equation (4.1.5) by the left-hand side of the above equation to obtain the second-degree equation

$$x^2 + x + 0.15 = 0.$$

The equation has two solutions

$$x_{1,2} = \frac{-1 \pm \sqrt{0.4}}{2},$$

which are the 2-periodic points. To check the stability of the cycle $\{x_1, x_2\}$, we calculate

$$|f'(x_1)||f'(x_2)| = 0.6 < 1.$$

The cycle is asymptotically stable.

We now introduce a method to find periodic points of a given function. This method is based on the following lemma.

Lemma 4.1. (Carvalho's lemma[4]) If $x(t)$ is k-periodic, then either

$$x(t) = c_0 + \sum_{j=1}^{m}\left[c_j \cos\left(\frac{2j'\pi}{k}\right) + d_j \sin\left(\frac{2j'\pi}{k}\right)\right], \tag{4.1.6}$$

if $k > 1$ is odd and $m = (k-1)/2$, or

---

[4] See Carvalho (1998).

$$x(t) = c_0 + (-1)^n c_1 + \sum_{j=1}^{m} \left[c_j \cos\left(\frac{2j'\pi}{k}\right) + d_j \sin\left(\frac{2j'\pi}{k}\right) \right], \tag{4.1.7}$$

if $k = 2m$ and $t \geq 1$.

Example Consider a population dynamics

$$x(t+1) = x(t)\exp(r(1 - x(t))). \tag{4.1.8}$$

The population tends to grow at low densities and to decrease at high densities. The nontrivial fixed point of the equation is $x^* = 1$. As

$$f'(1) = 1 - r,$$

$x^* = 1$ is asymptotically stable if $0 < r \leq 2$ (check $r = 2$). At $r = 2$, $x^* = 1$ loses its stability and gives rise to an asymptotically stable 2 -cycle. Carvalho's lemma implies

$$x(t) = a + (-1)^t b.$$

Plugging this into equation (4.18) yields

$$a - (-1)^t b = \left(a + (-1)^t b\right)\exp\left(r\left(1 - a - (-1)^t b\right)\right).$$

The shift $t \mapsto t + 1$ gives

$$a + (-1)^t b = \left(a - (-1)^t b\right)\exp\left(r\left(1 - a + (-1)^t b\right)\right).$$

We have

$$a^2 - b^2 = \left(a^2 - b^2\right)\exp(r(1 - a)).$$

Thus either $a = b$, which gives the trivial solution zero, or $a = 1$. A 2 -periodic solution has the form

$$x(t) = 1 + (-1)^t b.$$

Plugging this equation again into equation (4.1.8) yields

$$1 - (-1)^t b = (1 + (-1)^t b)\exp((-1)^{t+1} rb).$$

Introduce

$$y = (-1)^{t+1} b.$$

Then

$$r = \frac{1}{y}\ln\left(\frac{1+y}{1-y}\right) = g(y).$$

The function $g(y)$ has its minimum at 0, where $g(0) = 2$. For $r < 2$, $g(y) = r$ has no solution, and we have no periodic points. However, each $r > 2$ determines $\pm\, y_r$.

Exercise 4.1

1 Find a periodic orbit of period 2 of

$$x(t+1) = -2x^2(t) + 1.$$

2 Let

$$x(t+1) = -ax^2(t) + 1.$$

For what values of the parameter a does a function f have none or one periodic orbit of period 2 ? Are there any values of a for which there is more than one periodic orbit of period 2 ?

3 Assume that x_0 is a periodic point of point 2 for a dynamical system governed by a differentiable function $f: R \to R$. Prove that the derivatives of the second iterate of f at x_0 and at $x(1) = f(x_0)$ are the same.

4 Assume that x_0 is a periodic point of point 3 for a dynamical system governed by a differentiable function $f: R \to R$. Prove that the derivatives of the third iterate of f at x_0, $x(1) = f(x_0)$, $x(2) = f(x(1))$ are the same.

5 Show the Schwarzian derivative $Sf(1/2) < 0$ for the map

$$f(x) = \frac{xe^{\beta(1-2x)}}{xe^{\beta(1-2x)} + 1 - x}.$$

6 Let

$$f(x) = x\exp\left[\beta\frac{1-x}{1+x}\right], \quad \beta > 4, \quad x \in (0, \infty).$$

Let $\{x_1, x_2\}$ be a 2 -cycle of f. Show that this 2 -cycle is asymptotically stable.

4.2 Period-doubling bifurcations

Let I be an interval, $a \in I$, and $f(x, a)$ be a *one-parameter family* of scalar maps. Assume that f is differentiable with continuous derivatives with respect to x and a. Let $J \subset I$ be an interval with more than one point and $x: J \to R$ be a continuous function such that for every $a \in J$ and $x(a)$ has the property

$$f(x(a), a) = x(a),$$

i.e., $x(a)$ is a stationary state of $f(x, a)$ for every $a \in J$. Then the graph of the map $x(a)$, i.e., the curve $(a, x(a))$, $a \in J$ is called a *branch* of fixed points of the one-parameter family $f(x, a)$. For simplicity, $x(a)$ is sometimes called a branch. In a similar manner, we define branches of periodic points of period p.

Example Let

$$f(x, a) = ax(1 - x), \quad x \in [0, 1], \; a > 0.$$

From $f(x, a) = x$, we have

$$x_1(a) = 0, \quad x_2(a) = \frac{a - 1}{a}.$$

The branch of fixed point $x_1(a) = 0$ is independent of a; while that of

$$x_2(a) = \frac{a - 1}{a}$$

is dependent on a. As

$$f'(x, a) = a(1 - 2x),$$

we conclude that: (i) 0 is an asymptotically stable fixed point for $0 < a < 1$; and (ii) 0 is an unstable fixed point for $a > 1$. When $a = 1$, we have

$$f'(0) = 1, \quad f''(0) = -2.$$

We may conclude that 0 is unstable if we consider negative as well as positive initial points in the neighborhood of 0. Since negative initial points are not in the domain of f, and 0 is semiasymptotically stable from the right, we conclude that 0 is asymptotically stable in the domain $[0, 1]$.

For $x_2(a) \in (0, 1]$, we require $a > 1$. Applying

$$f'(x, a) = a(1 - 2x)$$

to

$$x_2(a) = \frac{a - 1}{a},$$

we conclude that: (i) x_2 is an asymptotically stable fixed point for $1 < a \leq 3$; and (i) x_2 is an unstable fixed point for $a > 3$.

We now show that the map does not have any period point of period 2 for $a \in (-1, 3)$. A point x_0 is periodic of period 2 if $x(2) = x(0)$ and $x(1) \neq x(0)$. We have

$$x(1) = ax(0) - ax^2(0),$$
$$x(0) = ax(1) - ax^2(1). \tag{4.2.1}$$

Subtracting the second equation of equation (4.2.1) from the first yields

$$x(1) - x(0) = -a(x(1) - x(0)) + a(x(1) - x(0))(x(1) + x(0)).$$

Since $x(1) \neq x(0)$, from the above equation we obtain

$$x(1) + x(0) = \frac{1+a}{a}. \tag{4.2.2}$$

Adding the two equations of equation (4.2.1) gives

$$x(1) + x(0) = a(x(1) + x(0)) - a(x(1) + x(0))^2 + 2ax(1)x(0).$$

Substituting equation (4.2.2) into the above equation yields

$$x(1)x(0) = \frac{1+a}{a^2}. \tag{4.2.3}$$

We can determine $x(1)$ and $x(0)$ by solving the quadratic equation

$$z^2 - \frac{1+a}{a}z + \frac{1+a}{a^2} = 0. \tag{4.2.4}$$

Hence

$$x_3(a) \equiv x(1) = \frac{a + 1 + \sqrt{a^2 - 2a - 3}}{2a},$$

$$x_4(a) \equiv x(0) = \frac{a + 1 - \sqrt{a^2 - 2a - 3}}{2a}.$$

Both $x(1)$ and $x(0)$ are fixed points of f. We see that no periodic orbit of period 2 exists for $a \le 3$, and exactly exists for $a > 3$. That is, we have two branches of periodic points of $f(x, a)$ when $a > 3$. At $a = 3$, we obtain

$$x_2(a) = x_3(a) = x_4(a).$$

It should be noted that we may directly get solution (4.2.4) by dividing the equation $x = f^2(x)$ with $x - f(x)$.

The 2 -cycle is asymptotically stable if

$$|f'(x_3)||f'(x_4)| < 1,$$

or

$$-1 < a^2(1 - 2x_3)(1 - 2x_4) < 1.$$

Substituting the solutions into the above inequalities, we have

$$3 < a < 1 + \sqrt{6} \approx 3.44949\cdots.$$

Hence, the 2 -cycle is attracting if $3 < a < 3.44949\cdots$.

At $a = 1 + \sqrt{6}$, we have

$$f'(x_3)f'(x_4) = -1.$$

We conclude that the 2 -cycle is attracting. If $a > 1 + \sqrt{6}$, the 2 -cycle becomes unstable.

As the parameter a changes its values, the qualitative behavior of the solutions also changes. Let us denote the value of the parameter a at the k -th point where the changes occurs as a_k. We have the following table.

Table 4.2.1: The behavior of $f = ax(1-x)$ as a changes

parameter interval	type of behavior	critical value of a
$0 < a < 1$	zero equilibrium point is asymptotically stable	$a_0 = 1$
$1 < a < 3$	positive equilibrium point is asymptotically stable	$a_1 = 3$
$1 < a < 1 + \sqrt{3}$	period-2 solution is asymptotically stable	$a_2 = 1 + \sqrt{6}$

Definition 4.2.1. A value $a_0 \in I$ is a *bifurcation point* of the one-parameter family f if there are intervals

$$J_i \subset I, \quad a_0 \in J_i, \quad i = 1, 2 \text{ or } i = 1, 2, 3,$$

and continuous functions $x_i : J_i \to R$, whose graphs are branches of fixed points or periodic points of f such that

$$x_i(a_0) = x_j(a_0), \quad x_i(a) \neq x_j(a), \quad a \neq a_0, \quad i \neq j.$$

The value a_0 is called the *bifurcation value* or *critical value* of the parameter.

We require J_i to have nonzero length, and allow a_0 to be one of its endpoints. Hence, J_i may have the form

$$(b, c), \quad [a_0, b), \text{ or } (b, a_0].$$

When $i = 1, 2,$ then $J_1 = J_2$ and all points $x_i(a)$, $i = 1, 2,$ have the same period. When $i = 1, 2, 3,$ then

$$J_1 = (b, c), \quad J_2 = J_3 = [a_0, b) \text{ or } (b, a_0]$$

and the points of $x_i(a)$, $i = 2, 3,$ may have period equal or double the period of $x_1(a)$. We depict $(a, x(a))$ on the plane, representing the parameter a along the horizontal axis. The pictorial display obtained by plotting all branches of stationary states and periodic points is called the bifurcation diagram of f on I. At a bifurcation point a_0 there is frequently an exchange of stability between the different branches. In particular, the bifurcation taking place at a value of a is called *period-doubling* if the period of the points of one of the branches is double the period of the points of another.

With regard to the logistical map, if we continue to increase the value of a, we will find the next critical value a_3 which corresponds to the appearance of a prime period-4 solution, as shown in figure 4.2.1.

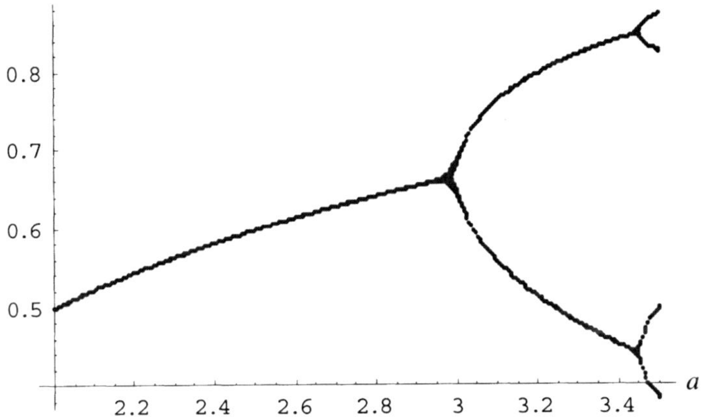

Figure 4.2.1: The bifurcation diagram for $2 \le a \le 3.5$

The next bifurcation value a_4 corresponds to a prime period-8 solution. In addition, it can be proved that for

$$a \in (a_2, a_3),$$

the prime period-4 solution is stable while the period-2 solution becomes unstable. Similarly, for

$$a \in (a_3, a_4),$$

the prime period-8 solution is stable while the prime period-4 solution becomes unstable. As this process continues, there is a sequence of bifurcation values of parameter $\{a_k\}_{k=0}^{\infty}$ with the following property: for

$$a \in (a_k, a_{k+1}),$$

the prime period-2^k is stable, while the periodic solutions of all periods

$$2, \cdots, 2^{k-1}$$

become unstable. This phenomenon is known as the *period-doubling bifurcation route to the chaos*. The entire sequence $\{a_k\}_{k=0}^{\infty}$ is called a *cascade of bifurcations*.

The sequence of period-doubling bifurcations ends at the value which is approximately $a = 3.56994\cdots$, where the map has the periodic solutions of all periods as well as some aperiodic solutions. The last situation is often described as *chaotic behavior* or *chaos*. The last period that can arise in this bifurcation process is period 3. There are three important features of this route to chaos. The periods finishes with 3; the order is known as Sharkovsky's order, which will be described below. There is only one periodic solution which is stable in each of the interval (a_k, a_{k+1}). The third is that the sequence $\{a_k\}_{k=0}^{\infty}$ has the remarkable property

$$\lim_{k \to \infty} \frac{b_k - b_{k-1}}{b_{k+1} - b_k} = \delta \approx 4.66920.$$

The constant δ is called *Myrberg's number* or *Feigenbaum's number*. It was first discovered by Myrberg and rediscovered by Feigenbaum.[5] Figure 4.2.2 is the bifurcation diagram of

$$f(x, a) = ax(1 - x).$$

Exercise 4.2

1 Determine the period-doubling bifurcation of the following maps $f(x, \mu)$ and investigate the stability of the periodic orbits of period 2.

(i) $\mu - x^2$;

(ii) $-\mu + x^2$;

---

[5] Myrberg (1958, 1959, 1963) and Feigenbaum (1978).

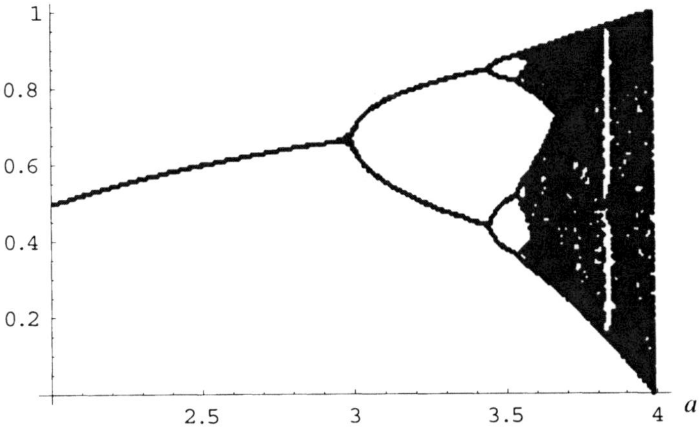

Figure 4.2.2: The bifurcation diagram of $f = ax(1-x)$

(iii) $\mu x + x^3$;

(iv) $\mu x^3 + (1-\mu)x$.

4.3 Aperiodic orbits

The orbit $O(x_0)$ of the dynamical system

$$x(t+1) = f(x(t))$$

has infinitely many states. These states may not be all distinct. For instance, they are all equal when x_0 is a fixed point and there are only p distinct points when the orbit $O(x_0)$ is periodic of period p.

Definition 4.3.1. A point $\widetilde{x}$ is said to be a *limit point* of $O(x_0)$ if there exists a subsequence

$$\{x(t_k) : k = 0, 1, \cdots\}$$

of $O(x_0)$ such that

$$\|x(t_k) - \tilde{x}\| \to 0 \text{ as } k \to \infty.$$

The *limit set* $L(x_0)$ of the orbit $O(x_0)$ is the set of all limit points of the orbit.

Definition 4.3.2. An orbit $O(x_0)$ is said to be *asymptotically stationary* if its limit set is a stationary state, and asymptotically periodic if its limit set is a periodic orbit. An orbit $O(x_0)$ such that

$$x(t + p) = x(t)$$

for some $t \geq 1$ and some $p \geq 1$ is said to be *eventually stationary* if $p = 1$ and *eventually periodic* if $p > 1$.

Every eventually stationary (eventually periodic orbit) is asymptotically stationary (asymptotically periodic). The converse is not true. For instance, for the difference equation

$$x(t + 1) = x(t)(1 - x(t)),$$

the orbit

$$x(0) = \frac{1}{2}, \quad x(1) = \frac{1}{4}, \quad x(2) = \frac{3}{16}, \quad \cdots$$

is asymptotically stationary (with the limit point 0), but not eventually stationary.

Example Let

$$x(t + 1) = 4x(t)(1 - x(t)).$$

The orbit $0(1/2)$

$$x(0) = \frac{1}{2}, \quad x(1) = 1, \quad x(2) = x(3) = \cdots = 0,$$

is eventually stationary ($t = 2,\ p = 1$).

It can be proved that whenever $L(x_0)$ has only finitely many points p, they actually constitute a periodic orbit with period p of the system.

Definition 4.3.3. The orbit $O(x_0)$ is *aperiodic* if its set $L(x_0)$ is not finite.

Example Figure 4.3.1 provides numerical evidence that the orbit of the dynamical system

$$f(x) = 4x(1 - x),$$

starting at $x_0 = 0.3$ is aperiodic. We will analyze aperiodic behavior later on.

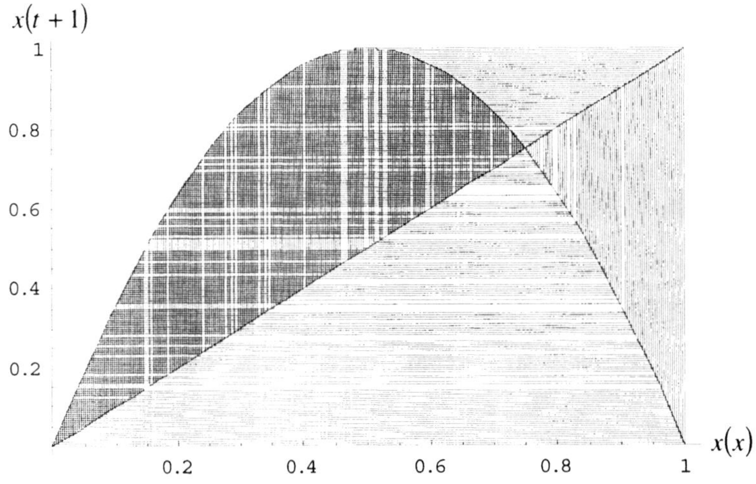

Figure 4.3.1: The orbit of $f(x) = 4x(1 - x)$ with $x_0 = 0.3$

It is difficult to establish theoretically the aperiodic character of an orbit. The following result ensures the existence of aperiodic orbits for dynamical systems in the real line.

Theorem 4.3.1. (Li-Yorke[6]) Let I be an interval and $f : I \to I$ be continuous. Assume that f has a periodic orbit of period 3. Then f has a periodic orbit of

[6] The proof is referred to Li and Yorke (1975), or Block and Coppel (1992).

every period and there is an infinite set S contained in S such that every orbit starting from a point of S is aperiodic.

Example Consider $f : [0, 1] \to [0, 1]$ defined by

$$f(x) = \begin{cases} 3x & \text{for } 0 \le x \le \frac{1}{3}, \\ \frac{17}{9} - \frac{8x}{3} & \text{for } \frac{1}{3} < x \le \frac{2}{3}, \\ 1/9 & \text{for } \frac{2}{3} < x \le 1. \end{cases}$$

The map is continuous and

$$f\left(\frac{1}{9}\right) = \frac{1}{3}, \quad f\left(\frac{1}{3}\right) = 1, \quad f(1) = \frac{1}{9}.$$

Hence, $x = 1/9$ is periodic of period 3. According to theorem 4.3.1, there is an infinite set S that the orbit of every point of S is aperiodic. Determining S is not easy.

Example Consider the difference equation generated by the tent function

$$T(x) = \begin{cases} 2x & \text{for } 0 \le x \le \frac{1}{2}, \\ 2(1 - x) & \text{for } \frac{1}{2} < x \le 1. \end{cases}$$

This may also be written in the compact form

$$T(x) = 1 - 2\left|x - \frac{1}{2}\right|.$$

We first observe that the periodic points of period 2 are the fixed points of T^2 which is given by

$$T^2(x)=\begin{cases} 4x & \text{for } 0 \le x < \dfrac{1}{4}, \\ 2(1-2x) & \text{for } \dfrac{1}{4} \le x < \dfrac{1}{2}, \\ 4\left(x-\dfrac{1}{2}\right) & \text{for } \dfrac{1}{2} \le x < \dfrac{3}{4}, \\ 4(1-x) & \text{for } \dfrac{1}{4} < x < \dfrac{1}{2}. \end{cases}$$

There are four fixed points, 0, 0.4, $2/3$, and 0.8, two of which, 0 and $2/3$ are fixed points of T. Hence, $\{0.4, 0.8\}$ is the only cycle of T. As shown in figure 4.3.2, $\{2/7, 4/7, 6/7\}$, is a 3-cycle.

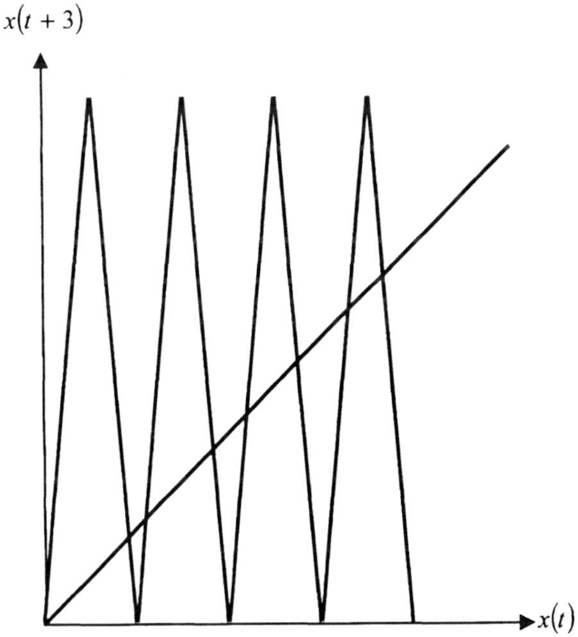

Figure 4.3.2: Fixed points of T^3

The Li-Yorke theorem is a special case of a theorem published in 1964 by the Ukranian mathematician A.N. Sharkovsky. Sharkovsky introduced a new ordering of the positive integers $\triangleright$ in which number 3 appears first. He proved that if $k \triangleright m$ and f has a k-periodic point, then it must have an m-periodic point. Here, the notation $k \triangleright m$ means that k appears before m in Sharkovsky's ordering. Sharkovsky's ordering is defined as follows

$$3 \triangleright 5 \triangleright 7 \triangleright \cdots \triangleright 2\times 3 \triangleright 2\times 5 \triangleright 2\times 7 \triangleright \cdots \triangleright 2^n \times 3 \triangleright 2^n \times 5 \triangleright 2^n \times 7 \cdots$$
$$\triangleright 2^n \cdots \triangleright 2^2 \triangleright 2 \triangleright 1$$

We first list all the odd integers except 1 and 2, then 2 times the old integers, 2^2 times the old integers, etc. This is followed by powers of 2 in descending order ending in 1.

Theorem 4.3.2. (Sharkovsky's theorem[7]) Let I be an interval in R (finite or infinite). Let

$$f : I \to I$$

be a continuous map. Let f has a periodic point of period k, then it must possess a periodic point of period m for all m with $k \triangleright m$.

This theorem gives a simply method for checking complicated behavior of the dynamical system. If we plot the *bisector* $y = x$ with the graph of the map $f(x)$ and graph of the third iterate $f^3(x)$, and if the graphs of the bisector and $f^3(x)$ intersect in some points that are not the equilibrium points, then there is at least one periodic point of period 3 and consequently there are periodic points of all other periods. Similarly, one can check virtually the existence of other periodic points of reasonably small periods. Figure 4.3.3 illustrates this method. The function f^3 is shown as a thick line; f in a thin line; and the identity function in a dashed line. Two intersections correspond to fixed points of f. The remaining six intersections indicate that there are two prime period-3 solutions.

It should be remarked that Sharkovsky's theorem does not extend to two or higher dimensions. It is known that one can construct a continuous map with a periodic point of 5 but not with periodic point 3, or with a periodic point of period 11 but not with period 5, etc.

---

[7] The proof is referred to Block and Coppel (1992).

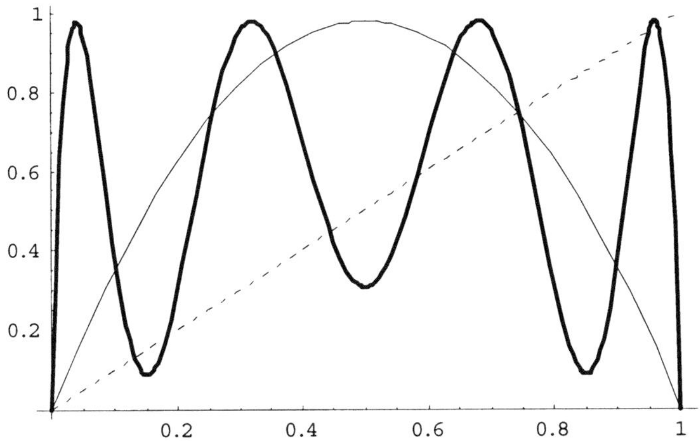

Figure 4.3.3: Plot of $f^3(x)$ for $f = 4.91x(1 - x)$

Consider a dynamical system: $f : I \to I$, where f is continuous and I is a bounded interval. We shall say that f is *chaotic* in I in the Li-Yorke sense if f has a periodic point in I of period 3. Notice that as a consequence of Sharkovsky's theorem f has a periodic orbit of period p for every integer p. Moreover, Li and Yorke proved that there is an infinite (actually, uncountable) set $S \subset I$ such that for every $x \in S$ is aperiodic and unstable.

Example Let

$$f = 2|x| - 1$$

and $[-1, 1]$. Then f has a periodic point of period 3,

$$\left\{-\frac{7}{9}, \frac{5}{9}, \frac{1}{9}\right\}.$$

Hence f is chaotic in the Li-Yorke sense.

Theorem 4.3.2. (a converse of Sharkovsky's theorem[8]) For any positive integer r, there exists a continuous map $f_r : I \to I$ on the closed interval I such that f_r has a point of minimal period r but no points of minimal periods s, for all positive integers s that precede r in the Sharkovsky's ordering, i.e., $s \rhd r$.

The following theorem by Singer tries to solve the problem of how many attracting periodic points a differentiable map can possess.

Theorem 4.3.3. (Singer's theorem[9]) Let $f : I \to I$ be a map defined on the closed interval I such that the Schwarizian derivative of f

$$Sf(x) < 0,$$

for all $x \in I$. If f has n critical points in I, then for every positive integer, k, the map f has at most $n + 2$ attracting period-k solutions.

Exercise 4.3

1 Show that for the map

$$f(x) = 2x - [2x],$$

where $[x]$ is the greatest integer less than or equal to x, the orbit $O(3/8)$ is eventually stationary and the orbit $O(1/24)$ is eventually periodic.

2 Let

$$f(x) = \begin{cases} 2x & \text{for } 0 \le x \le 0.5, \\ 2 - 2x & \text{for } 0.5 < x \le 1. \end{cases}$$

Find a periodic orbit of period 3 for f. Can the result of Li-Yorke be applied to f?

3 Let

---

[8] See Elaydi (1996).

[9] Singer (1978).

$$f(x) = 2|x| - 1.$$

Show that f has aperiodic orbits in $[-1, 1]$.

4.4 Some types of bifurcations

This section provides some typical examples of bifurcations.

Example Consider

$$f(x, \mu) = \mu x - x^3.$$

Fixed points of f are the solutions of

$$x = \mu x - x^3.$$

Hence, $x_1(\mu) = 0$ is always a fixed point. No other fixed point exists for $\mu \leq 1$. When μ crosses 1, two more stationary states appear, namely

$$x_2(\mu) = \sqrt{a - 1}, \quad x_3(\mu) = -\sqrt{a - 1}.$$

From

$$f_x = \mu - 3x^2,$$

we see that 0 is a repeller for $|\mu| > 1$ and a sink for $|\mu| < 1$. At $\mu = 1$, we have $f_x(0, 1) = 1$. Substituting $x_2(\mu)$ and $x_3(\mu)$ respectively into f_x, we conclude that for $\mu \in (1, 2)$, the points of the two branches are sinks for $\mu \in (1, 2)$ and source for $\mu > 2$. At $\mu = 1$, we have an exchange of stability between $x_1(\mu)$ and $x_2(\mu)$, $x_3(\mu)$. At $\mu = -1$, a periodic orbit of period 2 arises. Since the derivative of the second iterate of f is $(2a + 3)^2$, which is larger than 1 for $\mu > -1$, the periodic orbit is always a source. There is no exchange of stability at $\mu = -1$. Figure 4.4.1 is the bifurcation diagram for the map.

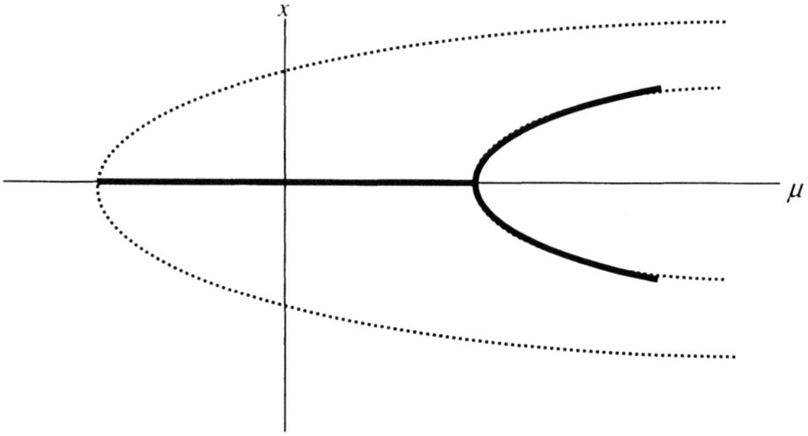

Figure 4.4.1: Part of the bifurcation diagram of $f = \mu x - x^3$

Example Let

$$f(x, \mu) = x^2 - \mu.$$

Fixed points of f are given by

$$x_{1,2}(\mu) = \frac{1}{2} \pm \sqrt{\frac{1}{4} + \mu}.$$

The fixed points exist for $\mu \geq -0.25$. At $\mu_0 = -0.25$,

$$x_1(-0.25) = x_2(-0.25).$$

Hence, μ_0 is a bifurcation point. The type of bifurcation taking place at μ_0 is called *supercritical fold*, as illustrated in figure 4.4.2. Along the branch $(x_1(\mu), \mu)$, $f_x > 1$. That is, $x_1(\mu)$ are sources for all $\mu \geq -0.25$. Along $(x_2(\mu), \mu)$

$$f_x = 1 - (1 + 4a)^2,$$

$x_2(\mu)$ are sinks for

$$\mu \in \left(-\frac{1}{4}, \frac{3}{4}\right),$$

and sources for $\mu > 3/4$.

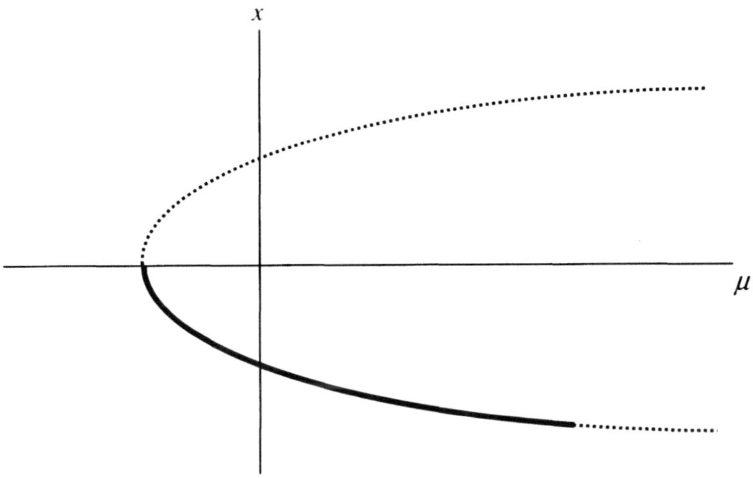

Figure 4.4.2: Supercritical fold bifurcation for $f = x^2 - \mu$

Example Let

$$f(x, \mu) = x^2 + \mu.$$

Fixed points of f are given by

$$x_{1,2}(\mu) = \frac{1}{2} \pm \sqrt{\frac{1}{4} - \mu}.$$

The fixed points exist for $\mu \leq 0.25$. At $\mu_0 = 0.25$,

$$x_1(0.25) = x_2(0.25).$$

Hence, μ_0 is a bifurcation point. The type of bifurcation taking place at μ_0 is called *subcritical fold*, as illustrated in figure 4.4.3. The fixed points of the upper branch are always sources. The fixed points of the lower branch are sinks for

$$\mu \in \left(-\frac{3}{4}, \frac{1}{4}\right),$$

and sources for $\mu < -3/4$.

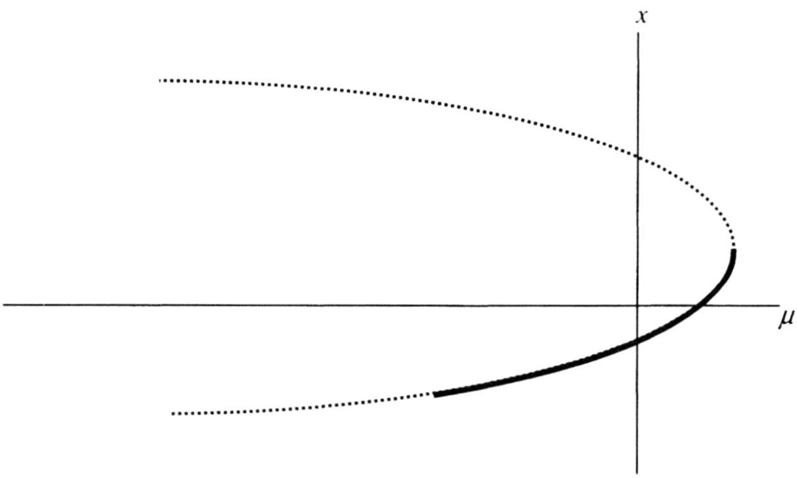

Figure 4.4.3: Subcritical fold bifurcation for $f = x^2 + \mu$

Example Let

$$f(x, \mu) = \mu x - x^3.$$

$x_1(\mu) = 0$ is a fixed point independent of μ. No other fixed point exists for $\mu \leq 1$. When μ crosses 1, two more fixed points appear for $\mu > 1$,

$$x_{2,3}(\mu) = \pm\sqrt{\mu - 1}.$$

Hence, $\mu = 1$ is a bifurcation point. The type of bifurcation taking place at $\mu = 1$ is called *supercritical pitchfork*. The *subcritical pitchfork* case is when $\mu \leq 1$.

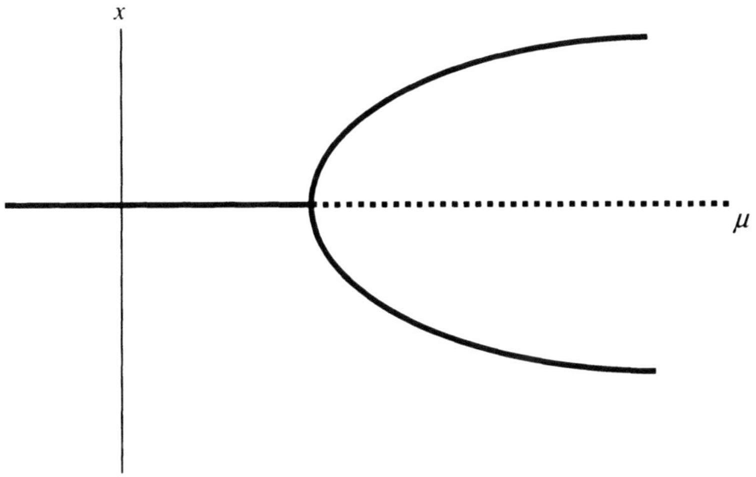

Figure 4.4.4: Supercritical pitchfork bifurcation for $f = \mu x - x^3$

Example Consider the logistic map again

$$f(x, \mu) = \mu x(1 - x).$$

We know that the two branches of fixed points

$$x_1(\mu) = 0,$$
$$x_2(\mu) = 1 - \frac{1}{\mu},$$

meet at $\mu = 1$. Hence, $\mu = 1$ is a bifurcation point. This type of bifurcation is called *transcritical* (see figure 4.4.5).

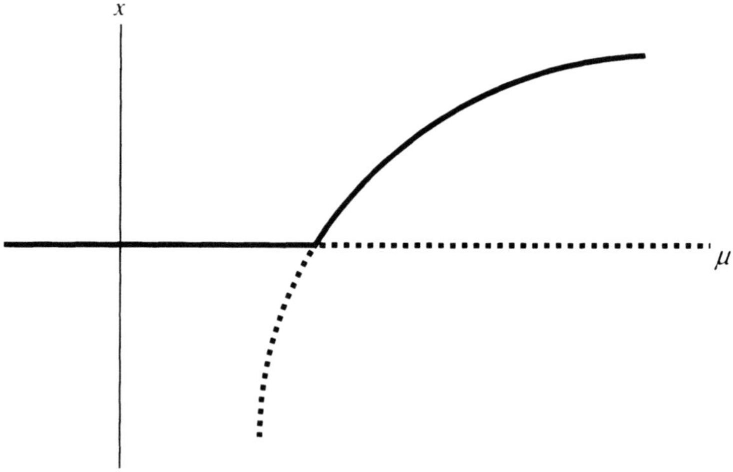

Figure 4.4.5: Transcritical bifurcation for $f = \mu x(1 - x)$

In summary, we see that transcritical bifurcation takes place at $\mu = \mu_0$ when two branches of fixed points (or periodic points of the same period) exist in an open interval J, $\mu_0 \in J$, and they meet at $\mu = \mu_0$. Fold bifurcation is observed when two branches of fixed points (or periodic points of the same period) exist only in an interval $[\mu_0, \mu_1)$ (supercritical case) or $(\mu_1, \mu_0]$ (subcritical case) and they equal at $\mu = \mu_0$. Pitchfork bifurcation requires three branches of fixed points (or periodic points of the same period). Two of them exist only in an interval $[\mu_0, \mu_1)$ (supercritical case) or $(\mu_1, \mu_0]$ (subcritical case). The third exists in an open interval J, $\mu_0 \in J$. The three branches meet at $\mu = \mu_0$. It can be shown that

$$f_x(x(\mu_0), \mu_0) = 1$$

is necessary for μ_0 to be a fold, transcritical, or pitchfork bifurcation. The condition is not sufficient. This is verified by

$$f(x, \mu) = (1 + \mu^2)x + x^3.$$

At $\mu = 0$, there is a unique fixed point $x_1(0) = 0$. We have

$$f'(0,0) = 1.$$

But $\mu = 0$ is not a bifurcation point.

The above examples are typical cases of bifurcation types. We provide the necessary conditions for the four types of bifurcations. We require the function $f(x, \mu)$, its first and second partial derivatives, and f_{xxx} to be continuous.

Table 4.4.1: The conditions for the four types of bifurcations[10]

	f_x	f_μ	f_{xx}	other at (x_0, μ_0)
fold	1	$\neq 0$	$\neq 0$	$f_{xx}f_\mu < 0$ supercritical
transcritical	1	0	$\neq 0$	$f_{x\mu}^2 - f_{xx}f_{\mu\mu} > 0$
pitchfork	1	0	0	$f_{x\mu} \neq 0$, $f_{xxx} \neq 0$, $f_{x\mu}f_{xxx} < 0$ supercritical
period-doubling	-1			$r = f_{xx}f_\mu + 2f_{x\mu} \neq 0$, $s = 2f_{xxx} + 3f_{xx}^2 \neq 0$, $rs < 0$, supercritical

Exercise 4.4

1 Study the bifurcation diagram of

$$f(x, \mu) = \mu - x^2.$$

In particular, determine the fold bifurcation and investigate the stability of the two branches of the fixed point.

2 Study the bifurcation diagram of

$$f(x, \mu) = \mu x - x^2.$$

In particular, determine the transcritical bifurcation and investigate the stability of the branches of the fixed point.

---

[10] The table is referred to Martelli (1999: 92). The examples of this section are also cited from the same source.

3 Study the bifurcation diagram of

$$f(x, \mu) = -\mu x + x^3.$$

In particular, determine the pitchfork bifurcation and investigate the stability of branches of the fixed point.

4 Study the bifurcation diagram of

$$f(x, \mu) = (x + \mu^2 - 1)(x^2 - 2x - \mu) + x.$$

Determine type of bifurcation point when $\mu = -1$.

4.5 Liapunov numbers

The Liapunov number is important for measuring the complexity of solutions behavior of difference equations. It is known that chaotic dynamics is characterized by an exponential divergence of initial close points. In the case of the one-dimensional discrete map of an interval (a, b) into itself

$$x(t + 1) = f(x(t)),$$

the Liapunov exponent is a measure of the divergence of two orbits starting with slightly different initial conditions x_0 and $x_0 \pm \delta_0$. If x_0 is a point of a period-k orbit and if we start the orbit from a nearby point $x_0 \pm \delta_0$, then after one iteration the distance between the two is approximated by

$$\delta_1 \approx |f'(x_0)|\delta_0 = M_0\delta_0.$$

where M_0 is the magnification factor for the first step. At the second step,

$$\delta_2 \approx |f'(x(1))|\delta_1 = M_1\delta_1 = M_1 M_0 \delta_0.$$

Continuing in this manner, we conclude that the total magnification factor over one cycle of the period-k orbit is the product, $M_0 M_1 \cdots M_{k-1}$. We are concerned with the geometric average of the factor

$$(M_0 M_1 \cdots M_{k-1})^{1/k},$$

which by taking logarithms leads to the arithmetic average

$$\lambda = \ln(M_0 M_1 \cdots M_{k-1})^{1/k} = \frac{1}{k}\sum_{t=0}^{k-1} \ln M_t = \frac{1}{k}\sum_{t=0}^{k-1} \ln|f'(x(t))|.$$

The condition for stability of a periodic orbit is that the average magnification factor is less than 1, which is equivalent to say that the orbit is stable (unstable) if $\lambda < (>)\ 0$. An interpretation of the Liapunov exponent is the measure of information loss during the process of iteration.

Definition 4.5.1. Let f be a smooth map on R and x_0 be a given initial point. The *Liapunov exponent* $\lambda(x_0)$ of a map f is given by

$$\lambda(x_0) = \lim_{k\to\infty} \frac{1}{k}\sum_{t=0}^{t-1} \ln|f'(x(t))|,$$

provided the limit exists. In the case when any of the derivatives is zero, set

$$\lambda(x_0) = -\infty.$$

The *Liapunov number* $L(x_0)$ is defined as the exponent of the Liapunov exponent, whenever the latter exists

$$L(x_0) \equiv \exp(\lambda(x_0)).$$

In fact, we may consider two orbits $\{\tilde{x}(t)\}$ and $\{x(t)\}$ originating from two nearby points $\tilde{x}_0$ and x_0. At the nth iteration they are separated by an amount

$$|\tilde{x}(t) - x(t)| = |f'(\tilde{x}_0) - f'(x_0)|.$$

Taking x_0 as a constant, expanding $f'(\tilde{x}_0)$ in a Taylor series around x_0 and retaining only the first-order term, we have

$$\tilde{x}(t) - x(t) \approx (f'(x_0) f'(x(1)) \cdots f'(x(t-1)))(\tilde{x}_0 - x_0),$$

and, providing that the derivative is different from zero, the approximation can be made very accurate by taking $|\tilde{x}_0 - x_0|$ sufficiently small. Asymptotically, we have

$$\lim_{t\to\infty}|\tilde{x}(t) - x(t)| \approx e^{n\lambda(x_0)}(\tilde{x}_0 - x_0).$$

We thus can interpret that $\lambda(x_0)$ is the (local) average, asymptotic, exponential rate of divergence of nearby orbits. Usually, it is difficult to exactly calculate Liapunov exponents. But there are exceptions.

Example Consider the tent map

$$T_p(x) = \begin{cases} px & \text{if } x \leq \frac{1}{2}, \\ p(1-x) & \text{if } x > \frac{1}{2}. \end{cases}$$

We have

$$T_p'(x) = \begin{cases} p & \text{if } x < \frac{1}{2}, \\ -p & \text{if } x > \frac{1}{2}, \end{cases}$$

and $T_p'(1/2)$ is undefined. For any orbit of the tent map that does not contain the point $1/2$, we have

$$\lambda(x_0) = \ln p.$$

This holds for all x_0 that are not eventually equal to $1/2$.

Definition 4.5.2. An orbit

$$\{x(1), x(2), x(3), \cdots\},$$

is called asymptotically periodic if there exists a periodic orbit

$$\{y(1), y(2), y(3), \cdots\},$$

such that

$$\lim_{t \to \infty} |x(t) - y(t)| = 0.$$

Definition 4.5.3. Let f be a map of R and let

$$O(x_0) = \{x(0), x(1), x(2), \cdots\}$$

be a bounded orbit of f. The orbit is *chaotic in the sense of Yorke*[11] if: (i) $O(x_0)$ is not asymptotically periodic; (ii) no Liapunov exponent is 0; and (iii) $\lambda(x_0) > 0$.

In general, there are very few examples where the Liapunov exponents can be computed exactly. In most cases one can just compute them numerically. Figure 4.5.1 depicts the Liapunov exponents of the logistic map $f = \mu x(1-x)$, for $\mu \in [2.8, 4]$. It can be seen that as μ is approaching 3.6, the exponents become positive. The system becomes chaotic as shown in figure 4.5.1.

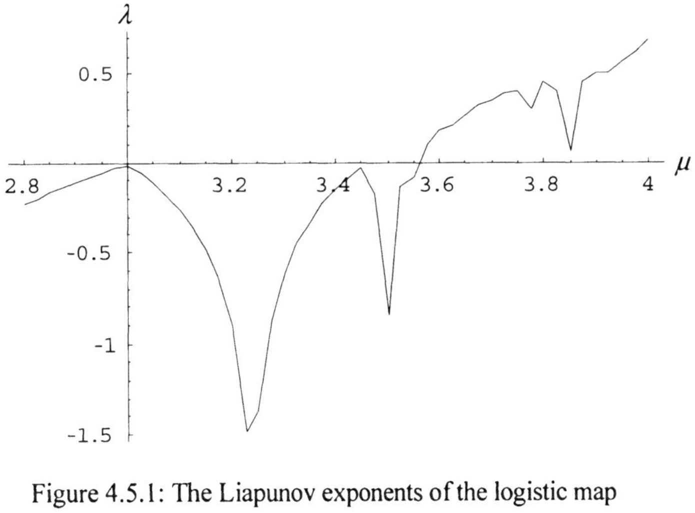

Figure 4.5.1: The Liapunov exponents of the logistic map

[11] See Alligood *et al.* (1997).

Example (a model of chaotic labor market) We now introduce another economic mechanism of economic chaos in labor market. The model is proposed by Bhaduri to demonstrate the existence of chaos when wage bargain takes place according to the usual specification of the Phillips curve while firms maximize profit.[12] The lack of coordination between profit-maximizing firms and wage-bargaining workers may result in an irregular dynamic path. In this model, the wage settlement takes place according to the usual specification of the Phillips curve. All agents are endowed with perfect foresight in the sense that they correctly anticipate the short-term rate of inflation. Neutrality of money is thus built into the model by assumption. There is no misperception of the real wage rate on the part of the concerned agent.

Consider the following linear Phillips curve[13]

$$\frac{w(t+1)-w(t)}{w(t)} = a\frac{L(t)}{N} - b, \quad a, b > 0, \tag{4.5.1}$$

where $L(t)$ is the level of employment, and N is the level of full employment. Perfect foresight resulting in consistently correct short-term anticipation of the inflation rate by the workers implies that workers bargain for real wages. The variable w in equation (4.5.1) is interpreted as the real wage. The unique rate of unemployment u^* that keeps the real wage rate at a stationary level is given by

$$u^* = 1 - \frac{L^*}{N} = \frac{b}{a}.$$

Perfect foresight on the part of the firms ensures that they also anticipate correctly the real wage rate and maximizes profit by equating the real wage rate to the marginal product of labor which is assumed to be diminishing linearly. That is

$$w(t) = \alpha - \beta L(t), \quad \alpha, \beta > 0. \tag{4.5.2}$$

Inserting equation (4.5.2) into equation (4.5.1) yields

$$w(t+1) = \left(1 + \frac{\alpha a}{\beta N} - b\right)w(t) - \frac{a}{\beta N}w^2(t).$$

---

[12] See Bhaduri (2002).

[13] See Phillips (1954). There are various forms of the so-called Phillips curve, see, for instance, Goodwin (1967), Friedman (1968), and Hahn and Solow (1985).

Introduce the following linear transformation

$$x(t) = \frac{kw(t)}{1 + \alpha k - b}, \quad k \equiv \frac{a}{\beta N},$$

then the difference equation for the wage rate is transformed into

$$x(t+1) = Ax(t)(1 - x(t)), \quad 1 > x(t) > 0.$$

This is the much studied logistic map. The reader is encouraged to interpret economic implications of the model.

Example (rational choice and erratic behavior) We now introduce a behavioral model of chaos by Benhabib and Day.[14] It is assumed that preference is changeable as people experience more. Let us begin describing optimal behavior of the consumer in t by maximizing the Cobb-Douglas utility function

$$U(t) = x^a(t) y^{1-a}(t),$$

subject to the budget constraint

$$p_x x + p_y y = m,$$

where x and y denote the consumption of two goods, p_x and p_y their prices, and m is the income. The demand functions from the optimal problem are given by

$$x(t) = a\frac{m}{p_x}, \quad y(t) = (1 - a)\frac{m}{p_y}.$$

Assume that the parameter a in the utility function depends on past choices in the following way

$$a(t+1) = bx(t)y(t).$$

Substituting the preference change equation into the demand functions yields

[14] This example is based on Benhabib and Day (1981).

$$x(t+1) = \frac{bm}{p_x} x(t)y(t),$$

$$y(t+1) = (1 - bx(t)y(t))\frac{m}{p_y}.$$

We now concentrate on x. For simplicity, we normalize the prices $p_x = p_y = 1$. From

$$x(t) + y(t) = m$$

and the demand equation for x, we get

$$x(t+1) = bmx(t)(m - x(t)) \equiv f(x(t)).$$

The unique nontrivial fixed point is

$$x^* = \frac{bm^2 - 1}{bm}.$$

We require $bm^2 > 1$ so that x^* is positive. It is straightforward to show that the map f has its maximum at $x = m/2$; the maximum consumption is

$$f\left(\frac{m}{2}\right) = \frac{bm^3}{4}.$$

On the other hand, as $x \le m$, we have $bm^3/4 \le m$, i.e., $bm^2 \le 4$. In summary, we should require

$$1 < bm^2 < 4.$$

to have meaningful solutions. Introduce

$$z(t) = \frac{x(t)}{m},$$

we have

$$z(t+1) = \alpha z(t)(1 - x(t)), \quad 1 < \alpha \leq 4,$$

where $\alpha \equiv bm^2$. This is the logistical map. Hence, the model exhibits chaos.

4.6 Chaos

This section gives another definition of chaos.

Definition 4.6.1. The map f on a metric space X is said to possess *sensitive dependence on initial conditions,* if there exists $\varepsilon > 0$ such that for any $x_0 \in X$ and any open set U containing x_0 there exists $y_0 \in U$ and $k \in Z^+$ such that

$$d(f^k(x_0), f^k(y_0)) > \varepsilon.$$

Example Consider

$$x(t+1) = cx(t), \quad c > 1.$$

Let $y_0 = x_0 + \delta$. Then

$$f^t(y_0) - f^t(x_0) = c^t \delta.$$

Hence

$$\left| f^t(y_0) - f^t(x_0) \right|$$

will increase to ∞ as t goes to ∞, regardless of how small δ is.

Definition 4.6.2. Let f be a map on a metric space (X, d).[15] Then f is said to be (topologically) *transitive* if for any pair of nonempty open sets U and V, there exists a positive integer k such that

[15] The concepts related to metric space are referred to appendix A.3.

$$f^k(U) \cap V \neq 0.$$

Under a transitive map a point wanders all over the space X where its orbit gets as close as we wish from every point in x.

Theorem 4.6.1. Let $f : X \to X$, where X is a metric space. Then the map f is transitive if it has a dense orbit. Furthermore, if X is closed interval in R, then f is transitive if and only if it has a dense orbit.

The following definition of chaos is due to Devaney.[16]

Definition 4.6.3. A map $f : X \to X$, where X is a metric space, is said to be *chaotic* if (i) f is transitive; (ii) the set of periodic points P is dense in X; and (iii) f has sensitive dependence on initial conditions.

It should be noted that it has been proved now that conditions (i) and (ii) in definition 4.6.3 imply condition (iii) of sensitive dependence on initial conditions; but no other two conditions imply the third. The proof of the following theorem is referred to Elaydi.[17]

Theorem 4.6.2. Let $f : X \to X$ be a continuous map on a metric space (X, d). If f is transitive and its set of periodic points is dense, then f possesses sensitive dependence on initial conditions, i.e., f is chaotic.

The following theorem demonstrates that for continuous maps on intervals in R, transitivity implies that the set of periodic points is dense.

Theorem 4.6.3. Let $f : J \to J$ be a continuous map on an interval J (not necessarily finite) in R. If f is transitive, the set of periodic points is dense, that is, f is chaotic.

Example (chaos in a demand and supply model) We now demonstrate chaos in a simple demand-supply model.[18] We consider that supply involves a time lag.

---

[16] Devaney (1989). See also Peitgen *et al.* (1992)

[17] Section 3.5 in Elaydi (2000).

[18] This example draws heavily on Hommes (1991: section 5.1). See also Shone (2002: section 8.1).

At low prices supply increases slowly, partly because of start-up costs and fixed costs of production. Supply might also increase only slowly at high prices, say, because of capacity constraints. This suggests a S-supply curve. The *arctan function* exhibits such a S-shape. Let us express supply $q^s(t)$ as function of expected price $p^e(t)$ by

$$q^s(t) = \arctan(\mu p^e(t)).$$

The origin is an inflection point. As shown in figure 4.6.1, the parameter μ determines the steepness of the S-shape. The higher the value of μ the steeper the curve.

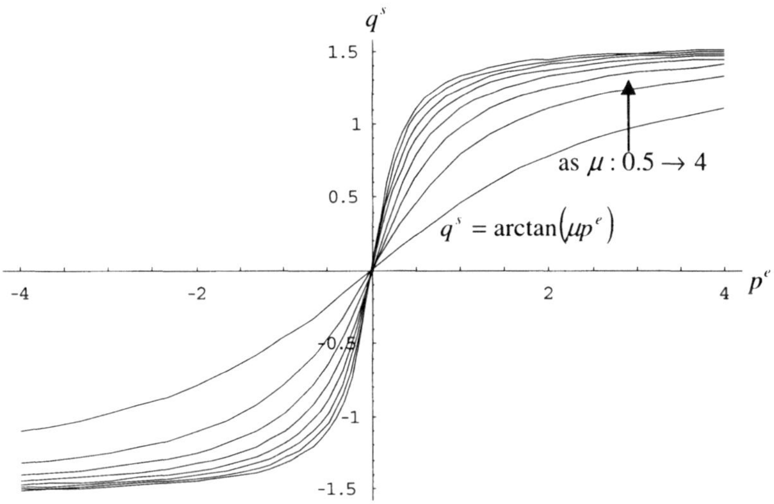

Figure 4.6.1: Specified relations between expected price and supply

For simplicity, assume that demand is a linear function of actual prices, i.e.

$$q^d(t) = a - bp(t), \quad b > 0.$$

We assume that price expectation is formed as follows

$$p^e(t+1) = \lambda p(t) + (1-\lambda)p^e(t).$$

It can be shown that under these specifications, the market condition that the demand equals supply is expressed by the following difference equation

$$p^e(t+1) = (1-\lambda)p^e(t) + \frac{a\lambda}{b} - \frac{\lambda \arctan(\mu p^e(t))}{b} \equiv f(p^e(t)). \tag{4.6.1}$$

We will demonstrate dynamic behavior of the model by simulation. In the remainder of this section, we fix: $\lambda = 0.3$, $b = 0.25$ and consider a as a bifurcation parameter for different values of μ. In the case of $\mu = 0.5$, figure 4.6.2 depicts the bifurcation diagram for

$$a \in [-1.25, 1.25].$$

We see that there is a unique fixed point for the map f.

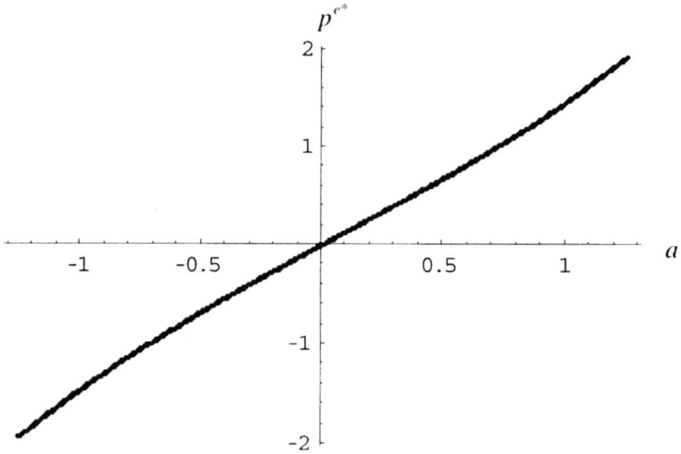

Figure 4.6.2: Bifurcation diagram for $\mu = 0.5$, $a \in [-1.25, 1.25]$

Let us rise the value of μ to 3 and depict the bifurcation diagram with a as the bifurcation parameter as in figure 4.6.3. For low values of a, there is a unique fixed point. Around $a = -0.9$ a period-doubling bifurcation occurs. The stable orbit remains until a reaches 0.9 and then a period-halving occurs. Thereafter the system settles down again to a unique stable equilibrium. The diagram is symmetrical about the origin because of the characteristic of the arctan function.

In figures 4.6.5 to 4.6.6, we depict respectively the bifurcation diagrams when

$$\mu = 4, \quad \mu = 4.5.$$

We see that within the period-four orbit chaos occurs.

Figure 4.6.4 depicts the bifurcation diagram when $\mu = 3.5$. The figure shows a doubling bifurcation into a period-four orbit, which then turns into a period-two orbit and finally a stable equilibrium.

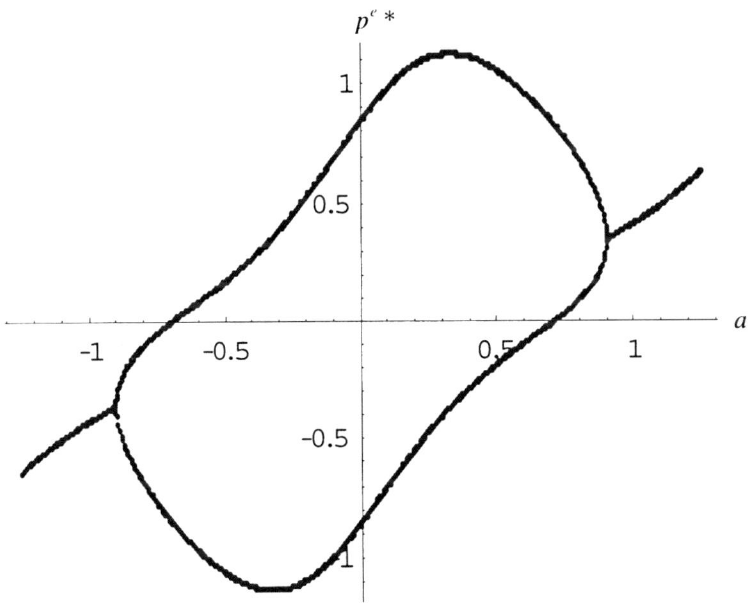

Figure 4.6.3: Bifurcation diagram for $\mu = 3$, $a \in [-1.25, 1.25]$

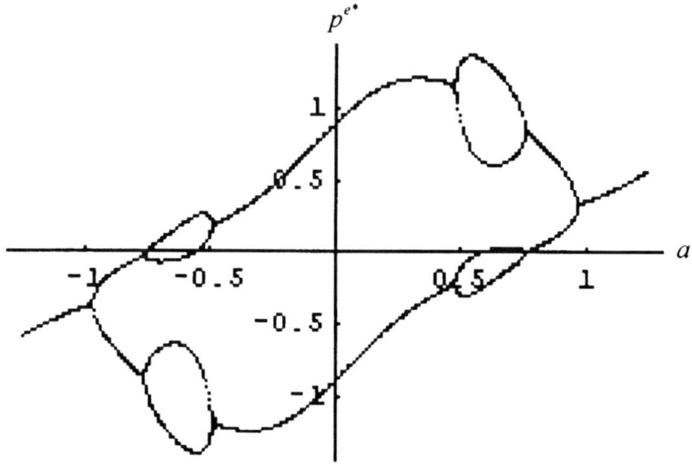

Figure 4.6.4: Bifurcation diagram for $\mu = 3.5$, $a \in [-1.25, 1.25]$

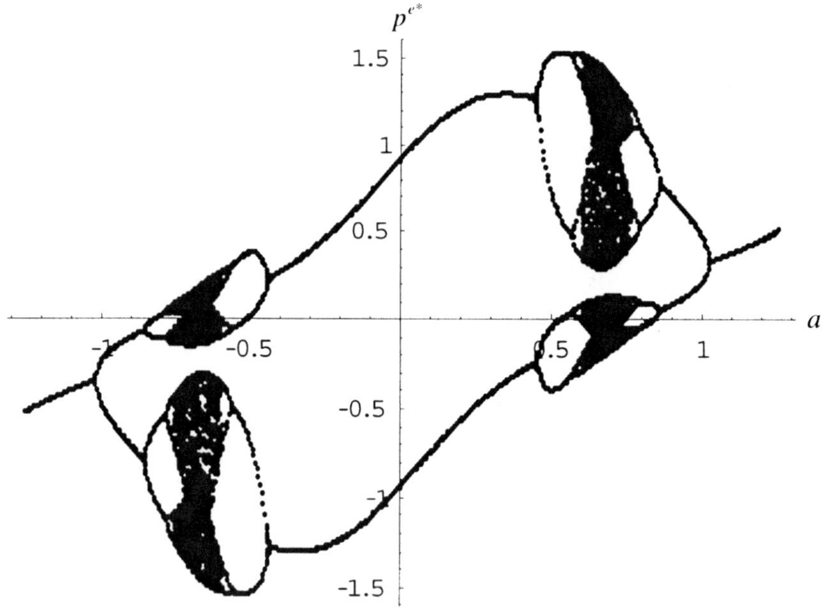

Figure 4.6.5: Bifurcation diagram for $\mu = 4$, $a \in [-1.25, 1.25]$

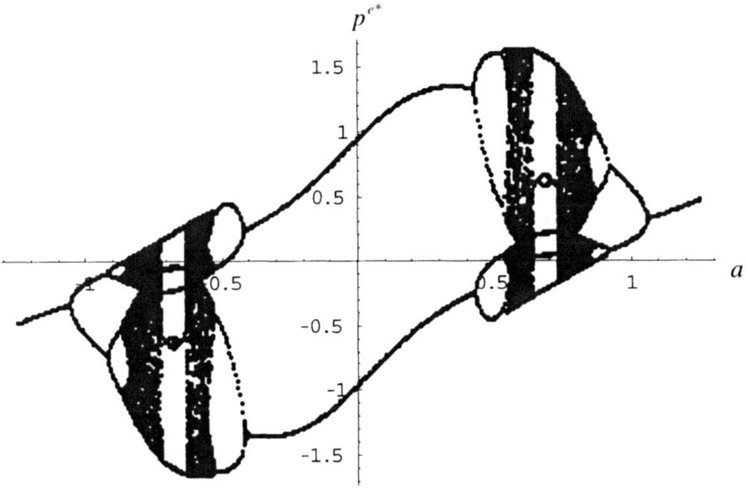

Figure 4.6.6: Bifurcation diagram for $\mu = 4.5$, $a \in [-1.25, 1.25]$

Chapter 5

Economic bifurcations and chaos

This chapter applies the concepts and theorems of the previous chapters to analyze different models in economics. The models in this chapter exhibit periodic, aperiodic, or chaotic behavior. Section 5.1 studies a model of endogenous business cycles in the presence of knowledge spillovers. Many economic indicators, such as GDP, exhibit asymmetry as if they repeatedly switch between different regimes. For instance, it has been found that (i) positive shocks are more persistent than negative shocks in the United States and France; (ii) negative shocks are more persistent than negative shocks in the United Kingdom and Canada; and (iii) there is almost no asymmetry in persistence in Italy, Japan, and (former) Germany. The model in this section provides some insights into the well-observed asymmetric nature of business cycles. Section 5.2 studies a nonlinear cobweb model with normal demand and supply, naïve expectations and adaptive production adjustment. The model exhibits horseshoes. Section 5.3 examines an inventory model with rational expectations. In section 5.4, we discuss an economic growth model with pollution. The model is an extension of the standard neoclassical growth model which has a unique stable equilibrium point. Chaos exists in the model because of the effects of pollution upon production. It is know that the neoclassical growth theory based on the Solow growth model focuses accumulation as an engine of growth, while the neo-Schumpeterian growth theory stresses innovation. Section 5.5 studies a model to capture these two mechanisms within the same framework. The model generates an unstable balanced growth path and the economy achieves sustainable growth cycles, moving back and forth between the two phases – one is characterized by higher output growth, higher investment, no innovation, and a competitive market structure; the other by lower output growth, lower investment, high innovation, and a more monopolistic structure. Section 5.6 identifies economic fluctuations in a monetary economy within the OLG framework. Section 5.7 shows chaos in a model of interaction of economic and population growth.

5.1 Business cycles with knowledge spillovers

We now introduce a model of endogenous business cycles in the presence of knowledge spillovers. [1] Many economic indicators, such as GDP, exhibit asymmetry as if they repeatedly switch between different regimes. For instance, in their study of changes in GDP, Hess and Iwata find that (i) positive shocks are more persistent than negative shocks in the United States and France; (ii) negative shocks are more persistent than negative shocks in the United Kingdom and Canada; and (iii) there is almost no asymmetry in persistence in Italy, Japan, and (former) Germany.[2] The model is supposed to provide some insights into well-observed asymmetric nature of business cycles.

Time discretely extends from zero to infinity. There is a continuum of firms, each indexed by i and the total population is normalized to 1. At the end of each period, a constant fraction $1 - \delta$, where $\delta \in (0, 1)$, of randomly chosen firms disappear and are replaced by new ones. Each firm is risk neutral and attempts to maximize the discounted sum of expected profits with $\gamma \in (0, 1)$ being the subjective discount factor. Define

$$\beta \equiv \gamma\delta,$$

as the effective discount factor. In any period, each firm is either active or inactive and engages in at most one project at a time. Each project is characterized by its quality, either high or low, that determines the cost of production. The cost of production is k if the project is of low quality and zero if it is of high quality. Upon entering the market, each new firm has a low-quality project at hand and must decide either to adopt the project as it is or to innovate it. If a firm decides to adopt, it immediately becomes active and produces until it disappears; if a firm decides to innovate (and each firm can innovate only once), it can upgrade the project to high quality in the next period while it must stay inactive for that period. Let

$$n(t) \in [0, 1 - \delta],$$

denote the fraction of firms that decide to innovate in period t, and

$$x(t) \in [0, 1],$$

[1] This section is based on Ishida and Yokoo (2004).

[2] Hess and Iwata (1997).

the fraction of firms with a high-quality project at the beginning of period t. The law of motion of $x(t)$ is

$$x(t+1) = \delta(x(t) + n(t)).$$

Assume that the productivity of each firm depends on the fraction of firms with a high-quality (innovated) project at the beginning of each period.[3]

Each firm is equally productive in any given period. Let $y_i(t)$ denote the output level of firm i in period t and

$$y_i(t) = h(x(t)),$$

where

$$h: [0, 1] \to [0, 1],$$

is continuous and strictly increasing in x and

$$h(0) = \theta \in [0, 1), \quad h(1) = 1.$$

Firm i 's profit is

$$\pi_i(t) = h(x(t)) - k,$$

if its project is of low quality and the profit is

$$\pi_i(t) = h(x(t)),$$

if it is of high quality. If a firm chooses to adopt, the expected gain is

$$E_t \sum_{s=0}^{\infty} \beta^s \pi_i(t+s) = E_t \sum_{s=0}^{\infty} \beta^s h(x(t+s)) - \frac{k}{1-\beta}.$$

If a firm chooses to innovate, the expected gain is

[3] This assumption is accepted by, for instance, Durlauf (1991, 1993), and Gale (1996).

$$E_t \sum_{s=0}^{\infty} \beta^s \pi_i(t+s) = E_t \sum_{s=1}^{\infty} \beta^s h(x(t+s)).$$

The firm chooses not to innovate if and only if

$$E_t \sum_{s=0}^{\infty} \beta^s h(x(t+s)) - \frac{k}{1-\beta} \geq E_t \sum_{s=1}^{\infty} \beta^s h(x(t+s)),$$

which can be rewritten as

$$h(x(t)) \geq \frac{k}{1-\beta}.$$

To examine $h(x(t)) \geq k/(1-\beta)$ under

$$\theta \leq h(x(t)) \leq 1$$

for any $x(t) \in [0, 1]$, we may consider three cases.

Case 1 $\theta > k/(1-\beta)$
It is never profitable to innovate, $n(t) = 0$. Hence,

$$x(t+1) = \delta x(t)$$

for $x_0 \in [0, 1]$, which implies $x(t) \to 0$ as $t \to \infty$. The project of every firm is of low quality in the end.

Case 2 $k/(1-\beta) > 1$
Each firm always chooses to innovate

$$n(t) = 1 - \delta,$$

for any t. Hence

$$x(t+1) = \delta x(t) + \delta(1-\delta),$$

for $x_0 \in [0, 1]$. The system has a globally asymptotically stable steady state.

Case 3 $\theta \leq k/(1-\beta) \leq 1$

This is a dynamically interesting case. First, we note that there exists a real number c, called the *threshold*, such that

$$h(c) = \frac{k}{1-\beta}.$$

Consider c as a function of k. It can be seen that $c(k)$ continuously increases with k for

$$\theta \leq \frac{k}{1-\beta} \leq 1.$$

From the definition of c and strictly increasing h, we see that each new firm in period t chooses to innovate if $x(t) < c$ (i.e., $n(t) = 1 - \delta$) and chooses not to innovate if $x(t) \geq c$ (i.e., $n(t) = 0$). The economy is said to be in the *contraction phase* if $x(t) < c$; in the *expansion phase* if $x(t) \geq c$. In summary, the law of motion of $x(t)$ is characterized by the following piecewise linear difference equation

$$x(t+1) = f(x(t)) = \begin{cases} \delta x(t) + \delta(1-\delta) \equiv f_L(x(t)), & \text{if } x(t) < c, \\ \delta x(t) \equiv f_R(x(t)), & \text{if } x(t) \geq c. \end{cases} \tag{5.1.1}$$

Equation (5.1.1) can lead to an asymmetric periodic cycle in which the expansion and the contradiction phase alternate with each other asymmetrically.[4] Let us first examine a type of periodic cycle

$$\{p(1), p(2), \cdots, p(m-1), p(m)\}$$

such that

$$p(1) < p(2) \cdots < p(m-1) < c \leq p(m), \tag{5.1.2}$$

[4] This section is only concerned with some simple cases. Methods for a comprehensive examination of this type of equations is referred to, for instance, Nagumo and Sato (1972).

for some natural number m. Such a cycle may be generally illustrated as in figure 5.1.2.

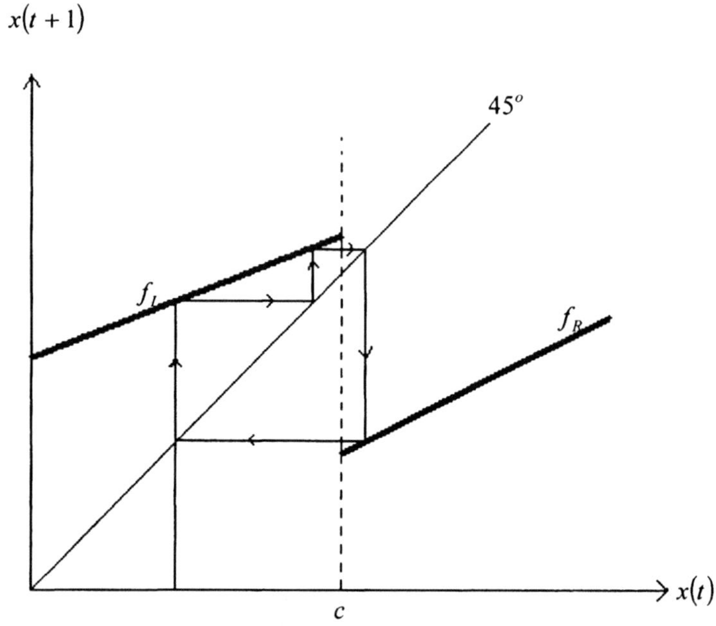

Fig. 5.1.1: A periodic cycle

To compute $p(1)$, we need to calculate

$$f^m(p(1)) = p_1,$$

under equation (5.1.2). As

$$f^m(p(1)) = f_R \circ f_L^{m-1}(p(1)),$$

we see that $f^m(p(1)) = p_1$ becomes

$$\delta^m p_1 + \delta^2(1 - \delta^{m-1}) = p_1 .$$

We thus solve

$$p_1 = \frac{\delta^2\left(1 - \delta^{m-1}\right)}{1 - \delta^{m-1}}. \qquad (5.1.3)$$

For this to be consistent, we should require p_1 in equation (5.1.3) to satisfy

$$f_L^{m-2}(p(1)) < c, \quad f_R(c) \le p_1.$$

From

$$f_L^{m-2}(p(1)) < c,$$

we have

$$c > f_L^{m-2}(p(1)) = \delta^{m-2} p_1 + \delta - \delta^{m-1} = \frac{\delta^m - \delta^{m+1} - \delta^{m-1} + \delta}{1 - \delta^m} \equiv L_m. \quad (5.1.4)$$

From

$$f_R(c) \le p_1,$$

we have

$$c \le \frac{p_1}{\delta} = \frac{\delta - \delta^m}{1 - \delta^m} \equiv R_m. \qquad (5.1.5)$$

The two conditions

$$f_L^{m-2}(p(1)) < c, \quad f_R(c) \le p_1,$$

are thus represented by conditions (5.1.4) and (5.1.5). That is

$$L_m < c \le R_m.$$

As $R_m - L_m > 0$, $(L_m, R_m]$ is well-defined. Conversely, suppose $L_m < c \le R_m$ be satisfied. Further, define $\hat{c}$ such that

$$f_L^{m-2}(\hat{c}) = c.$$

Solving this equation, we have

$$\hat{c} = \frac{c - \delta + \delta^{m-1}}{\delta^{m-2}}.$$

We can show that the mapping f^m restricted to the interval

$$T = [\delta c, \hat{c}],$$

is well-defined (that is, f^m maps T into itself) where $f^m|T$ is linear with a constant slope δ^m and a unique fixed point $p_1 \in T$. We see that

$$L_m < c \leq R_m$$

is a necessary and sufficient condition for the trajectory of any initial value to converge to a period-m cycle of formula (5.1.2). For instance, a period-2 cycle appears when

$$\frac{\delta^2}{1+\delta} < c \leq \frac{\delta}{1+\delta}.$$

Exercise 5.1

1 Similar to the procedures for examining formula (5.1.2), examine a type of periodic cycle

$$\{p(1), p(2), \cdots, p(m-1), p(m)\}$$

such that

$$p(1) > p(2) \cdots > p(m-1) \geq c > p(m).$$

5.2 A cobweb model with adaptive adjustment

A nonlinear cobweb model with normal demand and supply, naïve expectations and adaptive production adjustment is recently proposed by Onozaki *et al.*[5] Let us consider a market of a single commodity. In period t, a supplier decides his production $x(t+1)$. Nevertheless, this level may not be equal to the profit maximum $\tilde{x}(t+1)$ which he calculates and uses as a target of adjustment. Suppose that the calculation is made under the quadratic cost function $bx^2/2,\ b>0$ with the *naïve price expectation* (which means that his price expectation for the next period is equal to the current price $p(t)$). The profit maximum level of output is given by

$$\tilde{x}(t+1) = \frac{p(t)}{b}.$$

It is assumed that the producer will adjust his production according to the following hedging rule in the uncertain economy

$$x(t+1) = x(t) + \alpha(\tilde{x}(t+1) - x(t)),$$

where $\alpha \in (0, 1)$ is the speed of adjustment. Suppose that there are n identical suppliers in the market. The aggregate supply is thus given by

$$X(t) = nx(t).$$

Assume a monotonic demand function with constant price elasticity of $1/\beta\ (\beta > 0)$

$$p(t) = \frac{c}{Y^{\beta}(t)}.$$

Price clears the market in each period, i.e.

$$X(t) = Y(t).$$

---

[5] This section is based on Onozaki *et al.* (2000). See also Hommes (1994), Gallas and Nusse (1996), Goeree and Hommes (2000), for this type of nonlinear models. An extension of the model to two types of producers, cautious adapters and naïve optimizers is carried out by Onozaki *et al.* (2003).

It is straightforward to show that under the above specifications, the motion of the aggregate supply is given by

$$X(t+1) = (1-\alpha)X(t) + \frac{\alpha c n}{b X^{\beta}(t)}.$$

Introduce a linear transformation

$$z(t) \equiv \left(\frac{b}{cn}\right)^{1/(1+\beta)} X(t).$$

Then, the above equation is transformed into

$$z(t+1) = (1-\alpha)z(t) + \frac{\alpha}{z^{\beta}(t)} \equiv f(z(t), \alpha, \beta), \ \ (\alpha, \beta) \in (0, 1) \times (0, \infty). \tag{5.2.1}$$

The map f has a unique fixed point, $z^{*} = 1$. The first and second derivates are

$$f' = 1 - \alpha - \frac{\alpha\beta}{z^{1+\beta}}, \ \ f'' = \frac{\alpha\beta(1+\beta)}{z^{2+\beta}} > 0, \ \ z \in R^{++},$$

which implies that f is a strictly convex and unimodal function on R^{++} with its minimum at the critical point

$$\bar{z} = \theta \equiv \left[\frac{\alpha\beta}{1-\alpha}\right]^{1/(1+\beta)}.$$

The fixed point is a repeller if

$$f'(z^{*}) = 1 - \alpha - \alpha\beta < -1,$$

or equivalently

$$\frac{2-\alpha}{\alpha} < \beta. \tag{5.2.2}$$

For this map, it can be demonstrated that for sufficiently large β, the map f exhibits a *horseshore*. By a horseshoe it means here a compact invariant set on which some iterate of f is topologically conjugate to the one-sided full-shift on two symbols. The existence of a horseshoe is assured by that of a transverse homoclinic point. A map is said to exhibit *topological chaos* if it has a horseshoe or, alternatively, if the topological entropy of the map is positive. It should be remarked that a map restricted on horseshoes behaves in a complicated way, the existence of horseshoes itself does not assume complex dynamics in the long run; the system may eventually settle down to a periodic motion even if horseshoes are present.[6] In the following theorem proved by Onozaki *et al*, an attractor is said to be *strange* if it contains a dense orbit with positive Liapunov exponent.

Theorem 5.2.1. For any $\alpha \in (0, 1)$, there is generically a positive measure set of parameter values of β, $E \subset R^+$, such that for every $\beta \in E$, the map f exhibits a strange attractor.

The results can be demonstrated numerically. For instance, for $\alpha = 0.7$, figure 5.2.1 depicts the Liapunov exponent for different values of β. Figure 5.2.2 depicts the bifurcation diagram of the map with regard to β $(1{,}5 \le \beta \le 4.7)$ with $\alpha = 0.7$.

5.3 Inventory model with rational expectations[7]

This section introduces a disequilibrium inventory model. The actual labor employed, $L(t)$, is given by the short side of the labor market

$$L(t) = \min\{L^d(t), L^s(t)\},$$

where $L^d(t)$ and $L^s(t)$ are respectively demand and supply of labor. Assume

$$L^s(t) = c,$$

[6] See Block and Coppel (1992) and de Melo and van Strien (1993) for issues related to horseshoes and topological chaos.

[7] This example is given in Hommes (1991: chapter 28). See also Shone (2002: section 7.9).

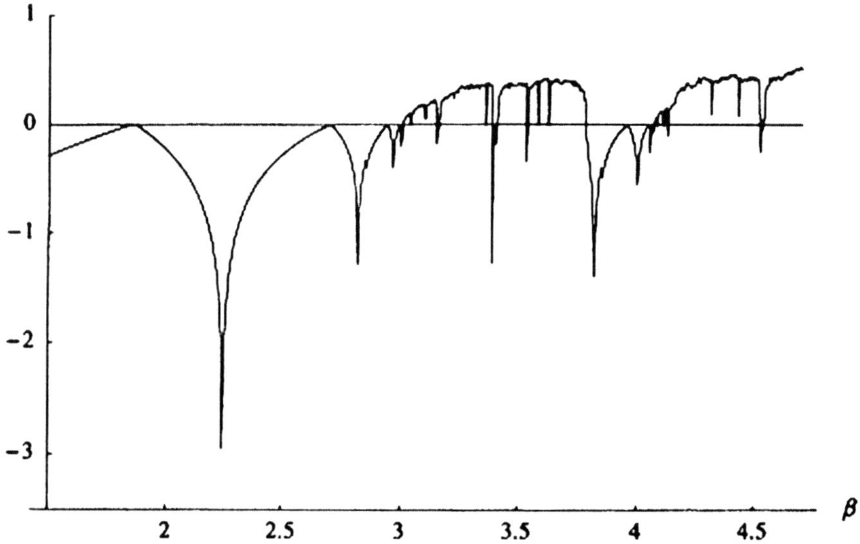

Figure 5.2.1: The Liapunov exponent for $1{,}5 \leq \beta \leq 4.7$ with $\alpha = 0.7$

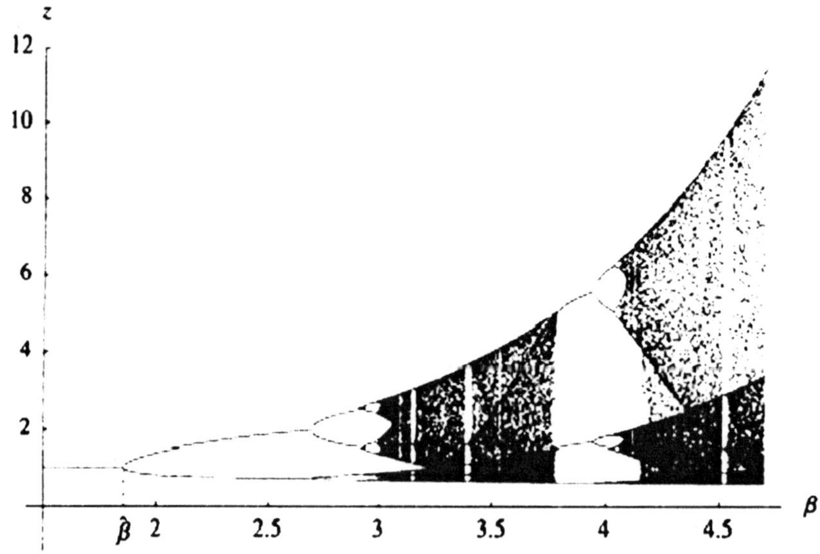

Figure 5.2.2: The bifurcation diagram for $1{,}5 \leq \beta \leq 4.7$ with $\alpha = 0.7$

where c is constant. Let $y^d(t)$ and $y^s(t)$ denote respectively aggregate demand and aggregate supply. The level of inventories $I(t)$ is positive when there is excess demand, otherwise it is zero

$$I(t) = \max\{0,\ y^s(t) - y^d(t)\}.$$

We denote expected aggregate demand by $E(y^d(t))$ and the desired level of inventories by $I^d(t)$. We assume perfect foresight, i.e.

$$E(y^d(t)) = y^d(t).$$

Suppose

$$I^d(t) = \beta E(y^d(t)),$$

where β is a parameter. Production is proportional to labor employed, $\delta L(t)$. We thus have

$$y^s(t) = I(t-1) + \delta L(t),$$
$$y^d(t) = E(y^d(t)) + I^d(t) = (1+\beta)E(y^d(t)).$$

Setting these two equations to equal each other yields

$$L(t) = \frac{(1+\beta)E(y^d(t)) - I(t-1)}{\delta}. \tag{5.3.1}$$

Hence labor demand is given by

$$L^d(t) = Max\left\{0,\ \frac{(1+\beta)E(y^d(t)) - I(t-1)}{\delta}\right\}. \tag{5.3.2}$$

Aggregate demand is assumed to be a linear function of labor employed

$$y^d(t) = a + bL(t).$$

We assume that the labor productivity is greater than the marginal propensity to consume, $\delta > b$. We have completed the model. We now show that the evolution of the system can be described by a difference equation for I.

First, we are concerned with $L(t) = L^d(t)$, which implies $L(t) \in (0, c)$. From equation (5.3.2), we have

$$L(t) = \frac{(1+\beta)y^d(t) - I(t-1)}{\delta} = \frac{(1+\beta)(a + bL(t)) - I(t-1)}{\delta}.$$

Solve the above equation for L

$$L(t) = \frac{(1+\beta)a - I(t-1)}{\delta - b(1+\beta)}.$$

Assume

$$\delta - b(1+\beta) > 0.$$

As $L(t) \in (0, c)$, then $L(t) < c$ is guaranteed by

$$\gamma_1 < I(t-1),$$
$$\gamma_1 \equiv (1+\beta)a - c(\delta - b(1+\beta)).$$

Similarly, the condition $L(t) > 0$ is guaranteed if

$$\gamma_2 > I(t-1),$$

where

$$\gamma_2 \equiv a(1+\beta).$$

Hence, $L(t) \in (0, c)$ is guaranteed if

$$\gamma_2 > I(t-1) > \gamma_1.$$

We need to consider the following three situations

(i) $I(t-1) \leq \gamma_1$;
(ii) $\gamma_2 > I(t-1) > \gamma_1$; and
(iii) $I(t-1) \geq \gamma_2$.

(i) $I(t-1) \leq \gamma_1$
In this case, $L(t) = c$. Then

$$I(t) = y^s(t) - y^d(t) = I(t-1) + \delta c - a - c. \tag{5.3.3}$$

(ii) $\gamma_2 > I(t-1) > \gamma_1$
In this case

$$I(t) = y^s(t) - y^d(t) = I(t-1) + \delta L(t) - a - bL(t)$$
$$= I(t-1) + \delta - b\left[\frac{(1+\beta)a - I(t-1)}{\delta - b(1+\beta)}\right] - a.$$

From the above equation, we obtain

$$I(t) = \frac{-b\beta I(t-1)}{\delta - b(1+\beta)} + \frac{a\delta\beta}{\delta - b(1+\beta)}. \tag{5.3.4}$$

(iii) $I(t-1) \geq \gamma_2$
Under these circumstances, we have

$$L(t) = L^d(t) = 0.$$

We thus have

$$I(t) = y^s(t) - y^d(t) = I(t-1) - a. \tag{5.3.5}$$

Combining all these three cases, then

$$I(t) = f(I(t-1)),$$

is a piecewise function

$$I(t) = \begin{cases} I(t-1) + \delta c - a - c, & I(t-1) \le \gamma_1, \\ \dfrac{-b\beta I(t-1)}{\delta - b(1+\beta)} + \dfrac{a\delta\beta}{\delta - b(1+\beta)}, & \gamma_2 > I(t-1) > \gamma_1, \\ I(t-1) - a, & I(t-1) \ge \gamma_2. \end{cases} \qquad (5.3.6)$$

The equilibrium investment is defined only for $\gamma_2 > I(t-1) > \gamma_1$, which is given by the following condition

$$I^* = \frac{-b\beta I^*}{\delta - b(1+\beta)} + \frac{a\delta\beta}{\delta - b(1+\beta)} \Rightarrow I^* = \frac{a\delta\beta}{\delta - b}.$$

Since $\delta > b$, we have $I^* > 0$.

We now provide two numerical examples. First, we specify

$$a = 0.2, \quad b = 0.75, \quad c = 1, \quad \delta = 1, \quad \beta = 0.2.$$

Under these specifications, equation (5.3.6) becomes

$$I(t) = \begin{cases} I(t-1) + 0.05, & I(t-1) \le 0.14, \\ -1.5I(t-1) + 0.4, & 0.24 < I(t-1) < 0.24, \\ I(t-1) - a, & I(t-1) \ge 0.24. \end{cases} \qquad (5.3.7)$$

We illustrate the behavior of map (5.3.7) as in figure 5.3.1. Although the equilibrium level exists ($I^* = 0.1.6$), from the figure it is not clear whether this fixed point will be approached. In fact, as shown in figure 5.3.2, the system is chaotic in the present case.

We still assume

$$a = 0.2, \quad b = 0.75, \quad c = 1, \quad \delta = 1,$$

but leave β unspecified. We consider β as a bifurcation parameter. We allow β to range over the interval $(0, 1/3)$, the value of $1/3$ is determined from the condition

$$\delta - b(1+\beta) = 0.$$

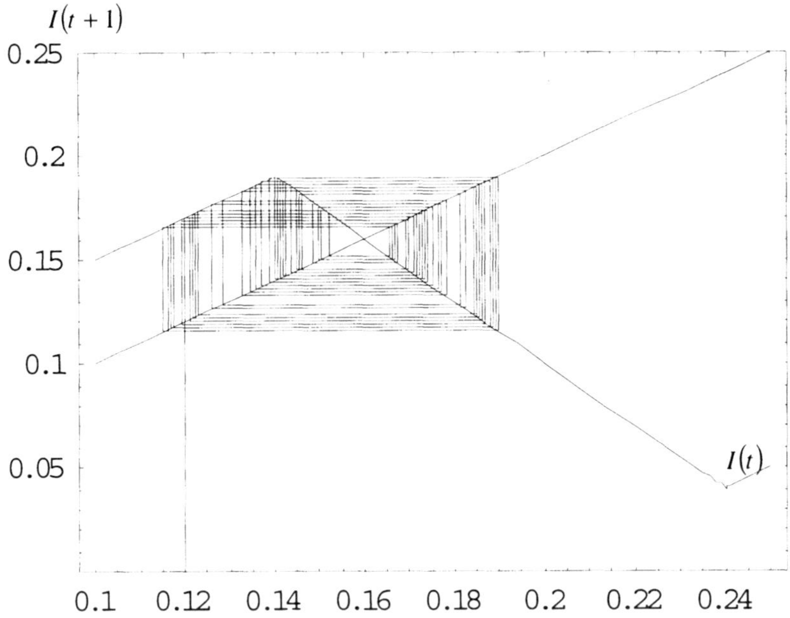

Figure 5.3.1: The dynamics with $\beta = 0.2$

In constructing the bifurcation diagram, we notice that with unspecified β, equation (5.3.6) becomes

$$I(t) = \begin{cases} I(t-1) + 0.05, & I(t-1) \le 0.2\beta_0 - \beta_1, \\ \dfrac{-0.75\beta I(t-1)}{\beta_1} + \dfrac{0.2\beta}{\beta_1}, & 0.2\beta_0 - \beta_1 < I(t-1) < 0.2\beta_0, \\ I(t-1) - 0.2, & I(t-1) \ge 0.2\beta_0. \end{cases}$$

where

$$\beta_0 \equiv 1 + \beta, \quad \beta_1 \equiv 1 - 0.75\beta_0.$$

Figure 5.3.2 depicts the bifurcation diagram. Although the system exhibits chaos over certain range for β, there are regular alternating behavior.

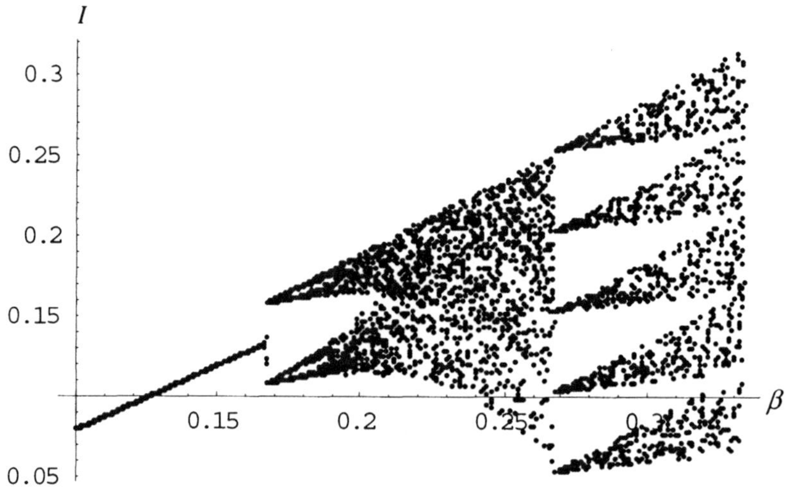

Figure 5.3.2: The bifurcation diagram with β as bifurcation parameter

5.4 Economic growth with pollution

The following model is due to Day.[8] First, we have the following relations

$$Y(t) = C(t) + I(t),$$
$$K(t+1) = I(t),$$
$$S(t) = Y(t) - C(t) = sY(t), \quad 1 > s > 0,$$
$$N(t) = (1+n)^t N_0, \quad n > 0,$$

where $Y(t)$ is output, $C(t)$ is consumption, $I(t)$ is investment, $S(t)$ is savings, $N(t)$ is the total labor force, and s and n are respectively the fixed rate of savings and the population growth rate. The production function $F(K(t), N(t))$ $(= Y(t))$ is linear-homogeneous. Introduce

$$k(t) \equiv \frac{K(t)}{N(t)}.$$

[8] See Day (1982).

Hence, we have

$$k(t+1) = \frac{s}{1+n} f(k(t)),$$

where $f(k) \equiv F(k, 1)$. The production function is taken on the following form

$$f(k) = Bk^{\beta}(t)(m - k(t))^{\gamma}, \quad k(t) \leq m,$$

where B, β, γ, and m are constant. Here, the term $(m - k(t))^{\gamma}$ reflects the influence of pollution on per-capita output. As the capita intensity increases, pollution increases as well. Suppose that resources have to be sacrificed in order to avoid this pollution. Substituting f into the equation for capital accumulation yields

$$k(t+1) = \tilde{n} k^{\beta}(t)(m - k(t))^{\gamma},$$

where

$$\tilde{n} \equiv \frac{sB}{1+n}.$$

If we choose

$$\beta = \gamma = m = 1,$$

then the equation becomes

$$k(t+1) = \tilde{n} k(t)(m - k(t)).$$

This is formally identical with the logistic equation. Hence, all the properties of the logistical map can be identified in Day's growth model.

We now consider the general equation

$$k(t+1) = \tilde{n} k^{\beta}(t)(m - k(t))^{\gamma}.$$

To apply the Li-Yorke theorem to this equation, we consider the following three different values of k. Let k^c be the critical point of the map, i.e., the value of $k(t)$ that implies the highest possible capital intensity in the next period. We determine k^c by solving the first-order condition

$$\frac{\partial k(t+1)}{\partial k(t)} = \tilde{n}\left[\beta k^{\beta-1}(t)(m-k(t))^{\gamma} - \gamma k^{\beta}(t)(m-k(t))^{\gamma-1}\right] = 0.$$

We have

$$k^c = \frac{\beta m}{\gamma+\beta}.$$

As shown in figure 5.4.1, if we make B sufficiently large, it is always possible to make k^c lower than the fixed point k^*. Next let k^b be the result of the backward iteration

$$k^b = f^{-1}(k^c).$$

When $k^b < k^*$, k^b will be smaller than k^c. Finally, let k^m denote the maximum attainable capital intensity, i.e., the intersection of the graph of the map with the abscissa. Variations in B eventually imply that the graph of $\tilde{n}k^{\beta}(m-k)^{\gamma}$ is stretched upward such that k^m is the forward iteration of k^c

$$f(k^c) = \frac{sB}{1+n}(k^c)^{\beta}(m-k^c)^{\gamma} = k^m.$$

As k^m is mapped to the origin, the following relations between these values of k hold

$$0 < k^b < k^c < k^m \Rightarrow f(k^m) < k^b < f(k^b) < f(k^c)$$
$$\Rightarrow f^3(k^b) < k^b < f(k^b) < f^2(k^b).$$

Hence, the requirements of the Li-Yorke theorem are satisfied. The model is chaotic in the Lie-Yorke sense for appropriate values of the parameters.

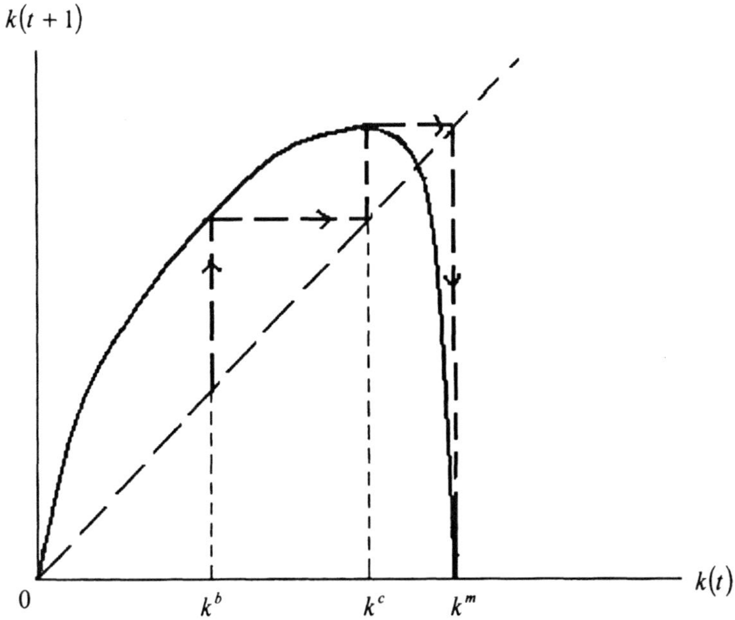

Figure 5.4.1: Day's growth model with pollution

5.5 The Solow-Schumpeter growth oscillations

As shown in chapter 3, the neoclassical growth theory based on the Solow growth model focuses accumulation as an engine of growth, while the neo-Schumpeterian growth theory stresses innovation. Recently, Matsuyama proposes a model to capture these two mechanisms within the same framework.[9] The model generates an unstable balanced growth path and the economy achieves sustainable growth cycles, moving back and forth between the two phases – one is characterized by higher output growth, higher investment, no

---

[9] This section is based on Matsuyama (1999). The model is based on the lab equipment model of Rivera-Batiz and Romer (1991) and the search-based model of technology evolution by Jovanovic and Rob (1990). It should be noted that innovations cycles are also generated by many other models, for instance, by Deneckere and Judd (1992) and Gale (1996). Matsuyama (2001) proposed another model to endogenously determine behavior of consumers within the framework of an infinitely lived agent economy. See also Mitra (2001) for discussing conditions for topological chaos with regard to the model by Matsuyama.

innovation, and a competitive market structure; the other by lower output growth, lower investment, high innovation, and a more monopolistic structure.

There is a single final good, taken as a numeraire; it is competitively produced and can either be consumed or invested. Let $K(t-1)$ be the final good left unconsumed in period $t-1$ and made available for use in production in t. Labor is supplied inelastically in amount N. Labor goes directly into final goods production. Capital $K(t-1)$ must be first converted into a variety of differentiated intermediate products; these intermediates are aggregated into the composite by a symmetric CES. The final goods production function is

$$Y(t) = A_0 N^{1/\sigma} \int_0^{Z(t)} x^{\alpha}(z, t) dz, \tag{5.5.1}$$

where $x(z, t)$ is the input of variety z in period t

$$\alpha = 1 - \frac{1}{\sigma}, \quad \sigma \in (1, \infty),$$

and $[0, Z(t)]$ is the range of intermediates available at t.

The market structure of the intermediate sector is described as follows. Prior to t, the economy developed all the intermediates in the range, $[0, Z(t-1)]$, with $Z(0) > 0$. These old intermediates are manufactured by converting a units of capital into one unit of an intermediate, and sold competitively in t. In addition, the intermediates of variety

$$z \in [Z(t-1), Z(t)],$$

may be introduced and sold exclusively by their innovators in t. Innovating new intermediates require F units of capital per variety. The process of manufacturing new intermediates requires a units of capital per output. Let $r(t)$ denote the price of capital. The marginal cost of manufacturing intermediates in t is equal to $ar(t)$. The old products are supplied at the marginal costs

$$p(z, t) = p^c(t) = ar(t), \quad z \in [0, Z(t-1)].$$

All the new products are sold at

$$p(z,t) = p^m(t) = \frac{a\sigma}{\sigma - 1} r(t), \quad z \in [Z(t-1), Z(t)].$$

Since all the intermediates enter symmetrically in the final goods production, we have

$$\frac{x^c(t)}{x^m(t)} = \left[\frac{p^c(t)}{p^m(t)}\right]^{-\sigma} = \alpha^{-\sigma}, \tag{5.5.2}$$

where

$$x(z,t) = p^c(t), \quad z \in [0, Z(t-1)],$$
$$p(z,t) = p^m(t), \quad z \in [Z(t-1), Z(t)].$$

The one-period monopoly enjoyed by the innovator yields the monopoly profit

$$\pi = p^m x^m - r(ax^m + F),$$

at t. The profit is negative if and only if

$$x^m(t) < (\sigma - 1)\frac{F}{a}.$$

The free entry conditions ensure

$$\begin{gathered} ax^m(t) \le (\sigma - 1)F, \\ Z(t) \ge Z(t-1), \\ [ax^m(t) - (\sigma - 1)F][Z(t) - Z(t-1)] = 0. \end{gathered} \tag{5.5.3}$$

The resource constraint on capital in t is given by

$$K(t-1) = ax^c(t)Z(t-1) + (ax^m(t) + F)(Z(t) - Z(t-1)).$$

Under equations (5.5.2) and (5.5.3), the above constraint becomes

$$ax^c(t) = a\sigma_0^{-\sigma} x^m(t) = \min\left\{\frac{K(t-1)}{Z(t-1)},\ \theta\sigma F\right\}, \tag{5.5.4}$$

$$Z(t) = Z(t-1) + \max\left\{0,\ \frac{K(t-1)}{\sigma F} - \theta Z(t-1)\right\}, \tag{5.5.5}$$

where $\theta \equiv \sigma^{1-\sigma}$. Under equations (5.5.3)-(5.7.5), the total output is given by

$$Y(t) = \begin{cases} A[\theta\sigma F Z(t-1)]^{1/\sigma} K^{\alpha}(t-1) & \text{if } K(t-1) \le \theta\sigma F Z(t-1), \\ AK(t-1) & \text{if } K(t-1) > \theta\sigma F Z(t-1), \end{cases} \tag{5.5.6}$$

where we use

$$A \equiv \frac{A_0}{a}\left(\frac{aN}{\theta\sigma F}\right),$$
$$\int_0^{Z(t)} x^{\alpha}(z, t)dz = \left(x^c(t)\right)^{\alpha} Z(t-1) + \left(x^m(t)\right)^{\alpha}\left(Z(t) - Z(t-1)\right).$$

Equations (5.5.5) and (5.5.6) tell that if

$$\frac{K(t-1)}{Z(t-1)} \le \theta\sigma F,$$

the resource base of the economy, K, is too small relative to the number of the existing products, Z, and there is no innovation. The economy in this situation is called the *Solow regime.* If

$$\frac{K(t-1)}{Z(t-1)} > \theta\sigma F,$$

some new products are introduced; the economy is said to be in the *Romer regime.*

To close the model, it is assumed that the economy carries over a constant fraction of its output to the next period, i.e.

$$K(t) = sY(t),$$

where s is a constant. The system has a unique solution for any positive initial condition, K_0 and Z_0. Set

$$k(t) \equiv \frac{K(t)}{\theta\sigma F Z(t)}.$$

Then the system can be expressed as the one-dimensional difference equation

$$k(t) = f(k(t-1)) = \begin{cases} sAk^{\alpha}(t-1), & \text{if } k(t-1) \leq 1, \\ \dfrac{sAk(t-1)}{1+\theta(k(t-1)-1)}, & \text{if } k(t-1) > 1. \end{cases}$$

The critical value of k, denoted by $\bar{k}$ $(=1)$, which separates the Solow and Romer regimes is equal to one. If

$$sA < 1,$$

the system has a unique steady state

$$k^{*} = (sA)^{\sigma} < \bar{k}.$$

Without innovation, all goods are competitively supplied and the economy does not grow: the steady is a *neoclassical stationary path.* If

$$sA > 1,$$

the steady state, given by

$$k^{**} = 1 + \frac{sA-1}{\theta} > \bar{k},$$

is in the Romer regime. New products are introduced steadily, and K and Z grow at the same rate. It is a *balanced growth path,* characterized by

$$K(t+1) = sAK(t).$$

Figures 5.5.1-3 depict the dynamics of

$$k(t) = f(k(t-1)).$$

Figure 5.5.1 depicts the case of $sA < 1$. For any initial conuition, after at most one period, innovation stops and the economy is trapped into the Solow regime.

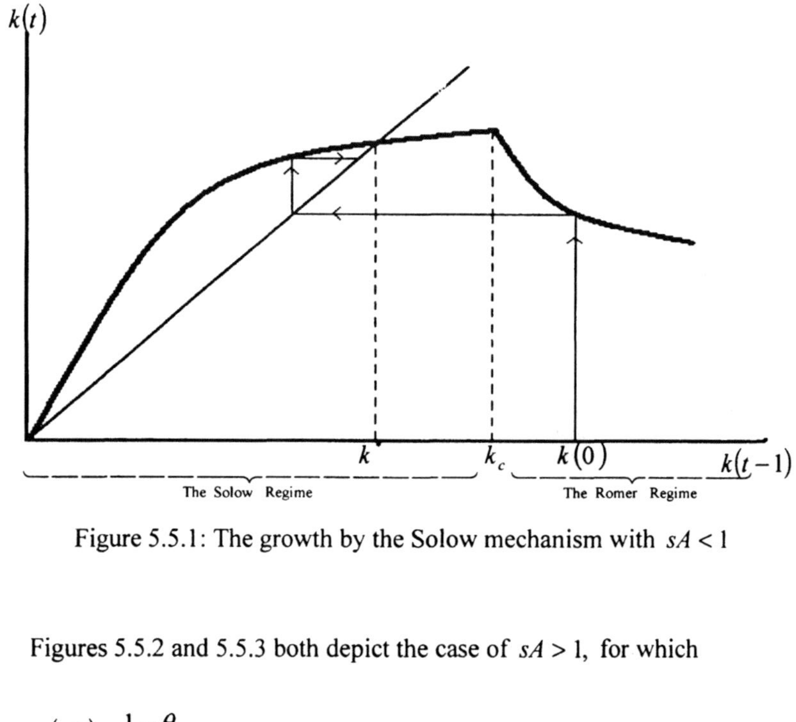

Figure 5.5.1: The growth by the Solow mechanism with $sA < 1$

Figures 5.5.2 and 5.5.3 both depict the case of $sA > 1$, for which

$$f'(k^{**}) = \frac{1-\theta}{sA} < 0.$$

They differ in the local stability of the steady state. Figure 5.5.2 illustrates the case of $1 < sA < \theta - 1$. The equilibrium point k^{**} is locally unstable because of

$$f'(k^{**}) < -1.$$

The interval $[f^2(k^{**}), f(k^{**})]$ represents the trapping region, i.e., the region that the economy enters eventually and once entered will never leave. The trapping region covers both the Solow and Romer regimes. If the economy starts with a small k_0, the economy may stay in the Solow regime for some periods, but eventually accumulates enough capital to enter the Romer regime, and innovation begins. The economy begins cycling between the two regimes. Formally, it can be proved that if

$$1 < sA < \theta - 1,$$

there are period $2-$cycles, $k(t)$ fluctuates forever between the Solow and Romer regimes for almost all initial conditions.

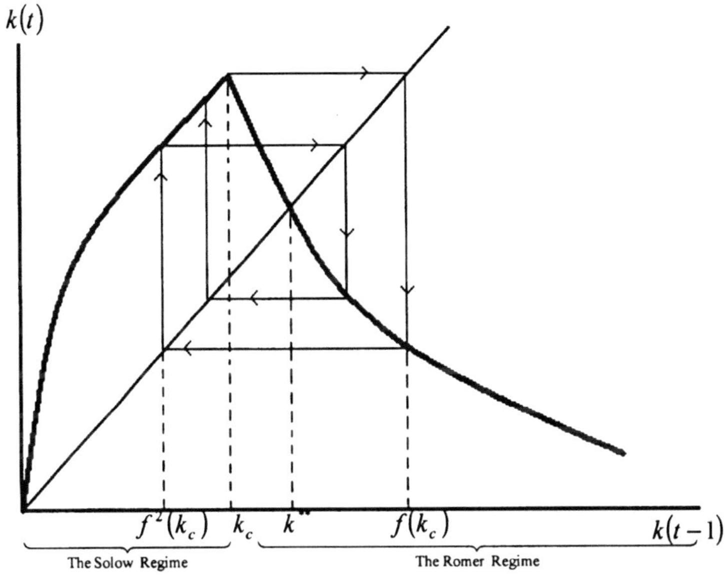

Figure 5.5.2: The Solow-Schumpeter mechanisms with $1 < sA < \theta - 1$

Figure 5.5.3 depicts the case of $sA > \theta - 1$. We have

$$-1 < f'(k^{**}) < 0.$$

The steady state is globally stable. In the long term, the economy will forever stay in the Romer regime.

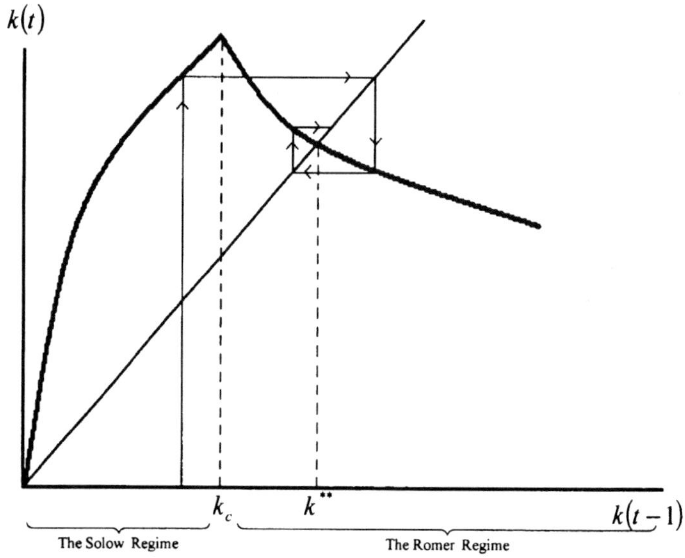

Figure 5.5.3: The Schumpeterian growth with $sA > \theta - 1$

Exercise 5.5

1 Consider the Matsuyama model presented in this section. Prove: (i) If

$$sA < 1,$$

then for any $k_0 \in R_+$, $k(t) < k_c$ for all t, and $\lim_{t \to \infty} k(t) = k^*$; (ii) If

$$1 < sA < \theta - 1,$$

there are period- 2 cycles, fluctuates forever between the Solow and Romer regimes for almost all initial conditions; and (iii) If

$$sA > \theta - 1,$$

then, for any $k_0 \in R_+$, there exists a t' such that

$$\{k(t);\, t \ge t'\} \subset [k_c,\, f(k_c)], \quad \lim_{t \to \infty} k(t) = k^{**}.$$

5.6 Money, growth and fluctuations

This section is to identify endogenous fluctuations in a monetary economy.[10] The economy is inhabited by an infinitely lived representative agent with perfect foresight, who maximizes

$$\sum_{t=0}^{\infty} \beta^t U(c(t), m(t)),$$

subject to the flow budget constraint

$$M(t) = P(t)(y - c(t)) + H(t) + M(t-1), \quad M(-1) \text{ given}$$

where $U : R^{+2} \to R$ is the one period utility function, which is strictly concave, increasing in both arguments, $\beta \in (0, 1)$ is the discount factor; y is the constant endowment of perishable consumption good; $c(t)$ is consumption level

$$m(t) = \frac{M(t)}{P(t)},$$

is real balance, $M(t)$ is nominal money holdings, and $P(t)$ is the price level. The agent considers $\{P(t)\}_t^{\infty}$ independent of his own money holdings. At the beginning of period t, the agent receives $H(t)$ units of paper money from the government though a "helicopter drop", also considered to be independent of his money holdings. The money supply grows at the rate $\mu > \beta$, which implies

$$H(t) = (\mu - 1)M(t - 1).$$

Thus the markets clear when

$$M(t) = \mu^t M_0, \quad c(t) = y,$$

for all t. An equilibrium point of the economy is given by a nonnegative sequence of real balances satisfying

$$\beta U_c(y, m(t+1))m(t+1) = \mu m(t)[U_c(y, m(t)) - U_m(y, m(t))], \tag{5.6.1}$$

[10] This section is based on Matsuyama (1991).

as well as the transversality condition

$$\lim_{t \to \infty} \beta^t U_c(y, m(t)) m(t) = 0.$$

The steady state is $m(t) = m^*$ or

$$P(t) = \mu^t \left(\frac{M_0}{m^*} \right)$$

for all t where $m^* > 0$ satisfies

$$(\mu - \beta) U_c(y, m^*) = \mu U_m(y, m^*),$$

exists uniquely.[11]

For simplicity, assume that the utility function is taken on the following form

$$U(c, m) = \begin{cases} -\dfrac{[g(c)m]^{-(1+\eta)}}{1+\eta}, & \text{if } \eta \neq -1, \\ \log g(c) + \log m, \end{cases} \tag{5.6.2}$$

where $g > 0,\ g' > 0$, and

$$\sup\left[\frac{gg''}{g'^2} \right] < 1,$$

and η is a parameter satisfying

$$(\eta + 2)\left[2 - \sup\left(\frac{gg''}{g'^2} \right) \right] > 1.$$

The elasticity of intertemporal substitution of real balances is equal to

---

[11] Strictly speaking, $m(t) = 0$ or $P(t) = \infty$ is another candidate for the steady state. Nevertheless, for the utility function used below, this case is ruled out.

$$\sigma = (\eta + 2)^{-1}.$$

Substituting equation (5.6.2) into equation (5.6.1) yields

$$p(t+1) = (1+\delta)^{1/\eta} p(t)(1-p(t))^{1/\eta} \equiv F(p(t)), \quad p(t) \in (0, 1), \tag{5.6.3}$$

for all t and

$$\delta \equiv \frac{\mu}{\beta} - 1 > 0, \quad p(t) \equiv \frac{g(y)}{g'(y)m(t)}.$$

For the new variable p, the transversality condition becomes

$$\lim_{t\to\infty} \beta^t p^\eta(t) = 0.$$

The unique steady state is

$$p^* = \frac{\delta}{1+\delta}.$$

We also rule out the case of $\eta = 0$. In the remaining of this section, we require: $\eta > 0$.[12] If $\eta = 1$, then

$$F = (1+\delta)p(t)(1-p(t)), \quad p(t) \in (0, 1).$$

This is again the well-studied logistic map. There are periodic, aperiodic and chaotic solutions to this system.

It is straightforward to check the following properties of F: (1)

$$F(0) = F(1) = 0;$$

(2) F has a single peak at

[12] Under this requirement, the transversality condition $\lim_{t\to\infty} \beta^t p^\eta(t) = 0$ is automatically satisfied for $p \in (0, 1)$.

$$\tilde{p} = \frac{\eta}{1+\eta};$$

F is strictly increasing on $[0, \tilde{p})$, and strictly decreasing on $(\tilde{p}, 1]$; (3)

$$F'(0) = (1 + \delta)^{1/\eta} > 1;$$

(4)

$$F'(p^*) = 1 - \frac{\delta}{\eta};$$

and (5) Let

$$\Delta(\eta) \equiv \eta^{-\eta}(1 + \eta)^{1+\eta} - 1.$$

If $\delta \leq (>) \Delta(\eta)$,

$$F(\tilde{p}) \leq (>) 1;$$

F maps $[0, 1]$ into itself if $\delta \leq \Delta(\eta)$; F maps $(0, 1)$ into itself if $\delta < \Delta(\eta)$. The function $\Delta(\eta)$ defined on $(0, \infty)$ is strictly increasing

$$\lim_{\eta \to \infty} \Delta(\eta) = 0, \quad \Delta(\eta) > 0$$

for all η.

Proposition 5.6.1. (i) If $0 < \delta \leq 2\eta$, then for all $p_0 \in (0, 1)$,

$$\lim_{t \to \infty} F^t(p_0) = p^*;$$

(ii) If $0 < 2\eta \leq \delta$, then a period-2 cycle exists, and the set $p_0 \in (0, 1)$ such that $F^t(p_0)$ converges, N, is at most countable. Furthermore, if

$$2\eta \leq \delta < \Delta(\eta),$$

the set of initial prices that lead to equilibrium points along which the price level fluctuates forever, N^{*}, is of full Lebesgue measure.

Proof: (i) Introduce

$$G(p) \equiv \{F^{2}(p)/p\}^{\eta}.$$

Clearly, $G(p^{*}) = 1$. If $G(\hat{p}) = 1$ for some

$$\hat{p} \in (0, 1)\backslash\{p^{*}\},$$

then there exists a period-2 cycle of F. From the definitions of F and G, we have

$$G(p) = (1+\delta)^{2}(1-p)(1-F(p)),$$

$$G'(p) = (1+\delta)^{2}\left[\{(1+\delta)(1-p)\}^{1/\eta}\left\{\left(2+\frac{1}{\eta}\right)p-1\right\}-1\right],$$

$$G''(p) = (1+\delta)^{2+1/\eta}\left(1+\frac{1}{\eta}\right)(1-p)^{1/\eta-1}\left\{2-\left(2+\frac{1}{\eta}\right)p\right\}.$$

Let

$$\overline{p} = \frac{2\eta}{1+2\eta}.$$

We see that G' is strictly increasing in $(0, \overline{p})$, strictly decreasing $(\overline{p}, 1]$, and $G'(\overline{p}) \leq 0$. Thus $G'(p) < 0$ in $(0, 1)|\backslash\{\overline{p}\}$, or G is strictly decreasing in $(0, 1)$. Therefore, p^{*} is the only solution of $G(p) = 1$. This implies that F has no period-2 cycle. From the theorem of Coppel,[13] which states that if f is a continuous map of a compact interval to itself that has no period-2, the sequence $\{f'(x)\}$ converges to a fixed point of f for every x in the interval. As no $p \in (0, 1)$ can approach 0 asymptotically, we have

[13] See Coppel (1955).

$$\lim_{t\to\infty} F^t(p) = p^*$$

for every p in $(0, 1)$.

(ii) For the existence of period-2 cycles, it is sufficient to show that $G(p) = 1$ has a solution in $(0, 1)\setminus\{p^*\}$. As

$$G(0^+) > 0, \quad G(1) = 0,$$

the intermediate value theorem implies that there exists $p_c \in (0, p^*)$ such that $G(p_c) = 1$ if

$$G'(p^*) = (1 + \delta)\left(\frac{\delta}{\eta} - 2\right) > 0.$$

For the proof of countability of N, let p be a point in N and p^∞ be the limit point of the sequence starting at p. From the continuity of F

$$p^\infty = \lim_{t\to\infty} F^{t+1}(p) = F\left(\lim_{t\to\infty} F^t(p)\right) = F(p^\infty),$$

we see

$$p^\infty = 0 \text{ or } p^\infty = p^*.$$

Since $0 < 2\eta < \delta$ implies that both 0 and p^* are locally unstable, $\lim_{t\to\infty} F^t(p)$ cannot approach them asymptotically. Therefore

$$N - \{q \in (0, 1) | F^T(q) = 0 \text{ or } p^* \text{ for a finite } T\},$$

which is at most countable, since property (2) of F implies that, for any z, there are at most two x's that solve $z = F(x)$. The second half follows from the fact that if

$$\delta < \Delta(\eta), \quad N^{*} = (0, 1) \setminus N.$$

Proposition 5.6.2. For any $\eta > 0$, there exists a value $\Delta^{*}(\eta)$ satisfying

$$2\eta < \Delta^{*}(\eta) < \Delta(\eta),$$

such that a period-3 cycle of F exists if $\delta > \Delta^{*}(\eta)$.

Proof: Introduce

$$G(p) \equiv \left\{\frac{F^{3}(p)}{p}\right\}^{\eta}.$$

Then $G(0^{+}) > 1$ and $G(p^{*}) = 1$. From the intermediate value theorem, it is sufficient to show that there is $p_c \in (0, p^{*})$ such that $G(p_c) < 1$. If $\delta > \Delta(\eta)$, then there exists $p_c \in (0, p^{*})$ such that $F(p_c) = 1$ so that

$$F^{2}(p_c) = F^{3}(p_c) = 0,$$

or $G(p_c) < 1$. From the continuity of G on δ, there exists a $\Delta^{*}(\eta) < \Delta(\eta)$ such that $G(p_c) < 1$. That $2\eta < \Delta^{*}(\eta)$ follows from proposition 5.6.1.

The proposition guarantees the existence of chaos.

5.7 Population and economic growth

The model in this section is based on section 6.2 in Zhang.[14] The model is constructed by Haavelmo in continuous form. Its discrete form was examined by Stutzer,[15] by applying modern mathematics for one-dimensional mappings. First, consider a macroeconomic growth model proposed by Haavelmo[16]

---

[14] Section 6.2 in Zhang (1991).

[15] Stutzer (1980).

[16] Haavelmo (1954).

$$\dot{N} = N\left(a - \frac{\beta N}{Y}\right), \quad a, \beta > 0, \quad Y = AN^{\alpha}, \ A > 0, \ 0 < \alpha < 1, \tag{5.7.1}$$

where N is the population, Y is the real output, and a, β, α and A are constant parameters. Substituting

$$Y = AN^{\alpha}$$

into the differential equation yields

$$\dot{N} = N\left(a - \frac{\beta N^{1-\alpha}}{A}\right). \tag{5.7.2}$$

We see that the growth law is a generalization of the familiar logistic form widely used in biological population and economic analysis. It is not difficult to see that the dynamics of this system are simple. If the initial condition satisfies

$$N(0) > (<) \left(\frac{aA}{\beta}\right)^{1/(2-\alpha)},$$

then both N and Y will decrease (increase) monotonically until approaching their unique equilibrium point, respectively. If we replace time derivatives by first differences and accept discrete time, then equation (5.7.2) becomes

$$N(t+1) = N(t)\left[(1+a) - \frac{\beta N(t)^{1-\alpha}}{A}\right],$$

which can be further simplified as

$$x(t+1) = (1+a)x(t)\left(1 - x^{1/2}(t)\right) = F(x(t); a, 0.5), \tag{5.7.3}$$

in which the new variable $x(t)$ is defined by the transformation

$$N(t) = \left[\frac{A(1+a)}{\beta}\right]^{1/(1-\alpha)} x(t),$$

with $\alpha = 1/2$. It can be shown that none of the qualitative properties of the system are affected by the particular choice of $0 < \alpha < 1$. The term chaotic dynamics refers to the dynamic behavior of certain equations F which possess: (a) a non-degenerate n-period point for each $n \geq 1$, and (b) an uncountable set

$$S \in J = [0, 1]$$

containing no periodic points and no asymptotically periodic points. The trajectories of such points wander around in J "randomly".

For each value of a, equilibrium points are given by the intersection of the graph of $F(x(t); a)$ with the 45-degree line. For each value of a, there are two equilibrium points

$$x_0 = 0, \quad x_0 = \left(\frac{a}{1+a}\right)^2.$$

The point $x_0 = 0$ is unstable and repels nearby points. The local stability of the other one can be determined by linearization at the equilibrium point. We have

$$F'(x_0; a) = 1 - \frac{a}{2} = \theta(a). \tag{5.7.4}$$

The eigenvalue $\theta(a)$ determines the local stability of x_0. When $0 < \theta < 1$, x_0 attracts nearby points in an exponential, monotonic fashion. When $0 > \theta > -1$, x_0 attracts nearby points in a damped oscillatory manner. When $\theta = 1$, x_0 is neither stable nor unstable. Finally, if $|\theta| > 1$, x_0 is unstable. These behaviors occur when

$$0 < a < 2, 2 < a < 4, \quad a = 4, \quad 4 < a < 5.57,$$

respectively. When the equilibrium point is stable, i.e., $a < 4$, the trajectory starting at any point always approaches it. In this region a traditional comparative statics analysis shows that an increase in the parameter a will increase $x(t)$ for sufficiently large t. If $4 < a < 5.57$, trajectories don't approach the equilibrium point, but bounded by 0 and 1. In fact, as the parameter a exceeds 4, the unstable equilibrium point bifurcates into two stable points of period two, i.e., into a stable periodic orbit of length 2. The 2-period cycle becomes unstable for values of a in excess of about 4.8, and each 2-period point bifurcates into two 4-period points, producing an stable cycle of length four denoted by

$$\{x_{01}^4, x_{02}^4, x_{03}^4, x_{04}^4\}.$$

Figure 5.7.1 illustrates the phenomenon.

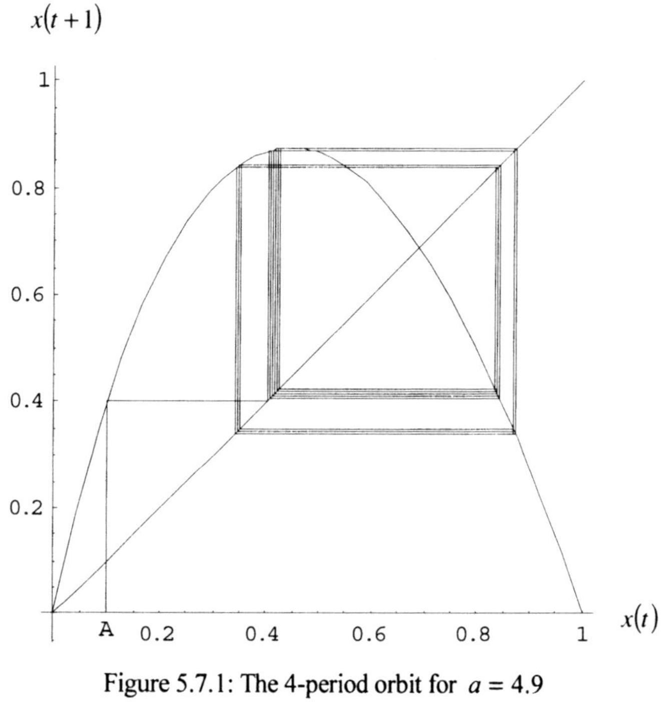

Figure 5.7.1: The 4-period orbit for $a = 4.9$

This pitchfork bifurcation process continues as the parameter a increases, producing non-degenerate orbits of length $2k$ ($k = 2,...$). These orbits are called harmonics of the 2-period orbit. It can be shown that all the harmonics occur prior to the parameter a reaching 5.54, although how much prior to this value is not known. Thus, the range of a, within which a stable orbit of length k first appears and later becomes unstable and bifurcates to a $2k$ k-period orbit, decreases in length as the parameter a increases to a limiting value $a_c < 5.54$. The range of $a_c < a \leq 5.75$ is termed the chaotic region. As the parameter a enters this region, even stranger behavior can occur. For example, a 3-period orbit exists at values of

a near 5.540. This, then gives rise to orbits of periods $3k$ ($k = 2, ...$) via the pitchfork process just described. In fact, if we can locate the 3-period orbit, the Li-Yorke theorem demonstrates that for any $F(x(t);\, a)$ in which a non-degenerate 3-period orbit arises, there must also exist non-degenerate points of all periods, as well as an uncountable set of periodic (not asymptotically periodic) points whose trajectories wander randomly throughout the domain of F.[17] Our dynamic economic system satisfies the requirements in the Li-Yorke theorem for some values of a. This guarantees the existence of chaotic behavior as illustrated in figure 5.7.2.

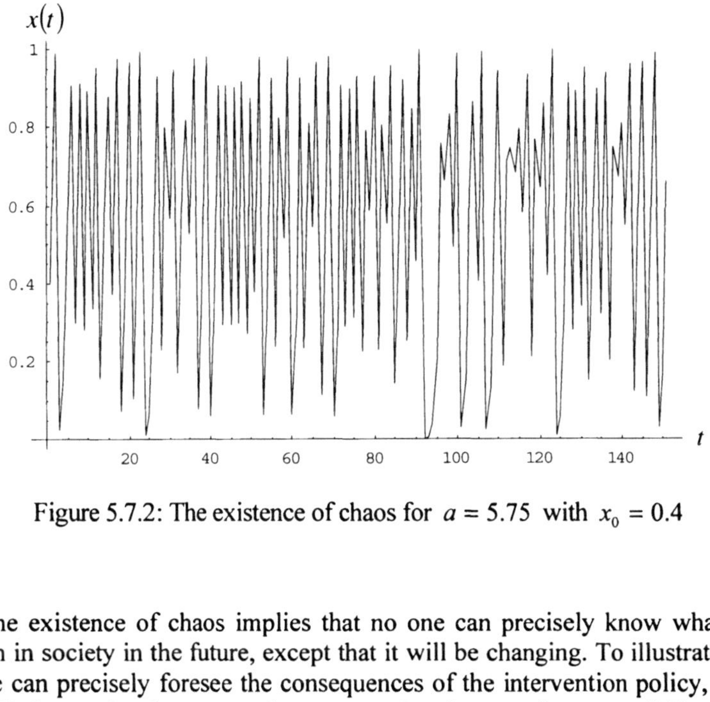

Figure 5.7.2: The existence of chaos for $a = 5.75$ with $x_0 = 0.4$

The existence of chaos implies that no one can precisely know what will happen in society in the future, except that it will be changing. To illustrate why no one can precisely foresee the consequences of the intervention policy, let us try to find out what happen to the system when it starts from two different but very near states. In figure 5.7.3, we consider the case of $a = 5.75$. Let us consider two cases of $x_0 = 0.400$ and $x_0 = 0.405$ over 100 years. It can be seen that the two behaviors are varied over time.

[17] Li and Yorke (1975).

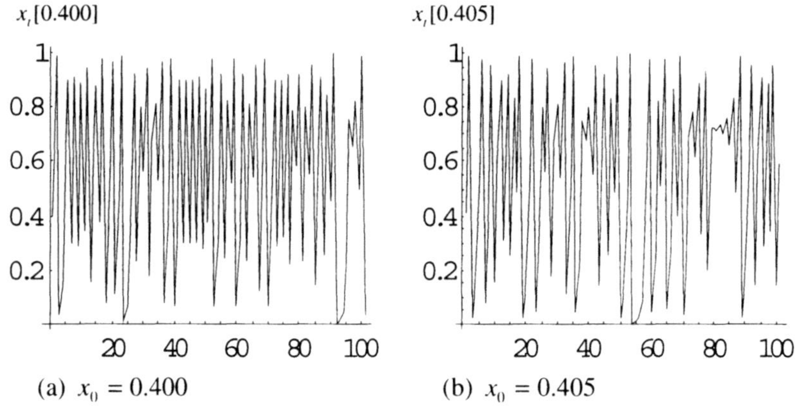

Figure 5.7.3: The dynamics with different initial conditions

We calculate the difference $x[t, 0.400] - x[t, 0.405]$ between the two paths started at $x_0 = 0.400$ and $x_0 = 0.405$ over 100 years.

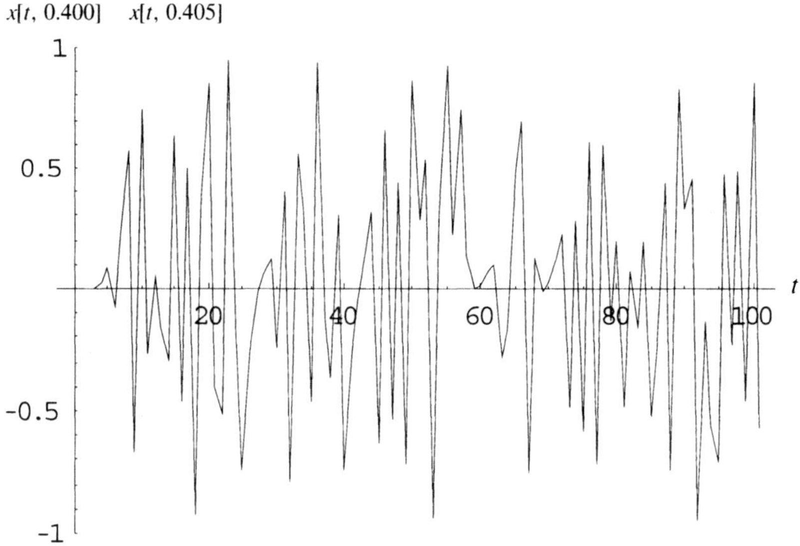

Figure 5.7.4: Small differences at the beginning signify much

In summary, as the autonomous growth rate a exceeds a certain value, the steady state ceases being approached monotonically, and an oscillatory approach occurs. If a is increased further, the steady state becomes unstable and repels nearby points. As a increases, one can find a value of a where the system possesses a cycle of period k for arbitrary k (see figure 5.7.5). Also, there exists an uncountable number of initial conditions from which emanate trajectories that fluctuate in a bounded and aperiodic fashion and are indistinguishable from a realization of some stochastic (chaotic) process.

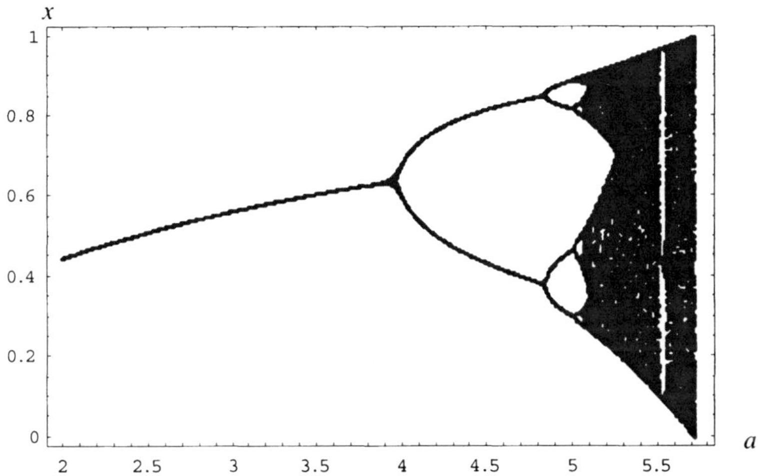

Figure 5.7.5: The map of bifurcations for $a \in [2, 5.75]$

Chapter 6

Higher dimensional difference equations

In this chapter, we study the dynamics of difference equations of the form

$$x_i(t+1) = f_i(x_1(t), x_2(t), \cdots, x_n(t)), \quad i = 1, 2, \cdots, n.$$

The system can be expressed in the vector form

$$x(t+1) = f(x(t)).$$

We organize the chapter as follows. Section 6.1 studies phase space analysis of planar linear difference equations. This section depicts dynamic behavior of the system when the characteristic equation has two distinct eigenvalues, or repeated eigenvalues, or complex conjugate eigenvalues. Section 6.2 studies autonomous linear difference equations. This section provides a procedure of finding general solutions of the system. Section 6.3 studies nonautonomous linear difference equations. This section provides a procedure of finding general solutions of the system. We also examine a few models of economic dynamics. They include a dynamic input-output model with time lag in production, a cobweb model in two interrelated markets, duopoly model, a model oligopoly with 3 firms, and a model of international trade between two countries. This section also shows how the one-dimensional difference equation of higher order can be expressed in multi-dimensional equations of first order. Section 6.4 defines concepts of stabilities and relations among these concepts. This section also provides conditions for stability or instability of difference equations. Section 6.5 studies Liapunov's second method or direct method. The theory of Liapunov functions is a global approach toward determining asymptotic behavior of solutions. Section 6.6 studies the theory of linearization of difference equations. There are two possible ways to simplify dynamical systems: one is to transform one complex system to another one which is much easier to analyze; and the other approach is to reduce higher dimensional problems to lower ones. The center manifold theorem helps us to reduce dimensions of dynamical problems. Section 6.7 defines the concept of conjugacy and shows how to apply the center manifold theorem. Section 6.8 studies the Hénon map, demonstrating bifurcations and chaos of planar difference equations. Section 6.9 studies the Neimark-Sacker (Hopf) bifurcation. This section identifies the Hopf bifurcation in the discrete Kaldor model. Section 6.10 introduces the Liapunov numbers and discusses chaos for planar dynamical systems.

6.1 Phase space analysis of planar linear systems

This section examines stability properties of second-order linear autonomous systems

$$x_1(t+1) = a_{11}x_1(t) + a_{12}x_2(t),$$
$$x_2(t+1) = a_{21}x_1(t) + a_{22}x_2(t),$$

or

$$x(t+1) = Ax(t), \tag{6.1.1}$$

If $Ax^* = x^*$ or

$$(A - I)x^* = 0,$$

x^* is an equilibrium point. If $(A - I)$ is nonsingular, then $x^* = 0$ is the only equilibrium point of system (6.1.1). On the other hand, if $(A - I)$ is singular, then there is a family of equilibrium points. In the latter case, let

$$y(t) = x(t) - x^*,$$

we obtain

$$y(t+1) = Ay(t).$$

Thus the stability properties of an equilibrium point $x^* \neq 0$ are the same as those of the equilibrium point $x^* = 0$. We can thus assume that $x^* = 0$ is the only equilibrium point of system (6.1.1). Let

$$J = P^{-1}AP$$

be the Jordan form of A. Introduce

$$y(t) = P^{-1}x(t).$$

Then system (6.1.1) is transformed into

$$y(t+1) = Jy(t), \quad y_0 \ (= P^{-1}x_0). \tag{6.1.2}$$

As the qualitative properties of systems (6.1.1) and (6.1.2) are identical, it is sufficient for us to be concerned only with equation (6.1.2). It can be seen that J has only three canonical forms. We examine these cases separately.

Case 1
When A has two distinct eigenvalues, $\rho_1 \neq \rho_2$

$$J = diag[\rho_1, \rho_2].$$

The solution of system (6.1.2) is given by

$$y_i(t) = \rho_i^t y_{i0}, \ i = 1, 2.$$

Thus

$$\frac{y_2(t)}{y_1(t)} = \left(\frac{\rho_2}{\rho_1}\right)^t \frac{y_{20}}{y_{10}}, \quad i = 1, 2.$$

We conclude

$$\lim_{t\to\infty}\frac{y_2(t)}{y_1(t)}=\begin{cases}0 & \text{if } |\rho_1|>|\rho_2|,\\ \infty & \text{if } |\rho_1|<|\rho_2|.\end{cases}$$

Figures 6.1.1 to 7.6.5 illustrate different combinations of ρ_1 and ρ_2.

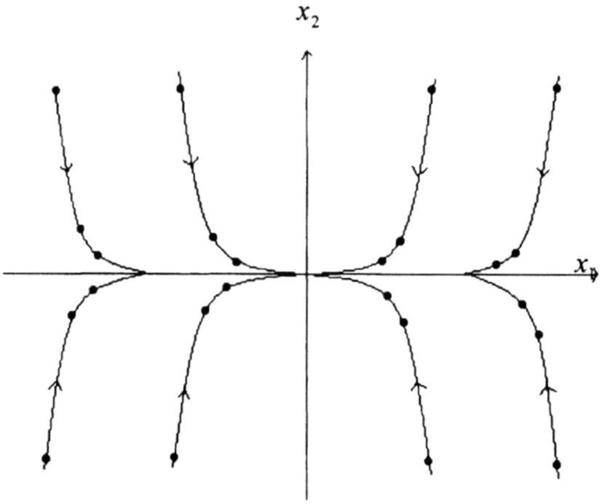

Figure 6.1.1: $\rho_1 < \rho_2 < 1$, asymptotically stable node

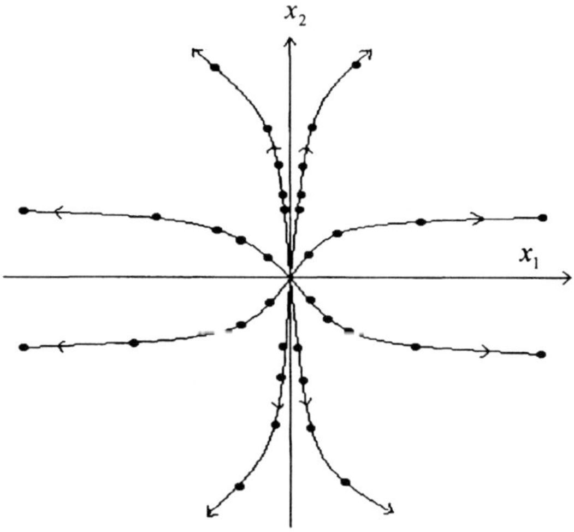

Figure 6.1.2: $\rho_1 > \rho_2 > 1$, unstable node

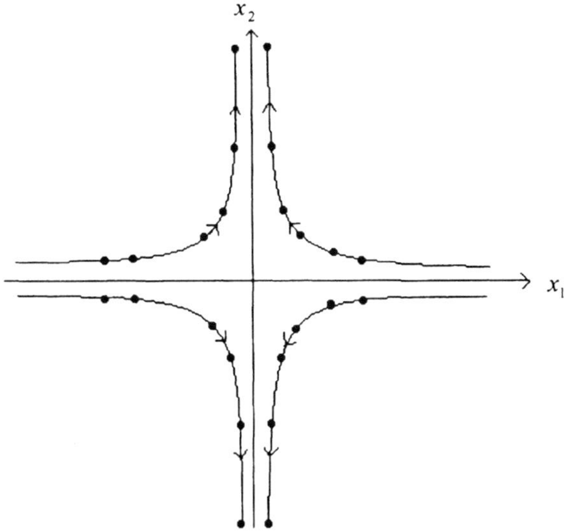

Figure 6.1.3: $0 < \rho_1 < 1$, $\rho_2 > 1$, saddle (unstable)

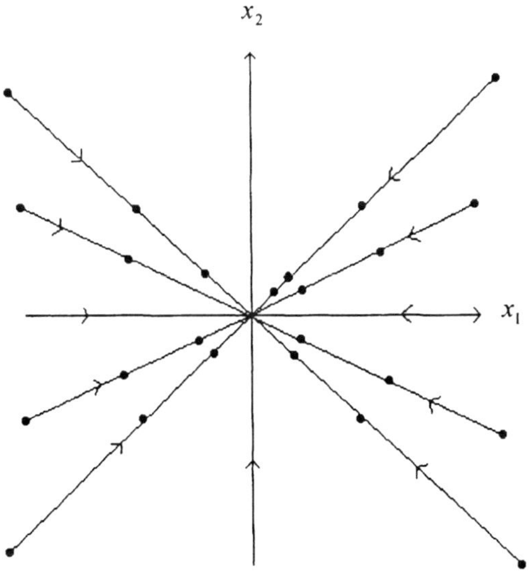

Figure 6.1.4: $0 < \rho_1 = \rho_2 < 1$, asymptotically stable node

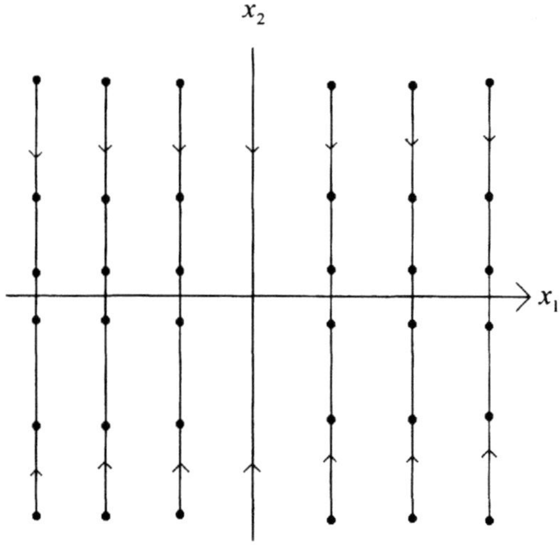

Figure 6.1.5: $\rho_1 = 1$, $\rho_2 < \rho_1$, degenerate node

Case 2
When A has repeated eigenvalues

$$\rho = \rho_1 = \rho_2,$$

we have

$$J = \begin{bmatrix} \rho & 1 \\ 0 & \rho \end{bmatrix}.$$

The solution of system (6.1.2) is given by

$$y_1(t) = \rho^t y_{10} + t\rho^t y_{20},$$
$$y_2(t) = \rho^t y_{20}.$$

Thus

$$\lim_{t\to\infty} \frac{y_2(t)}{y_1(t)} = 0.$$

We have two cases as illustrated in figures 6.1.6 and 7.1.7.

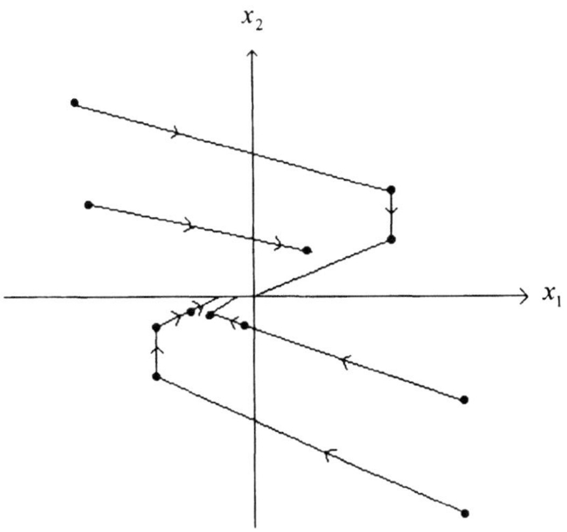

Figure 6.1.6: $\rho_1 = \rho_2 < 1$, asymptotically stable node

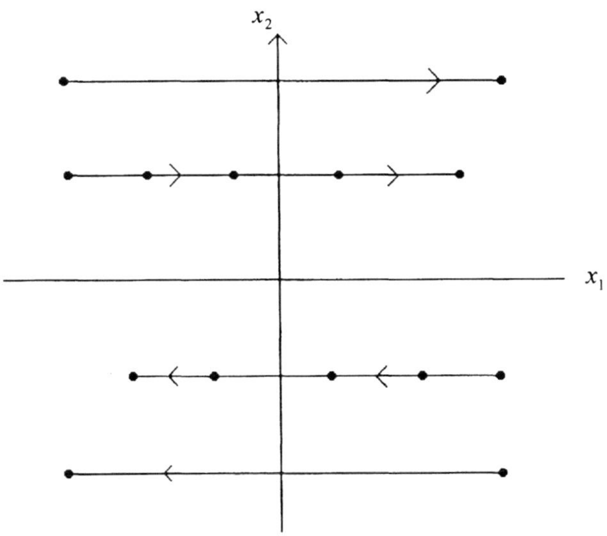

Figure 6.1.7: $\rho_1 = \rho_2 = 1$, degenerate case (unstable)

Case 3
When A has complex conjugate eigenvalues

$$\rho_{1,2} = \alpha \pm i\beta, \quad \beta \neq 0,$$

we have

$$J = \begin{bmatrix} \alpha & \beta \\ -\beta & \alpha \end{bmatrix}.$$

Suppose $\xi = [1 \quad i]^T$ is the eigenvector of A corresponding to

$$\rho_1 = \alpha + i\beta.$$

In this case, a solution of system (6.1.2) may then be given by

$$y(t) = (\alpha + i\beta)^t \begin{bmatrix} 1 \\ i \end{bmatrix}. \tag{6.1.3}$$

Introduce

$$r = \sqrt{\alpha^2 + \beta^2}, \quad \theta = \tan^{-1}\left(\frac{\beta}{\alpha}\right).$$

Then this solution can now be written as

$$\begin{aligned} y(t) &= \begin{bmatrix} 1 \\ i \end{bmatrix} [r(\cos\theta + i\sin\theta)]^t = \begin{bmatrix} 1 \\ i \end{bmatrix} r^t(\cos t\theta + i\sin t\theta) \\ &= r^t \begin{bmatrix} \cos t\theta \\ -\sin t\theta \end{bmatrix} + ir^t \begin{bmatrix} \sin t\theta \\ \cos t\theta \end{bmatrix} = u(t) + iv(t), \end{aligned} \tag{6.1.4}$$

where

$$u(t) = r^t \begin{bmatrix} \cos t\theta \\ -\sin t\theta \end{bmatrix}, \quad v(t) = r^t \begin{bmatrix} \sin t\theta \\ \cos t\theta \end{bmatrix}.$$

One can show that $u(t)$ and $v(t)$ are two linearly independent solutions of system (6.1.2). Hence we do not need to consider the solution generated by $\overline{\rho}$ and $\overline{\xi}$. A general solution may then be given by

$$\begin{bmatrix} y_1(t) \\ y_2(t) \end{bmatrix} = a_1 u(t) + a_2 v(t) = r^t \begin{bmatrix} a_1 \cos t\theta + a_2 \sin t\theta \\ -a_1 \sin t\theta + a_2 \cos t\theta \end{bmatrix}.$$

Given initial values y_{i0}, we obtain

$$\begin{bmatrix} y_1(t) \\ y_2(t) \end{bmatrix} = r^t \begin{bmatrix} y_{10} \cos t\theta + y_{20} \sin t\theta \\ -y_{10} \sin t\theta + y_{20} \cos t\theta \end{bmatrix}.$$

If we introduce

$$\cos\gamma = \frac{y_{10}}{r_0}, \quad \sin\gamma = \frac{y_{20}}{r_0},$$

where

$$r_0 = \left(y_{10}^2 + y_{20}^2\right)^{1/2},$$

then we have

$$y_1(t) = r^t r_0 \cos(t\theta - \gamma),$$
$$y_2(t) = -r^t r_0 \sin(t\theta - \gamma).$$

Using polar coordinates we can now write the solution as

$$R(t) = r^t r_0,$$
$$\omega(t) = -(t\theta - \gamma).$$

If $r < 1$, we have an asymptotically stable focus, as illustrated in figure 6.1.8. If $r > 1$, we have an unstable focus as shown in figure 6.1.9. When $r = 1$, we obtain a center where orbits are circles with radii r_0. This case is depicted in figure 6.1.10.

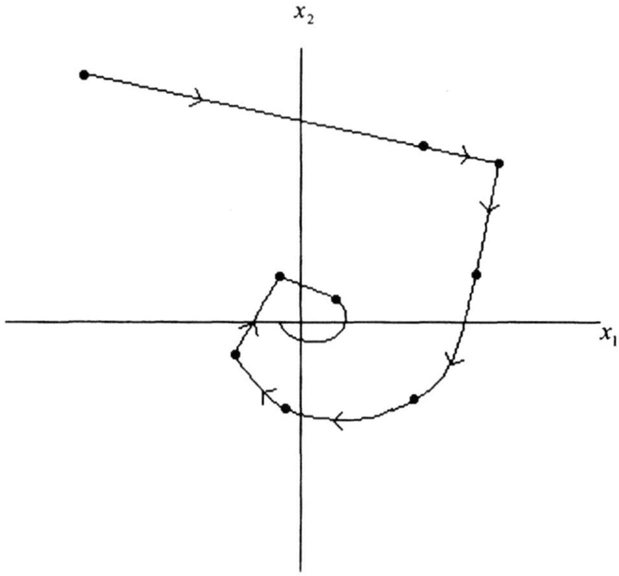

Figure 6.1.8: $|\rho| < 1$, asymptotically stable focus

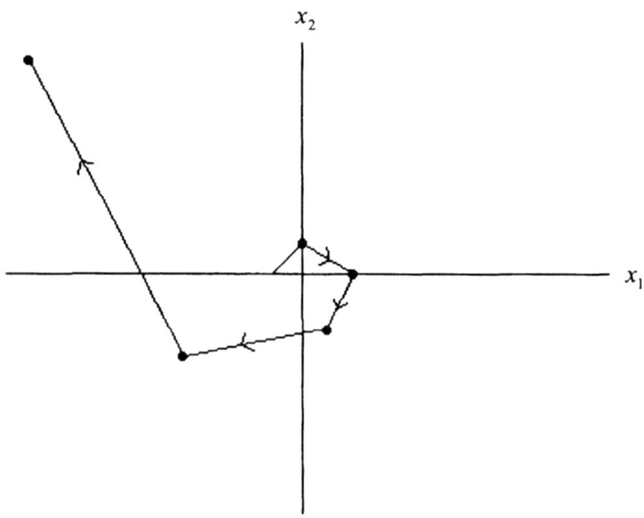

Figure 6.1.9: $|\rho| > 1$, unstable focus

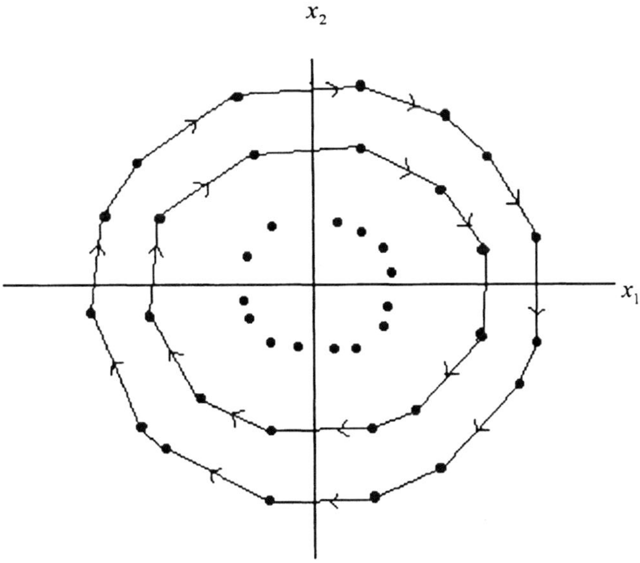

Figure 6.1.10: $|\rho| = 1$, center (stable)

Example Sketch the phase space portrait of

$$x(t+1) = Ax(t),$$

where

$$A = \begin{bmatrix} 1 & 1 \\ 0.25 & 1 \end{bmatrix}.$$

The eigenvalues of A are $\rho_1 = 1.5$ and $\rho_2 = 0.5$; the corresponding eigenvectors are

$$\xi_1 = (2 \quad 1)^T, \quad \xi_2 = (2 \quad -1)^T.$$

Hence

$$P = [\xi_1 \quad \xi_2],\ P^{-1}AP = J = \begin{bmatrix} 1.5 & 0 \\ 0 & 0.5 \end{bmatrix}.$$

Let $x(t) = Py(t)$. We obtain

$$y(t+1) = Jy(t).$$

The phase space portrait of $y(t+1) = Jy(t)$ is depicted as in figure 6.1.11. We may also describe the phase space portrait of the original system with

$$x(t) = Py(t),$$

as in figure 6.1.12.

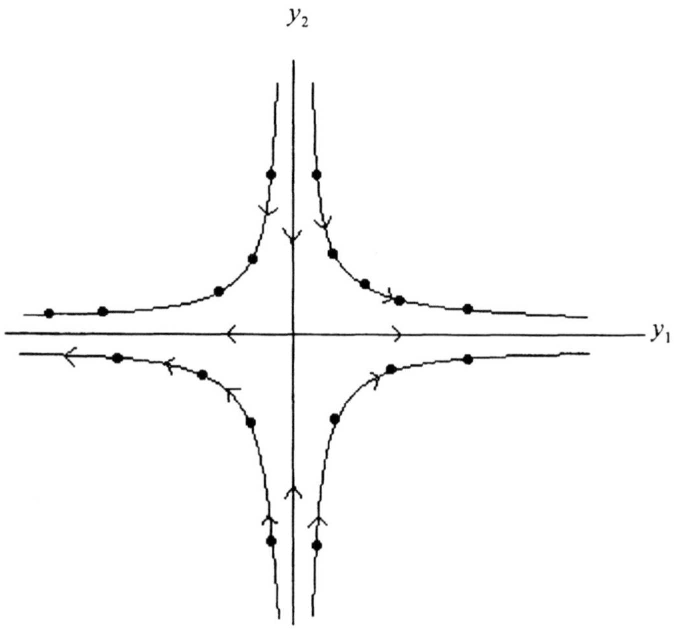

Figure 6.1.11: Saddle of the canonical form, $y(t+1) = Jy(t)$

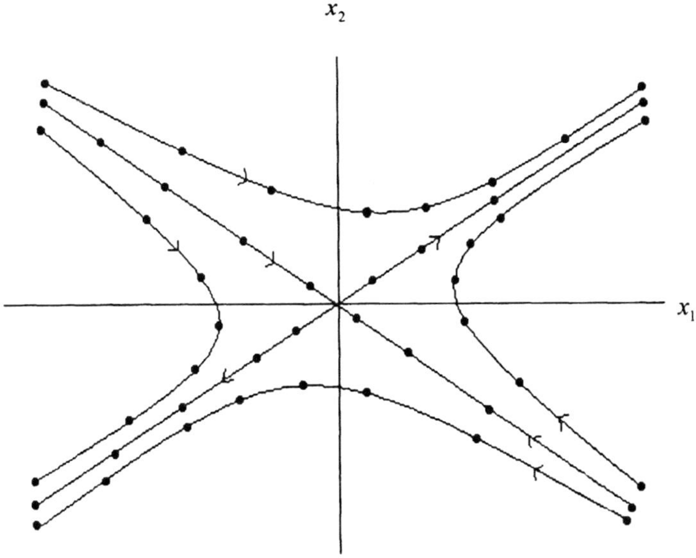

Figure 6.1.12: Saddle of the original system, $x(t+1) = Ax(t)$

Exercise 6.1

1 Depict the phase space diagram and determine the stability of the equation

$$x(t+1) = Ax(t),$$

where A is given by

(i) $\begin{bmatrix} 0.5 & 0 \\ 0 & 0.5 \end{bmatrix}$;

(ii) $\begin{bmatrix} -0.5 & 1 \\ 0 & -0.5 \end{bmatrix}$;

(iii) $\begin{bmatrix} 1 & 0.5 \\ -0.5 & 1 \end{bmatrix}$.

6.2 Autonomous linear difference equations

We now study the following linear homogenous difference equations

$$x(t+1) = Ax(t), \tag{6.2.1}$$

where

$$x(t) = (x_1(t), x_2(t), \cdots, x_n(t))^T,$$

and A is a $n \times n$ real nonsingular matrix. A *solution* of system (6.2.1) is an expression that satisfies this system for all $t \geq 0$. A *general solution* is a solution that contains all solutions of the system. A *particular solution* is one that satisfies an initial condition $x_0 = x(t_0)$. The problem of finding a particular solution with specified initial conditions is called an *initial value problem*. It can be seen that the solution of system (6.2.1) has the form

$$x(t) = A^{t-t_0} x_0.$$

Theorem 6.2.1. There exists a fundamental set, denoted by

$$\{X_1(t), X_2(t), \cdots, X_n(t)\},$$

of solutions for system (6.2.1). A general solution is given by

$$x(t) = \sum_{i=1}^{n} c_i X_i(t), \quad c_j \in R, \quad i = 1, \cdots, n.$$

Along with the homogeneous system (6.2.1), we consider the nonhomogeneous system

$$x(t+1) = Ax(t) + B(t), \quad x(0) = x_0, \quad t = 0, 1, \cdots. \tag{6.2.2}$$

The initial value problem (6.2.2) has a unique solution given by

$$x(t) = A^t x_0 + \sum_{i=0}^{t} A^{t-i} B(i), \quad t = 0, 1, \cdots.$$

We see that the main problem is to calculate A^t. There are some algorithms for computing A^t. Here, we introduce the *Putzer algorithm.*[1] Let the characteristic equation of A be

$$\sum_{i=0}^{n} a_i \rho^{n-i} = 0,$$

where $a_0 = 1$. Let

$$\rho_1, \rho_2, \cdots, \rho_n,$$

be the eigenvalues of A (some of them may be repeated). The following formula determines A^t

$$A^t = \sum_{j=1}^{n} u_j(t) M(j-1),$$

where

$$M(0) = I,$$
$$M(k) = \prod_{j=1}^{k} (A - \rho_j I), \quad k = 1, \cdots, n-1,$$
$$u_1(t) = \rho_1^t,$$
$$u_j(t) = \sum_{i=0}^{t-1} \rho_j^{t-1-i} u_{j-1}(i).$$

Example Solve

$$x(t+1) = Ax(t),$$

where

$$A = \begin{bmatrix} 4 & 1 & 2 \\ 0 & 2 & -4 \\ 0 & 1 & 6 \end{bmatrix}.$$

[1] The algorithm is referred to section 3.1 in Elaydi (1999).

The three eigenvalues of A are

$$\rho_1 = \rho_2 = \rho_3 = 4.$$

Hence, we have

$$M(0) = I,$$

$$M(1) = A - 4I = \begin{bmatrix} 0 & 1 & 2 \\ 0 & -2 & -4 \\ 0 & 1 & 2 \end{bmatrix},$$

$$M(2) = (A - 4I)M(1) = \begin{bmatrix} 0 & 0 & 0 \\ 0 & 0 & 0 \\ 0 & 0 & 0 \end{bmatrix},$$

$$u_1(t) = 4^t,$$

$$u_2(t) = \sum_{i=0}^{t-1} 4^{t-1-i}\left(4^i\right) = t4^{t-1},$$

$$u_3(t) = \sum_{i=0}^{t-1} 4^{t-1-i}\left(i4^i\right) = \frac{t(t-1)}{2}4^{t-2}.$$

Applying the above calculation results to

$$A^t = \sum_{j=1}^{n} u_j(t)M(j-1),$$

we get

$$A^t = \begin{bmatrix} 4^t & t4^{t-1} & 2t4^{t-1} \\ 0 & 4^t - 2t4^{t-1} & -t4^t \\ 0 & t4^{t-1} & 4^t + 2n4^{t-1} \end{bmatrix}.$$

Consequently

$$x(t) = A^t x_0$$

is the solution.

We now apply the Jordan to solve system (6.2.1). First, we consider the case that A is similar to the diagonal matrix

$$D = diag[\rho_i],$$

where ρ_i are the eigenvalues of A.[2] That is, there exists a non-singular matrix P such that

$$P^{-1}AP = D.$$

From $AP = PD$, we have

$$A\xi_i = \rho_i \xi_i,$$

where ξ_i is the ith column of P. We see that ξ_i is the eigenvector of A corresponding to the eigenvalue ρ_i. From

$$P^{-1}AP = D$$

and P being non-singular, we have

$$A = PDP^{-1}.$$

Consequently, we have

$$A^t = PD^tP^{-1} = Pdiag[\rho_i^t]P^{-1}. \qquad (6.2.3)$$

Substituting equation (6.2.3) into

$$x(t) = A^{t-t_0}x_0,$$

with $t_0 = 0$ yields the general solution

$$x(t) = Pdiag[\rho_i^t]P^{-1}x_0. \qquad (6.2.4)$$

[2] Refer to appendix A.1 for an introduction to matrix theory. It is known that a $n \times n$ matrix is diagonalizable if and only if it has n linearly independent eigenvectors.

As

$$P diag[\rho_i^t] = [\rho_1 \xi_1, \rho_2 \xi_2, \cdots, \rho_n \xi_n],$$

the general solution (6.2.4) can also be expressed by

$$x(t) = a_1 \rho_1^t \xi_1 + a_2 \rho_2^t \xi_2 + \cdots + a_n \rho_n^t \xi_n, \tag{6.2.5}$$

where

$$a \equiv P^{-1} x_0.$$

After having calculated the eigenvalues and eigenvectors, we may directly determine a by equation (6.2.5) through the initial conditions without calculating P^{-1}.

Example Find the general solution and the initial value problem of $x(t+1) = Ax(t)$

$$A = \begin{bmatrix} 2 & 2 & 1 \\ 1 & 3 & 1 \\ 1 & 2 & 2 \end{bmatrix}, \quad x(0) = \begin{bmatrix} 0 \\ 1 \\ 0 \end{bmatrix}.$$

The three eigenvalues of matrix A are

$$\rho_1 = 5, \ \rho_2 = \rho_3 = 1.$$

Correspondingly, we can find three linearly independent vectors[3]

$$\xi_1 = \begin{bmatrix} 1 \\ 1 \\ 1 \end{bmatrix}, \ \xi_2 = \begin{bmatrix} 1 \\ 0 \\ -1 \end{bmatrix}, \ \xi_3 = \begin{bmatrix} 0 \\ 1 \\ -2 \end{bmatrix}.$$

---

[3] It should be noted that if the matrix has repeated roots, then it is diagonalizable if it is *normal*, that is to say, if $A^T A = AA^T$.

It should be noted that there are infinite choices for ξ_2 and ξ_3 because of multiplicity of the corresponding eigenvalues. The general solution is

$$x(t) = a_1\rho_1^t\xi_1 + a_2\rho_2^t\xi_2 + a_3\rho_3^t\xi_3 = \begin{bmatrix} a_1 5^t + a_2 \\ a_1 5^t + a_3 \\ a_1 5^t - a_2 - 2a_3 \end{bmatrix}.$$

The solution of the initial value problem is solved by substituting the initial condition x_0 into the above equation and then solving a_i. We calculate

$$a_1 = \frac{1}{2}, \quad a_2 = -\frac{1}{2}, \quad a_3 = \frac{1}{2}.$$

Hence

$$x(t) = \frac{1}{2}\rho_1^t\xi_1 - \frac{1}{2}\rho_2^t\xi_2 + \frac{1}{2}\rho_3^t\xi_3 = \frac{1}{2}\begin{bmatrix} 5^t - 1 \\ 5^t + 1 \\ 5^t - 1 \end{bmatrix}.$$

We may also use $x(t) = A^t x_0$ and equation (6.2.3) to solve the initial value problem. We get the same solution by calculating

$$P = [\xi_1 \quad \xi_2 \quad \xi_3] = \begin{bmatrix} 1 & 1 & 0 \\ 1 & 0 & 1 \\ 1 & -1 & 2 \end{bmatrix} \Rightarrow P^{-1} = \begin{bmatrix} \frac{1}{4} & \frac{1}{2} & \frac{1}{4} \\ \frac{3}{4} & -\frac{1}{2} & -\frac{1}{4} \\ -\frac{1}{4} & \frac{1}{2} & -\frac{1}{4} \end{bmatrix}.$$

The reader is asked to check the result.

The matrix A may not be diagonalizable when A has repeated eigenvalues. There is something close to diagonal form called the *Jordan canonical form* of a square matrix. A *basic Jordan block* associated with a value ρ is expressed

$$J = \begin{bmatrix} \rho & 1 & 0 & \cdots & 0 & 0 \\ 0 & \rho & 1 & \cdots & 0 & 0 \\ \vdots & \vdots & \vdots & & \vdots & \vdots \\ 0 & 0 & 0 & \cdots & \rho & 1 \\ 0 & 0 & 0 & \cdots & 0 & \rho \end{bmatrix}.$$

The Jordan canonical form of a square matrix is compromised of such Jordan blocks.

Theorem 6.2.2. (the Jordan canonical form) Any $n \times n$ matrix A is similar to a Jordan form given by

$$J = diag[J_1, J_2, \cdots, J_k], \ 1 \le k \le n,$$

where each J_i is an $s_i \times s_i$ basic Jordan block and

$$\sum_{i=1}^{k} s_i = n.$$

Assume that A is similar to J under P, i.e., $P^{-1}AP = J$. We have

$$A = PJP^{-1}.$$

Hence,

$$A^t = PJ^tP^{-1}.$$

It can be seen that

$$J^t = diag[J_1^t, J_2^t, \cdots, J_k^t], \ 1 \le k \le n.$$

We can write J_i as

$$(J_i)_{s_i \times s_i} = \rho_i I + N_i,$$

where N_i is an $s_i \times s_i$ nilpotent matrix. Using $N_i^k = 0$ for all $k \ge s_i$, we have

$$
\left(J_i^t\right)_{s_i \times s_i} = (\rho_i I + N_i)^t = \begin{bmatrix} \rho_i^t & \binom{t}{1}\rho_i^{t-1} & \binom{t}{2}\rho_i^{t-2} & \cdots & \binom{t}{s_i - 1}\rho_i^{t-s_i+1} \\ 0 & \rho_i^t & \binom{t}{1}\rho_i^{t-1} & \cdots & \binom{t}{s_i - 2}\rho_i^{t-s_i+2} \\ \vdots & \vdots & & \ddots & \vdots \\ & & & & \binom{t}{1}\rho_i^{t-1} \\ 0 & 0 & & \cdots & \rho_i^t \end{bmatrix}.
$$

The general solution of equation (6.2.1) (for $t_0 = 0$) is now given by

$$
x(t) = A^t x_0 = PJ^t P^{-1} x_0 = PJ^t a, \quad a \equiv P^{-1} x_0.
$$

Corollary 6.2.1. Assume that A is any $n \times n$ matrix. Then

$$
\lim_{t \to \infty} A^t = 0,
$$

if and only if $|\rho| < 1$ for all eigenvalues ρ of A.

Exercise 6.2

1 Use the Putzer algorithm to evaluate A^t

(i) $A = \begin{bmatrix} 1 & 1 \\ -2 & 4 \end{bmatrix}$;

(ii) $A = \begin{bmatrix} 1 & 2 & -1 \\ 0 & 1 & 0 \\ 4 & -4 & 5 \end{bmatrix}$.

2 Solve the following systems with the Putzer algorithm

(i) $x_1(t+1) = -x_1(t) + x_2(t)$,
$x_2(t+1) = 2x_2(t), \; x_1(0) = 1, \; x_2(0) = 2$;

(ii) $x(t+1) = Ax(t)$ where

$$A = \begin{bmatrix} 1 & -2 & -2 \\ 0 & 0 & -1 \\ 0 & 2 & 3 \end{bmatrix}, \quad x_0 = \begin{bmatrix} 1 \\ 1 \\ 0 \end{bmatrix}.$$

3 Use formula (6.1.5) to find the solution of $x(t+1) = Ax(t)$

(i) $A = \begin{bmatrix} 2 & 3 & 0 \\ 4 & 3 & 0 \\ 0 & 0 & 6 \end{bmatrix}, \quad x(0) = \begin{bmatrix} 0 \\ 1 \\ 0 \end{bmatrix};$

(ii) $\begin{bmatrix} 1 & 0 & 1 \\ 1 & 2 & 3 \\ 0 & 0 & 3 \end{bmatrix}, \quad x(0) = \begin{bmatrix} 0 \\ 0 \\ 1 \end{bmatrix}.$

6.3 Nonautonomous linear difference equations

This section first examines the *nonautonomous homogenous difference equations*

$$x(t+1) = A(t)x(t), \tag{6.3.1}$$

where

$$x(t) = (x_1(t), x_2(t), \cdots, x_n(t))^T$$

and $A(t)$ is a $n \times n$ nonsingular matrix function.

Theorem 6.3.1. For each $x_0 \in R^n$ and $t_0 \in Z^+$ there exists a unique solution $x(t, x_0)$ of equation (6.1.1) satisfying the initial condition.

Let $X_1(t), X_2(t), \cdots, X_n(t)$ be n solutions of equation (6.3.1). They are said to be *linearly independent* if they are linearly independent for all $t \geq 0$. Otherwise, they are said to be *linearly dependent*. A set

$$\{X_1(t), X_2(t), \cdots, X_n(t)\},$$

of any n linearly independent solutions is said to be a *fundamental set of solutions*, and the matrix

$$X(t) = [X_1(t), X_2(t), \cdots, X_n(t)]$$

is called a *fundamental matrix*. We have

$$\det X(t) \neq 0$$

as $X_1(t), X_2(t), \cdots, X_n(t)$ are linearly independent. As each $X_i(t)$ satisfies equation (6.3.1), we have

$$X(t+1) = A(t)X(t). \tag{6.3.2}$$

In fact, the above equation has a unique solution.

Theorem 6.3.2. There is a unique solution $\widetilde{X}(t)$ of equation (6.3.2) with $\widetilde{X}(t_0) = I$.

The general solution of equation (6.3.1) is given by

$$x(t) = X(t)c,$$

where $c \in R^n$. The nonhomogeneous system corresponding to equation (6.3.1) is given by

$$x(t+1) = A(t)x(t) + g(t), \tag{6.3.3}$$

where $A(t)$ is a $n \times n$ nonsingular matrix function and $g(t) \in R^n$.

Theorem 6.3.3. Any solution $x(t)$ of equation (6.3.3) can be written as

$$x(t) = X(t)c + x_p(t),$$

where $X(t)$ is a fundamental matrix of the corresponding homogeneous system (6.3.1) and $x_p(t)$ is a particular solution of equation (6.3.3) for an appropriate choice of the constant vector c.

Example (a dynamic input-output with time lag in production) There are two industries in the economy. Each industry produces only one homogenous commodity; each industry uses a fixed input ratio for the production of its output; production in every industry is subject to constant return to scale. In order to produce each unit of the j th commodity, the input for the i th commodity is a fixed amount, denoted by a_{ij}. Let $d_i(t)$ indicate the final demand for the i th commodity in t. Assume that there is a one-period lag in production so that the amount demanded in period t determines not the current output but the output of period $(t+1)$. Let $x_i(t)$ denote the output of the i th industry. Then, the input-output model is represented by

$$x_i(t+1) = a_{i1}x_1(t) + a_{i2}x_2(t) + d_i(t).$$

In matrix form, we have

$$x(t+1) = Ax(t) + d(t),$$

where

$$x = [x_1 \quad x_2]^T, \quad A = \lfloor a_{ij} \rfloor_{2\times 2}.$$

Here, we specify

$$d(t) = [\delta^t \quad \delta^t],$$

where δ is a positive scalar. We now need to determine the particular integrals. Following the method of undetermined coefficients, we try solutions of the form

$$\begin{bmatrix} x_1(t) \\ x_2(t) \end{bmatrix} = \begin{bmatrix} b_1\delta^t \\ b_2\delta^t \end{bmatrix},$$

where b_i are unknown parameters to be determined. Substituting this into the input-output model yields

$$\begin{bmatrix} b_1 \\ b_2 \end{bmatrix} \delta^{t+1} = \begin{bmatrix} a_{11} & a_{12} \\ a_{21} & a_{22} \end{bmatrix} \begin{bmatrix} b_1 \\ b_2 \end{bmatrix} \delta^t + \begin{bmatrix} 1 \\ 1 \end{bmatrix} \delta^t .$$

If $I - A$ is nonsingular, we solve the particular integral as follows

$$\begin{bmatrix} x_1(t) \\ x_2(t) \end{bmatrix} = \begin{bmatrix} \delta - a_{11} & -a_{12} \\ -a_{21} & \delta - a_{22} \end{bmatrix}^{-1} \begin{bmatrix} 1 \\ 1 \end{bmatrix} \delta^t .$$

It is straightforward to find the general solution of $x(t+1) = Ax(t)$.

Example (A cobweb model in two interrelated markets) We now consider a model of two - corn and hog - markets with time lags.[4] Let subscripts 1 and 2 denote corn and hog, variables p_j the price of goods j, variables X_j and Y_j stand for the demand and supply of goods j, respectively. The corn market is described by

$$X_1(t) = a_1 - b_1 p_1(t), \ b_1 > 0,$$
$$Y_1(t) = c_1 + d_1 p_1(t-1), \ d_1 > 0,$$
$$X_1(t) = Y_1(t).$$

From the market equilibrium condition

$$X_1(t) = Y_1(t),$$

we have

$$p_1(t) = \frac{a_1 - c_1}{b_1} - \frac{d_1}{b_1} p_1(t-1).$$

The hog market is described by

$$X_2(t) = a_2 - b_2 p_2(t), \ b_2 > 0,$$
$$Y_2(t) = c_2 + d_2 p_2(t-1) + e p_1(t-1), \ d_2 > 0, \ e < 0,$$

[4] An early example of the corn-hog cycle was given by Ezekiel (1938) and Waugh (1964). This version of the model comes from Shone (2002: 346-348).

$$X_2(t) = Y_2(t).$$

From the market condition

$$X_2(t) = Y_2(t),$$

we have

$$p_2(t) = \frac{a_2 - c_2}{b_2} - \frac{d_2}{b_2} p_2(t-1) - \frac{e}{b_2} p_1(t-1).$$

The system consists of two-dimensional difference equations for $p_1(t)$ and $p_2(t)$. The fixed point is determined by

$$p_1^* = \frac{a_1 - c_1}{b_1} - \frac{d_1}{b_1} p_1^* \Rightarrow p_1^* = \frac{a_1 - c_1}{b_1 + d_1},$$

$$p_2^* = \frac{a_2 - c_2}{b_2} - \frac{d_2}{b_2} p_2^* - \frac{e}{b_2} p_1^* \Rightarrow p_2^* = \frac{a_2 - c_2}{b_2 + d_2} - \frac{e}{b_2 + d_2} p_1^*.$$

Introduce

$$x_j(t) = p_j(t) - p_j^*.$$

Then the dynamics are given by

$$\begin{bmatrix} x_1(t+1) \\ x_2(t+1) \end{bmatrix} = \begin{bmatrix} -\frac{d_1}{b_1} & 0 \\ -\frac{e}{b_2} & -\frac{d_2}{b_2} \end{bmatrix} \begin{bmatrix} x_1(t) \\ x_2(t) \end{bmatrix}.$$

The two eigenvalues are

$$\rho_j = -\frac{d_j}{b_j}, \quad j = 1, 2.$$

The condition for stability in the corn market is $|d_1 / b_1| < 1$. The condition for stability in the hog market is $|d_2 / b_2| < 1$.

Example (duopoly model) We consider a duopoly model. There are two firms, 1 and 2, producing a single commodity. The demand function in the market is

$$p(t) = 9 - Q(t),$$

where p is the price of the commodity and Q is the total supply

$$Q(t) = q_1(t) + q_2(t),$$

where $q_j(t)$ is firm j's output. Suppose that firm j's total cost is

$$TC_j(q_j) = 3q_j^2.$$

We assume that in time t a firm chooses its own output at time t so as to maximize its profits at t under the assumption that the rival of the firm will choose the same output level it chose in time $t - 1$. Profits (= revenue minus cost) are given by

$$\pi_1(t) = (9 - q_1(t) - q_2(t-1))q_1(t) - 3q_1^2(t),$$
$$\pi_2(t) = (9 - q_1(t-1) - q_2(t))q_2(t) - 3q_2^2(t).$$

Differentiate the profit functions

$$\frac{\partial \pi_1(t)}{\partial q_1(t)} = 9 - 2q_1(t) - q_2(t-1) - 6q_1(t) = 0,$$
$$\frac{\partial \pi_2(t)}{\partial q_2(t)} = 9 - q_1(t-1) - 2q_2(t) - 6q_2(t) = 0.$$

The Cournot-Nash solution is thus given by

$$q_1(t) = \frac{9}{8} - \frac{1}{8}q_2(t-1),$$

$$q_2(t) = = \frac{9}{8} - \frac{1}{8} q_1(t-1).$$

The unique fixed point is

$$(q_1^*, q_2^*) = (1, 1).$$

Solving the difference equations for initial values (q_{10}, q_{20}), we obtain

$$q_1(t) = 1 - \left(-\frac{1}{8}\right)^t + \frac{1}{2}\left(-\frac{1}{8}\right)^t (q_{10} + q_{20}) + \frac{1}{2}\left(\frac{1}{8}\right)^t (q_{10} - q_{20}),$$

$$q_2(t) = 1 - \left(-\frac{1}{8}\right)^t + \frac{1}{2}\left(-\frac{1}{8}\right)^t (q_{10} + q_{20}) + \frac{1}{2}\left(\frac{1}{8}\right)^t (-q_{10} + q_{20}).$$

The system converges to the equilibrium point.

Example (oligopoly with 3 firms) We consider a oligopoly model with three firms, 1, 2, and 3, producing a single commodity. The demand function in the market is

$$p(t) = 9 - Q(t),$$

where p is the price of the commodity and Q is the total supply

$$Q(t) = q_1(t) + q_2(t) + q_3(t),$$

where $q_j(t)$ is firm j 's output. Suppose firm j 's total cost is

$$TC_j(q_j) = 3q^2.$$

We assume that in time t a firm chooses its own output at time t so as to maximize its profits at t under the assumption that the rivals of the firm will choose the same output level it chose in time $t-1$. Profits (= revenue minus cost) are given

$$\pi_1(t) = (9 - q_1(t) - q_2(t-1) - q_3(t-1))q_1(t) - 3q_1(t),$$
$$\pi_2(t) = (9 - q_1(t-1) - q_2(t) - q_3(t-1))q_2(t) - 3q_2(t),$$
$$\pi_3(t) = (9 - q_1(t-1) - q_2(t-1) - q_3(t))q_3(t) - 3q_2(t).$$

Differentiate the profit functions

$$\frac{\partial \pi_1(t)}{\partial q_1(t)} = 9 - 2q_1(t) - 2q_2(t-1) - q_3(t-1) - 3 = 0,$$
$$\frac{\partial \pi_2(t)}{\partial q_2(t)} = 9 - q_1(t-1) - 2q_2(t) - q_3(t-1) - 3 = 0,$$
$$\frac{\partial \pi_3(t)}{\partial q_3(t)} = 9 - q_1(t-1) - q_2(t-1) - 2q_3(t) - 3 = 0.$$

The Cournot-Nash solution is thus given by

$$q_1(t) = -\frac{1}{2}q_2(t-1) - \frac{1}{2}q_3(t-1),$$
$$q_2(t) = 3 - \frac{1}{2}q_1(t-1) - \frac{1}{2}q_3(t-1),$$
$$q_3(t) = 3 - \frac{1}{2}q_1(t-1) - \frac{1}{2}q_2(t-1).$$

The unique fixed point is

$$\left(q_1^{\bullet}, q_2^{\bullet}, q_3^{\bullet}\right) = \left(\frac{3}{2}, \frac{3}{2}, \frac{3}{2}\right).$$

Solving the difference equations for initial values $\left(q_{10}, q_{20}, q_{30}\right)$, we obtain

$$q_1(t) = \frac{3}{2} - \frac{3}{2}(-1)^t + (-1)^t\left(\frac{q_{10} + q_{20} + q_{30}}{3}\right) + \left(\frac{1}{2}\right)^t\left(\frac{2q_{10} - q_{20} - q_{30}}{3}\right),$$
$$q_2(t) = \frac{3}{2} - \frac{3}{2}(-1)^t + (-1)^t\left(\frac{q_{10} + q_{20} + q_{30}}{3}\right) + \left(\frac{1}{2}\right)^t\left(\frac{-q_{10} + 2q_{20} - q_{30}}{3}\right),$$

$$q_3(t) = \frac{3}{2} - \frac{3}{2}(-1)^t + (-1)^t\left(\frac{q_{10} + q_{20} + q_{30}}{3}\right) + \left(\frac{1}{2}\right)^t\left(\frac{-q_{10} - q_{20} + 2q_{30}}{3}\right).$$

The system does not converge to the equilibrium point.

It is important to note that the duopoly dynamics with the same demand and total cost functions is stable; if the number of firms is larger than 3, then dynamics is unstable.

Theorem 6.3.4. (variation of constant formula) The unique solution of the initial problem

$$x(t+1) = A(t)x(t) + g(t), \quad x(t_0) = x_0,$$

is given by

$$x(t, x_0) = \left(\prod_{i=t_0}^{t-1} A(i)\right)x_0 + \sum_{r=t_0}^{t-1}\left(\prod_{i=r+1}^{t-1} A(i)\right)g(r).$$

In particular, when A is constant, we have

$$x(t, x_0) = A^{t-t_0}x_0 + \sum_{r=t_0}^{t-1} A^{t-r-1}g(r).$$

Example Apply theorem 6.3.4 to solve

$$x(t+1) = Ax(t) + g(t), \quad x(0) = x_0,$$

where

$$A = \begin{bmatrix} 2 & 1 \\ 0 & 2 \end{bmatrix}, \quad g(t) = \begin{bmatrix} t \\ 1 \end{bmatrix}, \quad x(0) = \begin{bmatrix} 1 \\ 0 \end{bmatrix}.$$

Using the Putzer algorithm, we obtain

$$A^t = \begin{bmatrix} 2^t & t2^{t-1} \\ 0 & 2^t \end{bmatrix}.$$

Hence

$$x(t)=\begin{bmatrix}2^t & t2^{t-1}\\ 0 & 2^t\end{bmatrix}\begin{bmatrix}t\\ 1\end{bmatrix}+\sum_{r=0}^{t-1}\begin{bmatrix}2^{t-r-1} & (t-r-1)2^{t--r-2}\\ 0 & 2^{t-r-1}\end{bmatrix}\begin{bmatrix}r\\ 1\end{bmatrix}$$

$$=\begin{bmatrix}2^t+t2^{t-1}-\frac{3}{4}t\\ 2^t-1\end{bmatrix}.$$

(check the last step).

Example Consider the trade between two countries, $i = 1, 2$. For country i, we have

$$Y_i(t)=C_i(t)+I_i+X_i(t)-M_i(t), \tag{6.3.4}$$

where $Y_i(t)$, $C_i(t)$, I_i, $X_i(t)$, $M_i(t)$ are respectively national income (in period t), total consumption, (fixed) net investment, exports, and imports. According to the definitions, country i 's consumption of domestic products

$$D_i(t)=C_i(t)-M_i(t).$$

Assume that the domestic consumption, $D_i(t)$, and imports, $M_i(t)$, of each country at period $t+1$ are proportional to the country's national income one time period earlier. That is

$$\begin{aligned}D_1(t)&=a_{11}Y_1(t),\\ M_1(t)&=a_{21}Y_1(t),\\ D_2(t)&=a_{22}Y_2(t),\\ M_2(t)&=a_{12}Y_2(t).\end{aligned} \tag{6.3.5}$$

As the world consists of the two countries, each country's exports equals the other country's imports, i.e., $X_1 = M_2$ and $X_2 = M_1$ in t. Substituting

$$D_i=C_i-M_i,\ \ X_1=M_2,\ \ X_2=M_1$$

and equation (6.3.5) into equation (6.3.4) yields the global economic model

$$\begin{bmatrix} Y_1(t+1) \\ Y_2(t+1) \end{bmatrix} = \begin{bmatrix} a_{11} & a_{12} \\ a_{21} & a_{22} \end{bmatrix} \begin{bmatrix} Y_1(t) \\ Y_2(t) \end{bmatrix} + \begin{bmatrix} I_1 \\ I_2 \end{bmatrix}.$$

By the variation of constant formula we solve

$$Y(t) = A^t Y_0 + \sum_{r=0}^{t-1} A^r I^*, \qquad (6.3.6)$$

where $I^* = (I_1, I_2)^T$. It can be shown that if

$$a_{11} + a_{21} < 1, \quad a_{12} + a_{22} < 1, \qquad (6.3.7)$$

then for all the eigenvalues ρ of A, $|\rho| < 1$.[5] Hence, under conditions (6.3.7)

$$\lim_{t \to \infty} A^t = 0.$$

On the other hand, according to the so-called Neumann's expansion, we have

$$\sum_{r=0}^{\infty} A^r = (I - A)^{-1}.$$

Hence

$$\lim_{t \to \infty} Y(t) = (I - A)^{-1} I^*.$$

We have examined one-dimensional difference equations of higher order. It can be shown that a one-dimensional difference equation of higher order can be expressed in multi-dimensional equations of first order.

Let us consider difference equations in forms

---

[5] This condition implies that the sum of domestic consumption $D_i(t+1)$ and imports $M_i(t+1)$ in period $t+1$ is less than the national product $Y_i(t)$ of period t, i.e.

$$D_i(t+1) + M_i(t+1) < Y_i(t).$$

$$x(t+1) = f(x(t), x(t-1)). \tag{6.3.8}$$

In this equation the state of the orbit at time $t + 1$ depends directly from its state at time t and $t - 1$. We call equation (6.3.8) a *discrete dynamical system with a delay of a one-unit unit*. We can eliminate the delay by increasing the dimension of the system. Set $y(t+1) = x(t)$. We can rewrite equation (6.3.8) as follows

$$\begin{aligned} x(t+1) &= f(x(t), y(t)), \\ y(t+1) &= x(t). \end{aligned}$$

We might express the above system in vector form

$$z(t+1) = g(z(t)),$$

where

$$z = (x, y), \quad g = (f(x, y), x).$$

More complicated cases are also possible; but we can make similar transformations. For instance, consider

$$x(t+1) = ax(t)(1 - x(t-2)).$$

The dynamical system contains a delay of two-time units. We can replace it with a three-dimensional system with no delay. Set

$$y(t+1) = x(t), \quad z(t+1) = y(t).$$

We obtain

$$\begin{aligned} x(t+1) &= ax(t)(1 - z(t)), \\ y(t+1) &= x(t), \\ z(t+1) &= y(t). \end{aligned}$$

The above example can be generalized. Assume

$$x(t+1) = f(x(t), x(t-1), x(t-2)).$$

Set $y(t+1) = x(t)$ and $z(t+1) = y(t)$. We obtain

$$
\begin{aligned}
x(t+1) &= f(x(t), y(t), z(t)), \\
y(t+1) &= x(t), \\
z(t+1) &= y(t).
\end{aligned}
$$

We can also write the system in vector form

$$w(t+1) = g(w(t)).$$

Hence, we can always assume that $x(t+1)$ depends directly only on $x(t)$. Systems with delay, i.e., when $x(t+1)$ depends directly on one or more states of the form $x(t-k)$, $k \geq 1$, are replaced by higher-dimensional systems with no delay.

Exercise 6.3

1 Solve the difference equations

$$
\begin{aligned}
x_1(t+1) &= 2x_1(t) + 3x_2(t) + 1, \\
x_2(t+1) &= x_1(t) + 4x_2(t), \quad x_1(0) = 0, \quad x_2(0) = -1.
\end{aligned}
$$

2 Let

$$x(t+1) = ax(t-1)(1 - x(t-2)).$$

Replace this one-dimensional system with a three-dimensional system with no delay.

3 Develop a mathematical model for a foreign trade among three countries using an argument similar to that used in the example of trade model between two countries in section 6.2.

4 Let

$$
\begin{aligned}
x(t+1) &= 2x(t) - 0.2x(t-1)y(t-1), \\
y(t+1) &= y(t) + 0.1x(t-1)y(t-1).
\end{aligned}
$$

Replace the two-dimensional system having a delay of a one-time unit with a four-dimensional system with no delay.

5 Apply theorem 2.6.4 to solve the following difference equations

(i) $x(t+2)+8x(t+1)+12x(t)=0;$

(ii) $x(t+2)-5x(t+1)+4x(t)=4^t.$

6.4 Stabilities

We now study the vector difference equation

$$x(t+1)=f(x(t),t),\quad x(t_0)=x_0, \tag{6.4.1}$$

where

$$x(t)\in R^n,\quad f: R^n\times Z^+\to R^n.$$

We assume f is continuous in x. Equation (6.4.1) is said to be *autonomous* or *time-invariant* if the variable t does not appear explicitly in the right-hand side of the equation

$$f(x(t),t)=f(x(t)).$$

It is said to be *periodic* if for all $n\in Z$,

$$f(x,t+N)=f(x),$$

for some positive integer. A point x^* in R^n is said to be an *equilibrium point* of equation (6.4.1) if

$$f(x^*,t)=x^*,$$

for all $t\geq t_0$. We can always assume x^* to be the origin. If x^* is not the origin, let

$$y(t) = x(t) - x^*.$$

Then equation (6.4.1) becomes

$$y(t+1) = f(y(t) + x^*, t) - x^* = g(y(t), t). \tag{6.4.2}$$

Notice that $y = 0$ corresponds to $x = x^*$. We now introduce some stability notations of x^*.[6]

Definition 6.4.1. (i) The equilibrium point x^* of equation (6.4.1) is said to be *stable* if given $\varepsilon > 0$ and $t_0 \geq 0$ there exists $\delta = \delta(\varepsilon, t_0) > 0$ such that

$$\|x_0 - x^*\| < \delta$$

implies

$$\|x(t, x_0) - x^*\| < \varepsilon$$

for all $t \geq t_0$, *uniformly stable* if δ may be chosen independent of t_0, *unstable* if it is not stable.[7]
(ii) x^* is said to be *attractive* if there exists $\mu = \mu(t_0)$ such that $\|x_0 - x^*\| < \mu$ implies

$$\lim_{t \to \infty} x(t, x_0) = x^*,$$

uniformly attractive if the choice of μ is independent of t_0.

---

[6] It should be noted that some notations are defined in section 2.3.

[7] The symbol $\|\ \|$ means the Euclidean norm in this book. For a vector x, $\|x\| = \left(\sum_i x_i^2\right)^{1/2}$, for a matrix A, $\|A\| = \left[\rho(A^T A)\right]^{1/2}$, where

$$\rho(A) \equiv Max\{|\lambda|, \lambda \text{ is an eigenvalue of } A\}.$$

(iii) x^* is said to be *asymptotically stable* if it is stable and attractive, and *uniformly asymptotically stable* if it is uniformly stable and uniformly attractive.
(iv) x^* is said to be *exponentially stable* if there exists $\delta > 0$, $M > 0$, and $\eta \in (0, 1)$ such that

$$\left\|x(t, t_0) - x^*\right\| < M\left\|x_0 - x^*\right\|\eta^{t-t_0},$$

whenever

$$\left\|x_0 - x^*\right\| < \delta.$$

(v) A solution $x(t, x_0)$ is *bounded* if for some positive constant M,

$$\left\|x(t, x_0)\right\| \le M$$

for all $t \ge t_0$, where M may depend on each solution.

If in (ii) and (iii) $\mu = \infty$ or in (iv) $\delta = \infty$, the corresponding stability property is said to be *global*. We illustrate the concept of stability in phase space as in figure 6.4.1.

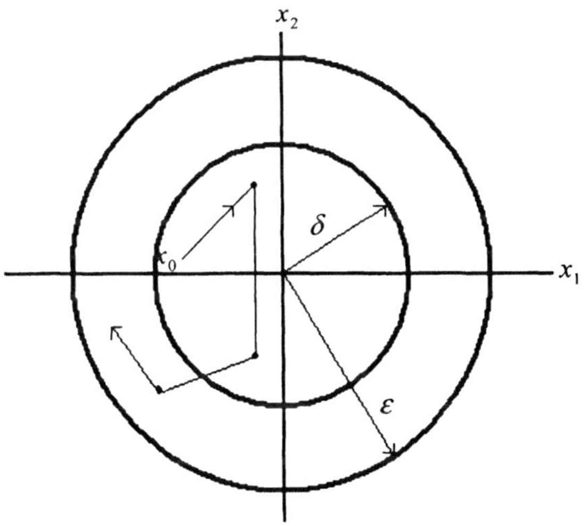

Figure 6.4.1: Stable equilibrium

Example Consider

$$f(x_1, x_2) = (x_1 \cos\theta - x_2 \sin\theta,\ x_1 \sin\theta + x_2 \cos\theta),\ \ \theta \in (0, 2\pi).$$

This is a counterclockwise rotation of the plane of an angle θ. Since $\theta \in (0, 2\pi)$, every point x_0 of the plane is moved from its initial position except the origin 0, which is the only fixed point of f. Setting

$$x_0 = (x_1(0), x_2(0)) \neq 0,$$

we obtain $x(1) = f(x(0))$ as

$$x_1(1) = x_1(0)\cos\theta - x_2(0)\sin\theta,$$
$$x_2(1) = x_1(0)\sin\theta + x_2(0)\cos\theta.$$

The new state $x(2) = f(x(1))$ can be obtained from $x(1)$ with a rotation of θ, or from x_0 with a rotation of 2θ. Hence

$$x_1(1) = x_1(0)\cos 2\theta - x_2(0)\sin 2\theta,$$
$$x_2(1) = x_1(0)\sin 2\theta + x_2(0)\cos 2\theta.$$

In general, $x(t)$ is obtained from x_0 with a rotation of $t\theta$

$$x_1(t) = x_1(0)\cos t\theta - x_2(0)\sin t\theta,$$
$$x_2(t) = x_1(0)\sin t\theta + x_2(0)\cos t\theta.$$

We thus have

$$\left\|x(t) - x^*\right\| = \sqrt{x_1^2(t) + x_2^2(t)} =$$
$$\sqrt{\{x_1(0)\cos t\theta - x_2(0)\sin t\theta\}^2 + \{x_1(0)\sin t\theta + x_2(0)\cos t\theta\}}$$
$$= \sqrt{x_1^2(0) + x_0^2(0)} = \left\|x_0 - x^*\right\|,$$

where $x^* = 0$ is the unique fixed point. Hence, all states $x(t)$, $t = 0, 1, \cdots$ of the orbit $O(x_0)$ are at the same distance from the fixed point x^*. The origin is thus stable.

In the above definitions, some of the stability properties automatically imply one of more of the others. Figure 6.4.2 shows the relations among these concepts. In general, none of the arrows can be reversed. For instance, the solution of $x(t+1) = x(t)$ is given by $x(t, x_0) = x_0$. This solution is uniformly stable but not asymptotically stable.

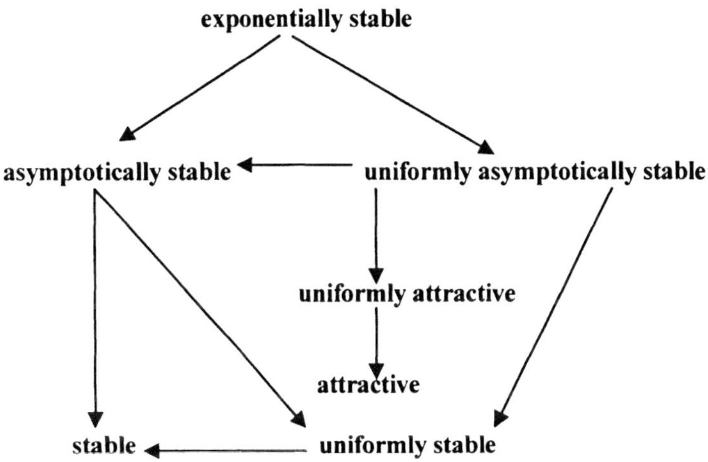

Figure 6.4.2: Relations of stabilities

Example Consider[8]

$$x(t+1) = \left(\frac{t+1}{2}\right)x^2(t). \tag{6.4.3}$$

The solution of this equation is given by

[8] This and the following examples are according to Elaydi (1999: 160-61).

$$x(t, x_0) = \left(\frac{t}{2}\right)\left(\frac{t-1}{2}\right)^2\left(\frac{t-4}{2}\right)^4 \cdots \left(\frac{t_0+1}{2}\right)^{2^{t-t_0-1}} x_0^{2^{t-t_0}}.$$

If $|x_0|$ is sufficiently small, then

$$\lim_{t\to\infty} x(t) = 0.$$

Thus the zero solution is attractive. However, it is not uniformly attractive. For if $\delta > 0$ is given and t_0 is so chosen that

$$(t_0 + 1)\delta^2 \geq 2,$$

then for $|x_0| = \delta$

$$|x(t_0 + 1, x_0)| = \left(\frac{t_0+1}{2}\right)|x_0|^2 \geq 1.$$

The zero solution is stable. For given $\varepsilon > 0$ and $t_0 \geq 0$, choose

$$\delta = \frac{\varepsilon}{1 + t_0}.$$

If $|x_0| < \delta$, then

$$|x(t_0 + 1, x_0)| < \varepsilon,$$

for all $t \geq t_0$. Since δ depends on the choice of t_0, the zero solution is stable but not uniformly stable.

Example The equilibrium point $(1, 0)$ of

$$\begin{aligned} r(t+1) &= \sqrt{r(t)}, \quad r > 0, \\ \theta(t+1) &= \sqrt{2\pi\theta(t)}, \quad 0 \leq \theta \leq 2\pi, \end{aligned} \tag{6.4.4}$$

is attractive but not stable. The solution of the equation is

$$r(t) = r_0^{2^{-t}},$$
$$\theta(t) = (2\pi)^{1-2^{-t}} \theta_0^{2^{-t}}.$$

We have

$$\lim_{t\to\infty} r(t) = 1, \quad \lim_{t\to\infty} \theta(t) = 2\pi,$$

with $\theta_0 \neq 0$. Now if $r_0 \neq 0$, $\theta_0 = 0$, then

$$r(t) = r_0^{2^{-t}}, \quad \theta(t) = 0,$$

which converges to $(1, 0)$. However, if

$$\theta_0 = a\pi, \quad 0 < a < 1,$$

then the orbit of (r_0, θ_0) will spiral around the circle counterclockwise to converge to $(1, 0)$. Hence, $(1, 0)$ is attractive but not stable. The behavior is illustrated as in figure 6.4.4.

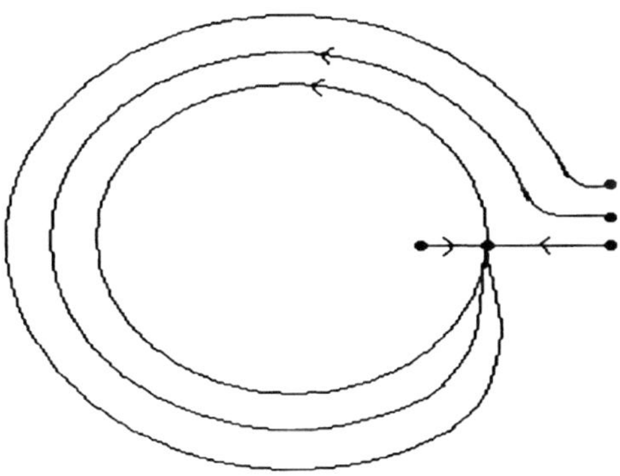

Figure 6.4.4: Attractive but not stable equilibrium

Under some additional conditions, the arrows in figure 6.4.2 may be reversed. For instance, for linear systems, uniformly asymptotically stable ↔ uniformly stable; for autonomous systems, stable ↔ uniformly stable, asymptotically stable ↔ uniformly asymptotically stable, and attractive ↔ uniformly attractive.

We provide some results about stability of the following linear autonomous system[9]

$$x(t+1) = A(t)x(t), \quad t \geq t_0 \geq 0. \tag{6.4.5}$$

We always assume $A(t)$ to be non-singular for any $t \geq t_0 \geq 0$. It should be remarked that for equation (6.4.5), every local stability property of the zero solution implies the corresponding global stability property.

Theorem 6.4.1. (i) If

$$\sum_{i=1}^{k} \left|a_{ij}(t)\right| \leq 1, \quad 1 \leq j \leq k, \quad t \geq t_0,$$

then the zero solution of system (6.4.5) is uniformly stable; and (ii) if

$$\sum_{i=1}^{k} \left|a_{ij}(t)\right| - \upsilon \leq 1, \quad 1 \leq j \leq k, \quad t \geq t_0,$$

for some $\upsilon > 0$, then the zero solution of system (6.4.5) is uniformly asymptotically stable.

Theorem 6.4.2. Consider

$$x(t+1) = Ax(t), \quad n \geq t_0 \geq 0, \tag{6.4.6}$$

where A is constant. The following statements hold:

(i) The zero solution of system (6.4.6) is stable if and only if $\rho(A) \leq 1$ and the eigenvalues of unit modulus are semisimple;[10] and

(ii) The zero solution of system (6.4.6) is asymptotically stable if and only if $\rho(A) < 1$.

---

[9] The stability properties in the remainder of this section is referred to Elaydi (1999: 168-172).

[10] An eigenvalue is said to be *semisimple* if the corresponding Jordan block is a diagonal.

Theorem 6.4.3. Consider

$$x(t+1) = A_{2\times 2}x(t), \quad n \ge t_0 \ge 0, \tag{6.4.7}$$

where A is a constant 2×2 real matrix. The zero solution of system (6.4.7) is asymptotically stable if and only if

$$|trA| < 1 + \det A < 2.$$

Exercise 6.4

1 Find fixed points of the system

$$x_1(t+1) = ax_1(t)(1 - x_1(t) - x_2(t)),$$
$$x_2(t+1) = bx_1(t)x_2(t).$$

2 Show that the system

$$x_1(t+1) = x_1(t)\cos 1 - x_2(t)\sin 1 + 1,$$
$$x_2(t+1) = x_1(t)\sin 1 + x_2(t)\cos 1 + 2.$$

has one and only one fixed point.

3 Find fixed points of the systems and discuss their stability

(a) $f(x_1, x_2) = (0.5x_1 + 1, 0.5x_2 + 2)$;
(b) $f(x_1, x_2) = (2x_1 + 1, 2x_2 + 2)$.

4 Discuss stability properties of the zero solution of $x(t+1) = Ax(t)$, where

(i) $A(t) = \begin{bmatrix} -1 & 5 \\ -0.5 & 2 \end{bmatrix}$;

(ii) $A(t) = \begin{bmatrix} 1.5 & 1 & -1 \\ -1.5 & -0.5 & 1.5 \\ 0.5 & 1 & 0 \end{bmatrix}$.

5 Using theorem 6.4.2 to determine whether or not the zero solution of $x(t+1) = A(t)x(t)$ is uniformly stable or uniformly asymptotically stable when:

(i) $A(t) = \begin{bmatrix} \frac{t}{1+t} & 0 \\ -1 & 1 \end{bmatrix}$;

(ii) $A(t) = \begin{bmatrix} \frac{1}{t+1} & 0 & \frac{1}{2}\sin(t) \\ \frac{1}{4} & \frac{1}{2}\sin(t) & \frac{1}{4}\cos(t) \\ \frac{1}{5} & 0 & 0 \end{bmatrix}$;

(iii) $A(t) = \begin{bmatrix} \frac{t+2}{t+1} & 0 & 0 \\ 0 & 1 & 0 \\ \frac{1}{t+1} & 0 & 1 \end{bmatrix}$.

6.5 Liapunov's direct method

Since the theory of linearization is a local theory, it does not address global issues. In this section, we discuss another approach, known as *Liapunov's second method* or *direct method.*[11] The method determines the stability or instability of a critical point by constructing a suitable auxiliary function. The theory of Liapunov functions is a global approach toward determining asymptotic behavior of solutions. Basically, the method is a generalization of two physical principles for conservative systems, namely, (i) a rest position is stable if the potential energy is a local minimum, otherwise it is unstable, and (ii) the total energy is a constant during any motion. The Liapunov function shows that initial values from a large region converge to an equilibrium point.

Consider the autonomous difference equation

$$x(t+1) = f(x(t)), \tag{6.5.1}$$

where

[11] The method is referred to as a direct method because no knowledge of the solution of the system of difference equations is required.

$$f: G \to R^n, \quad G \subset R^n,$$

is continuous. We assume that x^* is a fixed point of equation (6.5.1). That is

$$f(x^*) = x^*.$$

Let $V : R^n \to R$ be a real-valued function. The variation of V relative to equation (6.5.1) is defined as

$$\Delta V(x) = V(f(x)) - V(x),$$

and

$$\Delta V(x(t)) = V(f(x(t))) - V(x(t)) = V(x(t+1)) - V(x(t)).$$

If $\Delta V(x) \leq 0$, then V is nonincreasing along solutions of equation (6.5.1). The function V is said to be a *Liapunov function* on a subset H of R^n if
(i) V is continuous on H, and
(ii) $\Delta V(x) \leq 0$, whenever x and $f(x)$ belong to H.

Let $B(x, r)$ denote the open ball in R^n of radius r and center x defined by

$$B(x, r) \equiv \{y \in R^n \| \|y - x\| < r\}.$$

We denote $B(0, r)$ with $B(r)$. The real-valued function V is said to be *positive definite* at x^* if
(i) $V(x^*) = 0$, and
(ii) $V(x) > 0$ for all $x \in B(x^*, r)$, $x \neq x^*$ for some $r > 0$.

Theorem 6.5.1. (the Liapunov stability theorem)[12] If V is a Liapunov function for equation (6.5.1) in a neighborhood H of the equilibrium point, x^*, and V is positive definite with respect to x^*, then x^* is stable. If, in addition

[12] The proof is referred to section 4.5 in Elaydi (1999).

$$\Delta V(x) < 0$$

whenever $x,\ f(x) \in H$ and $x \neq x^*$, then x^* is asymptotically stable. Moreover, if $G = H = R^n$ and

$$V(x) \to \infty \text{ as } \|x\| \to \infty, \tag{6.5.2}$$

then x^* is globally asymptotically stable.

For illustration, let us consider a planar system with $x^* = 0$ as the equilibrium point. Suppose that equation (6.5.1) has a positive definite Liapunov function V defined on $B(r)$. Figure 6.5.1 illustrates the graph of V in a 3-dimensional coordinate system. If we now let $\varepsilon > 0$, $B(\varepsilon)$ then contains one of the level curves of V, say

$$V(x) = c.$$

The level curve $V(x) = c$ contains the ball $B(\delta)$ for some δ with $0 < \delta \le \varepsilon$. If a solution $x(t, x_0)$ starts at $x_0 \in B(\delta)$, then $V(x_0) \le c$. Since $\Delta V(x) \le 0$, V is a monotonic nonincreasing function along solutions of equation (6.5.1). Hence

$$V(x(t)) \le V(x_0) \le c,$$

for all $t \ge 0$. Thus, the solution $x(t, x_0)$ will stay forever in the ball $B(\varepsilon)$. Consequently, the zero solution is stable.

Theorem 6.5.2. If V is a Liapunov function on the set $\{x \in R^n \,|\, \|x\| > \alpha\}$ for some $\alpha > 0$, and if condition (6.5.2) holds, then all solutions of equation (6.5.1) are bounded.

Example Consider

$$x_1(t+1) = x_2(t),$$
$$x_2(t+1) = \frac{\alpha x_1(t)}{1 + \beta x_2^2(t)}, \quad \beta > 0.$$

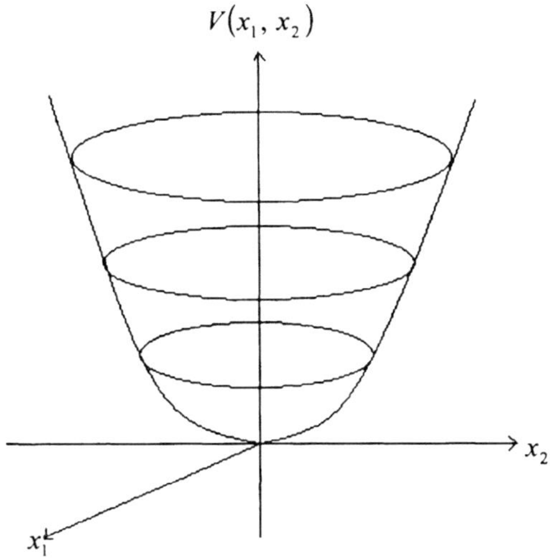

Figure 6.5.1: A Liapunov function

There are three equilibrium points,

$$(0, 0), \quad (\pm \beta^*, \pm \beta^*),$$

if $\alpha > 1$, where $\beta^* \equiv \sqrt{(\alpha - 1)/\beta}$. Consider the stability of the equilibrium point $(0, 0)$. Let

$$V(x(t)) = x_1^2(t) + x_2^2(t).$$

This is continuous and positive definite on R^2

$$\Delta V(x(t)) = \left(\frac{\alpha^2}{[1 + \beta x_2^2(t)]^2} - 1 \right) x_1^2(t) \leq (\alpha^2 - 1) x_1^2(t).$$

If $\alpha^2 \leq 1$, according to theorem 6.5.1, the unique equilibrium point $(0, 0)$ is stable. Since

$$V(x) \to \infty \text{ as } \|x\| \to \infty,$$

according to theorem 6.5.2, all solutions are bounded. Since

$$\Delta V(x(t)) = 0$$

for all points on the x_1 -axis, theorem 6.5.1 fails to determine asymptotical stability for this equation. In fact, it can be shown that in the case of $\alpha^2 = 1$, the zero solution is not asymptotically stable; in the case of $\alpha^2 < 1$, the origin is asymptotically stable; in the case of $\alpha^2 > 1$, the stability of origin is indeterminable.[13]

Example Consider

$$x_1(t+1) = 2x_2(t) - 2x_2(t)x_1^2(t),$$
$$x_2(t+1) = \frac{1}{2}x_1(t) + x_1(t)x_2^2(t).$$

There are three equilibrium points

$$(0, 0), \quad \left(\pm\sqrt{2}, \pm\sqrt{2}\right).$$

Let

$$V(x(t)) = x_1^2(t) + 4x_2^2(t).$$

This is continuous and positive definite on R^2

$$\Delta V(x_1(t), x_2(t)) = 4x_2^2(t)x_1^2(t)\left[x_1^2(t) + x_2^2(t) - 1\right].$$

If

$$x_1^2(t) + x_2^2(t) \leq 1,$$

then $\Delta V(x) \leq 0$. We conclude that the zero is stable.

[13] The results are based on LaSalle's invariance principle. See section 4.5 in Elaydi (1999).

For any real a, the solution with an initial value of $x_0 = (a, 0)^T$ is periodic with period 2 and with orbit

$$\{[a \quad 0]^T, [a/2]^T\},$$

and a solution with an initial value of $x_0 = (0, a)^T$ is also periodic with period 2. Hence, the zero solution cannot be asymptotically stable.

Theorem 6.5.3. If $\Delta V(x)$ is positive definite in a neighborhood of the origin and there exists a sequence $a_i \to 0$ with $V(a_i) > 0$, then the zero solution of equation (6.5.1) is unstable.[14]

Example Consider

$$x_1(t+1) = 4x_2(t) - 2x_2(t)x_1^2(t),$$
$$x_2(t+1) = \frac{1}{2}x_1(t) + x_1(t)x_2^2(t).$$

Define

$$V(x(t)) = x_1^2(t) + 16x_2^2(t).$$

Then

$$\Delta V(x_1(t), x_2(t)) = 3x_1^2(t) + 16x_1^2(t)x_2^2(t) + 4x_1^4(t)x_2^2(t) > 0, \quad \text{if } x_1(t) \neq 0.$$

From theorem 6.5.3, we see that the zero is unstable.

In section 6.2, we mentioned that the condition for asymptotical stability of the linear system is that $\rho(A) < 1$.

Exercise 6.5

1 For

[14] The conclusion of the theorem is also true if $\Delta V(x)$ is negative definite and $V(a_i) < 0$.

$$x_1(t+1) = x_1(t) + x_2^2(t) + x_1^2(t),$$
$$x_2(t+1) = x_2(t),$$

introduce the Liapunov function

$$V(x(t)) = x_1(t) + x_2(t).$$

Show that the origin is unstable.

2 For

$$x_1(t+1) = x_1(t) - x_1^3(t)x_2^2(t),$$
$$x_2(t+1) = x_2(t),$$

introduce the Liapunov function

$$V(x(t)) = x_1^2(t) + x_2^2(t).$$

Show that the origin is stable.

3 Consider the planar system

$$x_1(t+1) = \frac{x_2(t)}{1 + x_1^2(t)},$$
$$x_2(t+1) = \frac{x_1(t)}{1 + x_2^2(t)}.$$

Find the equilibrium points and determine their stability.

6.6 Linearization of difference equations

We consider the nonlinear systems of difference equations

$$x(t+1) = A(t)x(t) + g(x(t), t), \tag{6.6.1}$$

where $A(t)$ is a $n\times n$ nonsingular matrix for all $t\in Z^+$ and

$$g: Z^+\times G\to R^n,\quad G\subset R^n,$$

is a continuous function. Its corresponding linear system is

$$x(t+1)=A(t)x(t). \tag{6.6.2}$$

We may consider system (6.6.1) as a perturbation of system (6.6.2). System (6.6.1) may arise from the linearization of nonlinear equations of the form

$$x(t+1)=f(x(t),t), \tag{6.6.3}$$

where

$$f: Z^+\times G\to R^n,\quad G\subset R^n,$$

is continuously differentiable at an equilibrium point x^* $(=0)$ in this section. We require $f(0,t)=0$ for all $t\in Z^+$. The Jacobina matrix of f is defined as

$$\frac{\partial f(x,t)}{\partial x}\Big|_{x=0}=f'(0,t)=\begin{bmatrix}\frac{\partial f_1(x,t)}{\partial x_1} & \frac{\partial f_1(x,t)}{\partial x_2} & \cdots & \frac{\partial f_1(x,t)}{\partial x_n}\\ \frac{\partial f_2(x,t)}{\partial x_1} & \frac{\partial f_2(x,t)}{\partial x_2} & \cdots & \frac{\partial f_2(x,t)}{\partial x_n}\\ \vdots & \vdots & & \vdots\\ \frac{\partial f_n(x,t)}{\partial x_1} & \frac{\partial f_n(x,t)}{\partial x_2} & \cdots & \frac{\partial f_n(x,t)}{\partial x_n}\end{bmatrix}.$$

Let $\partial f(0,t)/\partial x=A(t)$ and

$$g(x(t),t)=f(x(t),t)-A(t)x(t).$$

Then equation (6.6.3) may be written in the form of system (6.6.1). In the case that equation (6.6.3) is autonomous, $x(t+1)=f(x(t))$, we have

$$x(t+1) = Ax(t) + g(x(t)), \tag{6.6.4}$$

where $A = f'(0)$ and

$$g = f - Ax.$$

Since f is differentiable at 0, it follows that

$$g(x) = o(x) \text{ as } \|x\| \to 0,$$

i.e.

$$\frac{\|g(x)\|}{\|x\|} \to 0 \text{ as } \|x\| \to 0.$$

Theorem 6.6.1.[15] Assume that

$$g(x(t), t) = o(\|x\|)$$

uniformly as $\|x\| \to 0$. If the zero solution of the homogeneous system (6.6.2) is uniformly asymptotically stable, the zero solution of the nonlinear system (6.6.1) is exponentially stable.

Corollary 6.6.1. If $\rho(A) < 1$, then the zero solution of equation (6.6.4) is exponentially stable.

Example Examine the stability of the zero solution of the planar system

$$x_1(t+1) = \frac{ax_2(t)}{1 + x_1^2(t)} \equiv f_1(x_1(t), x_2(t)),$$
$$x_2(t+1) = \frac{bx_1(t)}{1 + x_2^2(t)} \equiv f_2(x_1(t), x_2(t)).$$

As

---

[15] The proof of the theorem is referred to section 4.6 in Elaydi (1999).

$$\left.\frac{\partial f}{\partial x}\right|_{(0,0)} = \begin{bmatrix} \frac{\partial f_1}{\partial x_1} & \frac{\partial f_1}{\partial x_2} \\ \frac{\partial f_2}{\partial x_1} & \frac{\partial f_2}{\partial x_2} \end{bmatrix} = \begin{bmatrix} 0 & a \\ b & 0 \end{bmatrix},$$

the system may be rewritten as

$$\begin{bmatrix} x_1(t+1) \\ x_2(t+2) \end{bmatrix} = \begin{bmatrix} 0 & a \\ b & 0 \end{bmatrix} \begin{bmatrix} x_1(t) \\ x_2(t) \end{bmatrix} + \begin{bmatrix} \frac{-ax_2(t)x_1^2(t)}{1+x_1^2(t)} \\ \frac{-bx_1(t)x_2^2(t)}{1+x_2^2(t)} \end{bmatrix}, \tag{6.6.5}$$

or

$$x(t+1) = Ax(t) + g(x(t)).$$

The eigenvalues of A are $\pm\sqrt{ab}$. If $\sqrt{ab} < 1$, the zero solution of the linear part is asymptotically stable. Since $g(x(t))$ is continuously differentiable at the origin, the zero solution of equation (6.6.5) is exponentially stable.

Example Introducing a delay of time 1 into the *Pielou logistic equation*

$$x(t+1) = \frac{\alpha x(t)}{1+\beta x(t)},$$

yields the following difference delay equation

$$x(t+1) = \frac{\alpha x(t)}{1+\beta x(t-1)}, \quad \alpha > 0, \quad \beta > 0.$$

Let

$$y(t) = x(t) - \frac{\alpha - 1}{\beta}.$$

Then the above equation becomes

$$y(t+1) = \frac{\alpha y(t) - (\alpha - 1)y(t-1)}{\alpha + (\alpha - 1)y(t-1)}.$$

To change the above equation into a planar system, introduce

$$X_1(t) = y(t-1), \quad X_2(t) = y(t).$$

Then

$$\begin{bmatrix} X_1(t+1) \\ X_2(t+2) \end{bmatrix} = \begin{bmatrix} X_2(t) \\ \dfrac{\alpha X_2(t) - (\alpha - 1)X_1(t)}{\alpha + (\alpha - 1)X_1(t)} \end{bmatrix}. \quad (6.6.6)$$

By linearizing equation (6.6.6) around $(0, 0)$, we have

$$X(t+1) = AX(t) + g(X(t)),$$

where

$$A = \begin{bmatrix} 0 & 1 \\ \dfrac{1-\alpha}{\alpha} & 1 \end{bmatrix},$$

$$g(X) = \begin{bmatrix} 0 \\ \dfrac{\beta(\alpha - 1)X_1^2(t) - \alpha\beta X_1(t)X_2(t)}{\alpha(\alpha + \beta X_1(t))} \end{bmatrix}.$$

The eigenvalues of A are inside the unit disk if and only if

$$0 < \frac{\alpha - 1}{\alpha} < 1,$$

which is always held since $\alpha > 1$. Consequently, the zero solution of equation (6.6.6) is asymptotically stable.

Theorem 6.6.2. The following statements hold
(i) If $\rho(A) = 1$, then the zero solution of equation (6.6.4) may be stable or unstable;

(i) If $\rho(A) > 1$ and $g(x)$ is $o(x)$ as $\|x\| \to \infty$, then the zero solution of equation (6.6.4) is unstable.

Exercise 6.6

1 Determine the stability of the zero solution of the equation

$$x(t+2) - \frac{1}{2}x(t+1) + x(t+1)x(t) + \frac{13}{16}x(t) = 0.$$

2 Find the equilibrium points and determine their stability of the system

$$\begin{aligned} x_1(t+1) &= \frac{1}{2}x_1(t) - x_2^2(t) + x_3(t), \\ x_2(t+1) &= x_1(t) - x_2(t) + x_3(t), \\ x_3(t+1) &= x_1(t) - x_2(t) + \frac{1}{2}x_3(t). \end{aligned}$$

3 Determine conditions for the asymptotical stability of the zero solution of the system

$$\begin{aligned} x_1(t+1) &= \frac{ax_1(t)}{1 + x_2(t)}, \\ x_2(t+1) &= [bx_2(t) - x_1(t)][1 + x_1(t)]. \end{aligned}$$

6.7 Conjugacy and center manifolds

There are two possible ways to simplify dynamical systems: one is to transform one complex system to another one which is much easier to analyze; and the other is to reduce higher dimensional problems to lower ones. The center manifold theorem helps us to reduce dimensions of dynamical problems. A change of variable may reduce a system to a simple one. This idea is made more precise with the concept of conjugate maps. The concept of conjugacy arises in many subjects of mathematics. In linear algebra, the natural concept is linear conjugacy. Thus, if $x_1 = Ax$ is a linear map and $x = Cy$ is a linear change of coordinates for which C has an inverse, then the map on the y-variables is given as follows

$$y_1 = C^{-1}x_1 = C^{-1}Ax = C^{-1}ACy.$$

Thus, the matrix for the map in the y-variables is $C^{-1}AC.$ As long as the maps are defined on the same space, a conjugacy can be considered a change of coordinates of the variables on the space on which the function acts.

Definition 6.7.1. Let I and J be two intervals and $\phi : I \to J$ be continuous, one-to-one, and onto. We say that ϕ is a *conjugacy* between $G : I \to J$ and $F : J \to I$ if

$$\phi(G(x)) = F(\phi(x)),$$

for all $x \in I.$ The maps F and G are said to be *conjugate* by $\phi.$

Let F and G be *conjugate* by $\phi.$ Then x^* is a fixed point of G if and only if

$$y^* = \phi(x^*)$$

is a fixed point of $F.$ Similarly, x_p is a fixed point of G if and only if

$$y_p = \phi(x_p)$$

is a fixed point of $F.$

Example Let

$$G(x) = 4x(1 - x), \quad x \in [0, 1],$$
$$F(y) = 2y^2 - 1, \quad y \in [-1, 1],$$
$$y = \phi(x) = 1 - 2x.$$

We see that ϕ is continuous, one-to-one, and onto from $[0, 1]$ to $[-1, 1].$ We have

$$\phi(G(x)) = 8x^2 - 8x + 1.$$

We also have

$$F(\phi(x)) = 8x^2 - 8x + 1.$$

Hence the two maps F and G are conjugate in the given intervals by the map ϕ.

Since ϕ is invertible, we also have

$$G = \phi^{-1} \circ F \circ \phi,$$

where the symbol $\circ$ stands for the composition of functions.

We now introduce a way to transforming a higher dimensional problem to a lower one. Consider the m-parameter map $f(x, \lambda)$

$$f : R^n \times R^m \to R^n, \text{ with } x \in R^n \text{ and } \lambda \in R^m$$

where f is C^r ($r \geq 3$) on some sufficiently large open set in $R^n \times R^m$. Let (x_0, λ_0) be a fixed point of f, i.e.

$$f(x_0, \lambda_0) = x_0.$$

We know that the stability of hyperbolic fixed points of f is determined from the stability of the fixed points under the linear map

$$J = D_x f(x_0, \lambda_0).$$

However, the situation is different if one of the eigenvalues ρ of J lies on the unit cycle, that is, $|\rho| = 1$. There are three cases in which the fixed point (x_0, λ_0) is nonhyperbolic: (i) J has one real eigenvalue equal to 1 and the other eigenvalues are off the unit circle; (ii) J has one real eigenvalue equal to -1 and the other eigenvalues are off the unit circle; and (iii) J has two complex conjugate eigenvalues modulus 1 and the other eigenvalues are off the unit circle. To analyze behavior of the system in these cases, we need to introduce a rudiment of center manifold theory.

By a change of variables, we may assume without loss of generality that $x_0 = 0$. Let us temporarily suppress the parameter μ. Then the map $f(x, \lambda)$ can be written as

$$\begin{aligned} y &\mapsto A_{p\times p}y + f(y, z), \\ z &\mapsto B_{q\times q}z + g(y, z), \end{aligned} \tag{6.7.1}$$

where

$$p + q = n, \quad f(0, 0) = g(0, 0) = 0, \quad Df(0, 0) = Dg(0, 0) = 0,$$

and J has the form

$$J_{n\times n} = \begin{bmatrix} A_{p\times p} & 0 \\ 0 & B_{q\times q} \end{bmatrix},$$

where all eigenvalues of A lie on the unit circle and all of the eigenvalues of B are off the unit circle. Observe that system (6.7.1) corresponds to the system of difference equations

$$\begin{aligned} y(t+1) &= Ay(t) + f(y(t), z(t)), \\ z(t+1) &= Bz(t) + g(y(t), z(t)). \end{aligned} \tag{6.7.2}$$

The following theorems are referred to Carr.[16]

Theorem 6.7.1. There is a C^r center manifold for system (6.7.1) that can be represented locally as

$$M_c = \{(y, z) \in R^p \times R^q : \ z = h(y), \ |y| < \delta, \ h(0) = 0, \ Dh(0) = 0\},$$

for a sufficiently small δ. Furthermore, the dynamics restricted to M_c are given locally by the map

$$y \mapsto Ay + f(y, h(y)), \quad y \in R^p. \tag{6.7.3}$$

This theorem asserts the existence of a center manifold, i.e., a curve $z = h(y)$ on which the dynamics of system (6.7.1) is given by system (6.7.3). The next

[16] See Carr (1981).

theorem shows that the dynamics on the center manifold determines completely the dynamics of system (6.7.1).

Theorem 6.7.2. If the fixed point $(0, 0)$ of system (6.7.3) is stable, asymptotically stable, or unstable, then the fixed point $(0, 0)$ of system (6.7.1) is correspondingly stable, asymptotically stable, or unstable.

Theorem 6.7.3. For any solution $(y(t), z(t))$ of system (6.7.1) with an initial point (y_0, z_0) in a small neighborhood around the origin, there exists a solution $w(t)$ of system (6.7.3) and positive constants $L, \beta > 1$ such that

$$|y(t) - w(t)| \le L\beta^t,$$
$$|z(t) - h(w(t))| \le L\beta^t, \quad \text{for all } t \in Z^+.$$

The question now is how to calculate the center manifold, $z = h(y)$. Substituting $z = h(y)$ into system (6.7.1) yields

$$\begin{aligned} y(t+1) &= Ay(t) + f(y(t), h(y(t))), \\ z(t+1) &= Bh(y(t)) + g(y(t), h(y(t))). \end{aligned} \tag{6.7.4}$$

We also have

$$z(t+1) = h(y(t+1)) = h(Ay(t) + f(y(t), h(y(t)))),$$

where we use the first equation in system (6.7.4). Equating this equation and the second in system (6.7.4) we have

$$\Phi(h(y)) = h[Ay + f(y, h(y))] - Bh(y) - g(y, h(y)) = 0. \tag{6.7.5}$$

We may approximate this solution via power series. It can be shown that although the choice of h is not unique, any two center manifolds of a given fixed point may differ only in transcendentally small terms. The Taylor series expansion of any two center manifolds must agree in all orders. The following theorem justifies the approach.[17]

[17] See Carr 1981).

Theorem 6.7.3. Let $\Psi : R^p \to R^q$ be a C^1 map with

$$\Psi(0) = \Psi'(0) = 0.$$

Suppose that

$$\Phi(\Psi(y)) = O(|y|^r) \text{ as } x \to 0$$

for some $r > 1$. Then

$$h(y) = \Psi(y) + q(y) \text{ as } y \to 0,$$

where

$$q(y) = O(|y|^r) \text{ as } y \to 0,$$

if there is a positive number K such that

$$|q(y)| \le K|x|^r,$$

in a small neighborhood of zero.

Example Consider

$$\begin{bmatrix} x_1(t+1) \\ x_2(t+1) \end{bmatrix} = \begin{bmatrix} -1 & 0 \\ 0 & -\frac{1}{2} \end{bmatrix} \begin{bmatrix} x_1(t) \\ x_2(t) \end{bmatrix} + \begin{bmatrix} x_1(t)x_2(t) \\ x_1^2(t) \end{bmatrix}. \tag{6.7.6}$$

The center manifold is

$$M_c = \{(x_1, x_2) \in R \times R : x_2 = h(x_1), |x_1| < \delta, h(0) = h'(0) = 0\}.$$

The function h satisfies

$$h[Ax_1 + f(x_1, h(x_1))] - Bh(x_1) - g(x_1, h(x_1)) = 0$$

where

$$f = x_1(t)x_2(t), \quad g = x_1^2(t).$$

With equation (6.7.5), the above equation becomes

$$h[-x_1 + x_1 h(x_1)] + \frac{1}{2}h(x_1) - x_1^2 = 0. \tag{6.7.7}$$

Let us assume that $h(x_1)$ takes the form

$$h(x_1) = c_1 x_1^2 + c_2 x_1^3 + O(x_1^4).$$

Substituting this equation into equation (6.7.7) yields

$$c_1 x_1^2 - c_2 x_1^3 + \frac{1}{2}c_1 x_1^2 + \frac{1}{2}c_2 x_1^3 - x_1^2 + O(x_1^4) = 0.$$

Hence

$$c_1 = \frac{2}{3}, \quad c_2 = 0.$$

We thus obtain

$$h(x_1) = \frac{2x_1^2}{3} + O(x_1^4).$$

The map on the center manifold is given by

$$x_1(t+1) = -x_1(t) + \frac{2}{3}x_1^3(t) + O(x_1^5). \tag{6.7.8}$$

The Schwarzian derivative at zero is given by

$$S(f)(0) = \frac{f'''(0)}{f'(0)} - \frac{2}{3}\left[\frac{f''(0)}{f'(0)}\right]^2 = -4 < 0.$$

Hence the origin is asymptotically stable under equation (6.7.8). This implies that the origin is asymptotically stable under equation (6.7.5). We illustrate the local dynamics as in figure 6.7.1.

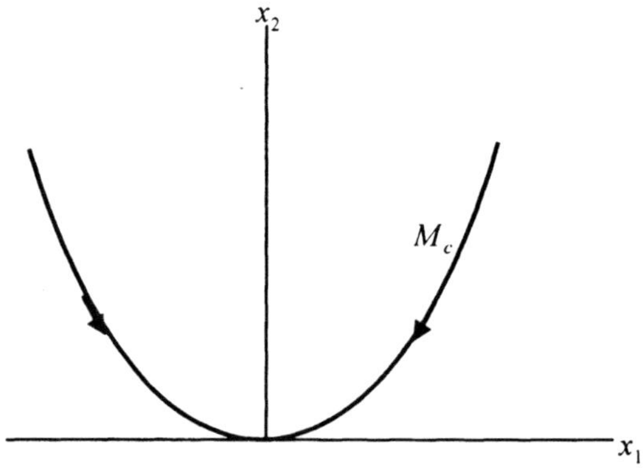

Figure 6.7.1: An asymptotically stable center manifold

We now illustrate the case that the system depends on a vector of parameters $\lambda \in R^m$. Then system (6.7.2) takes the form

$$\begin{aligned} y(t+1) &= Ay(t) + f(y(t), z(t), \lambda), \\ z(t+1) &= Bz(t) + g(y(t), z(t), \lambda), \end{aligned} \tag{6.7.9}$$

where f and g are C^r functions ($r \geq 3$), in some neighborhood of

$$(0, 0, 0), \quad f(0, 0, 0) = g(0, 0, 0) = 0, \quad Df(0, 0, 0) = Dg(0, 0, 0) = 0.$$

To find the center manifold of system (6.7.9), we consider λ as a function of time and rewrite system (6.7.9) as

$$\begin{aligned} y(t+1) &= Ay(t) + f(y(t), z(t), \lambda(t)), \\ \lambda(t+1) &= \lambda(t), \\ z(t+1) &= Bz(t) + g(y(t), z(t), \lambda(t)). \end{aligned} \tag{6.7.10}$$

The center manifold now takes the form of

$$M_c = \{(y,\ \lambda,\ z): z = h(y,\ \lambda),\ |y| < \delta_1,\ |\lambda| < \delta_2,\ h(0,\ 0) = 0,\ Dh(0,\ 0) = 0\}. \tag{6.7.11}$$

Substituting for $z = h(y,\ \lambda)$ into system (6.7.10) yields

$$\begin{aligned} y(t+1) &= Ay(t) + f(y(t),\ h(y(t),\ \lambda(t)),\ \lambda(t)), \\ z(t+1) &= h[Ay(t) + f(y(t),\ h(y(t),\ \lambda(t)),\ \lambda(t)),\ \lambda(t)] \\ &= Bh(y(t),\ \lambda(t)) + g(y(t),\ h(y(t),\ \lambda(t)),\ \lambda(t)). \end{aligned} \tag{6.7.12}$$

The latter equations lead to the function

$$\begin{aligned} \Phi(h(y,\ \lambda)) = h[Ay + f(y,\ h(y,\ \lambda),\ \lambda),\ \lambda] - Bh(y,\ \lambda) \\ + g(y,\ h(y,\ \lambda),\ \lambda) = 0. \end{aligned}$$

For instance, if y and λ are one-dimensional, we may take

$$h(y,\ \lambda) = c_1 y^2 + c_2 \lambda y + c_3 \lambda^2 +$$

to approach h.

Exercise 6.7

1 Calculate center manifolds near the origin and describe the bifurcations of the origin:

(i)
$$\begin{aligned} x_1(t+1) &= -\frac{1}{2}x_1(t) - x_2(t) - x_1(t)x_2^2(t), \\ x_2(t+1) &= -\frac{1}{2}x_1(t) + \lambda x_2(t) + x_1^2(t); \end{aligned}$$

(ii)
$$\begin{aligned} x_1(t+1) &= x_1^2(t) + \lambda x_1(t), \\ x_2(t+1) &= x_2(t) + x_1(t)x_2(t). \end{aligned}$$

6.8 The Hénon map and bifurcations

Consider a two-dimensional map

$$f : R^2 \to R^2, \quad f(x_1, x_2) \in C^r, \quad r \geq 5.$$

Let $x^* = (x_1^*, x_2^*)$ be a fixed point and the Jacobian matrix

$$J = D_x f(x^*).$$

Using the center manifold theorem we deduce the following:

(i) If J has an eigenvalue equal to 1, then we have a saddle node bifurcation;
(ii) If J has an eigenvalue equal to -1, then we have period-doubling bifurcation; and
(iii) If J has two complex conjugate eigenvalues of modulus 1, then we have a new type of bifurcation, called Neimark-Sacker (Hopf) bifurcation.

We use the Hénon map to illustrate the saddle node and the period bifurcation. In 1976, the French Astronomer Michael Hénon suggested a simplified model for the dynamics of the Lorenz system.[18] The Hénon map is given by

$$\begin{aligned} x(t+1) &= 1 - ax^2(t) + y(t), \\ y(t+1) &= bx(t). \end{aligned} \tag{6.8.1}$$

We explain a few properties of the Hénon map. We also denote the Hénon map by

$$X(t+1) = HX(t),$$
$$X(t) \equiv \begin{bmatrix} x(t) \\ y(t) \end{bmatrix},$$
$$H \equiv \begin{bmatrix} 1 - ax^2 + y \\ bx \end{bmatrix}.$$

It is known that the Hénon map contracts areas for $|b| < 1$. If $b = 0$, we get the quadratic map

$$x(t+1) = 1 - ax^2(t).$$

[18] Hénon (1976).

To see this, we find the determinant of the Jacobian matrix of H. If $|\det DH| < 1$ for all (x, y), the map is area contracting. From vector calculus we know that H maps an infinitesimal rectangle at (x, y) with area $dxdy$ into an infinitesimal parallelogram with area

$$|\det DH(x, y)dx\,dy|.$$

Thus, if

$$|\det DH(x, y)| < 1,$$

then H is area contracting. As

$$DH(x, y) = \begin{bmatrix} -2ax & 1 \\ b & 0 \end{bmatrix},$$

we have

$$\det DH(x, y) = -b.$$

Hence, if $|b| < 1$, the Hénon map is area contracting.

We show that the Hénon map is invertible. To see this, one may decompose H into three simple maps T_j as follows

$$T_1\begin{bmatrix} x \\ y \end{bmatrix} = \begin{bmatrix} x \\ 1 - ax^2 + y \end{bmatrix},$$
$$T_2\begin{bmatrix} x \\ y \end{bmatrix} = \begin{bmatrix} bx \\ y \end{bmatrix},$$
$$T_3\begin{bmatrix} x \\ y \end{bmatrix} = \begin{bmatrix} y \\ x \end{bmatrix}.$$

As shown in figure 6.8.1, T_1 is an area-preserving bending map, T_2 contracts in the x direction, and T_3 rotates by 90°. The composite transformation $T_3 \circ T_2 \circ T_1$ yields the Hénon map.

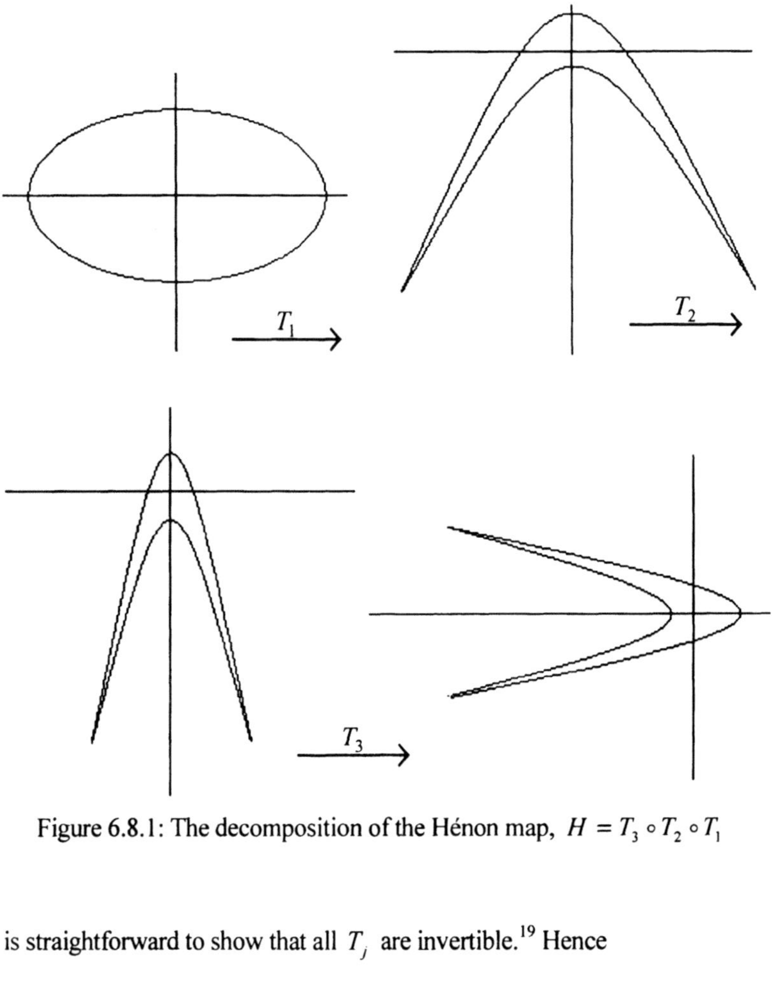

Figure 6.8.1: The decomposition of the Hénon map, $H = T_3 \circ T_2 \circ T_1$

It is straightforward to show that all T_j are invertible.[19] Hence

$$H^{-1} = T_1^{-1} \circ T_2^{-1} \circ T_3^{-1},$$

i.e.

---

19 $T_1^{-1}\begin{bmatrix} x \\ y \end{bmatrix} = \begin{bmatrix} x \\ y - 1 + ax^2 \end{bmatrix}$, $T_2^{-1}\begin{bmatrix} x \\ y \end{bmatrix} = \begin{bmatrix} x/b \\ y \end{bmatrix}$, and $T_3^{-1}\begin{bmatrix} x \\ y \end{bmatrix} = \begin{bmatrix} y \\ x \end{bmatrix}$.

$$H^{-}\begin{bmatrix} x \\ y \end{bmatrix} = \begin{bmatrix} \dfrac{y}{b} \\ -1 + \dfrac{a}{b^2} y^2 + x \end{bmatrix}.$$

As all T_j are one-to-one, H is also one-to-one.

If $a \neq 0$, H has a fixed point X^* if

$$a \geq -\left(\frac{1-b}{2}\right)^2.$$

A fixed point is determined by

$$1 - ax^2 + y = x,$$
$$bx = y.$$

We see that x is given by

$$ax^2 + (1-b)x - 1 = 0.$$

If

$$a \geq -\left(\frac{1-b}{2}\right)^2,$$

we see that H has two fixed points

$$X_1^* = \begin{bmatrix} \dfrac{b-1+b_0}{2a} \\ \dfrac{b-1+b_0}{2a} b \end{bmatrix}, \quad X_2^* = \begin{bmatrix} \dfrac{b-1-b_0}{2a} \\ \dfrac{b-1-b_0}{2a} b \end{bmatrix},$$

where

$$b_0 \equiv \sqrt{(1-b)^2 + 4a}.$$

It can be shown that if $a \neq 0$ and

$$a \in \left(-\left(\frac{1-b}{2}\right)^2, 3\left(\frac{1-b}{2}\right)^2 \right),$$

then X_1^* is asymptotically stable and X_2^* is a saddle. At

$$a = -\left(\frac{1-b}{2}\right)^2,$$

we have

$$X_1^* = X_2^* = \begin{bmatrix} \dfrac{b-1}{2a} \\ \dfrac{b-1}{2a}b \end{bmatrix}.$$

Moreover, the Jacobian matrix at this equilibrium point has an eigenvalue equal to 1. By the center manifold theorem, there is a saddle node bifurcation at X_1^*.

For a fixed value of the parameter $b \in (0, 1)$, X_1^* loses its stability and becomes a saddle point at

$$a = 3\left(\frac{1-b}{4}\right)^2$$

and a new stable 2-cycle appears. The reason is that one of the eigenvalues of $DH(X_1^*)$ will decrease and pass -1. In the case of $b = 0.3$, a period-doubling cascade starts at $a = 0.3675$ (see figure 6.8.2) and ends at $a \approx 1.06$.

Beyond $a \approx 1.06$, a strange attractor appears. Figure 6.8.3 shows a strange attractor of the Hénon map when $a = 1.4$ and $b = 0.3$. Zooming into the strange attractor, we can see that there are six parallel curves. If we zoom further on the top three curves, we can see that they are really six curves grouped the same way as the first batch. This "self-similarity" continues to arbitrarily small scales.[20]

[20] Benedick and Carleson (1991) demonstrate that this strange attractor is the closure of a branch of the unstable manifold.

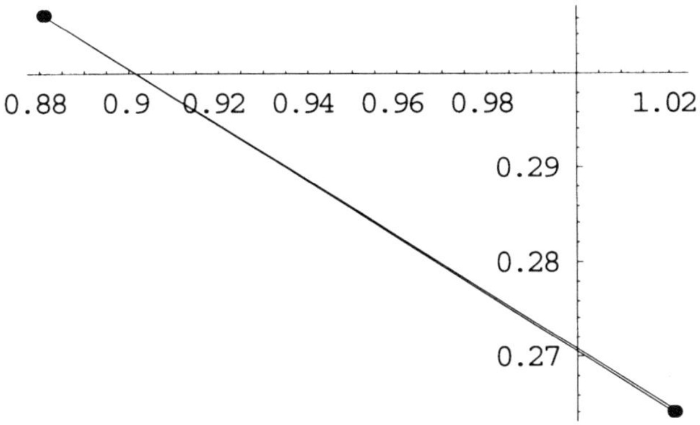

Figure 6.8.2: A 2 -cycle for $a = 0.3675$ and $b = 0.3$

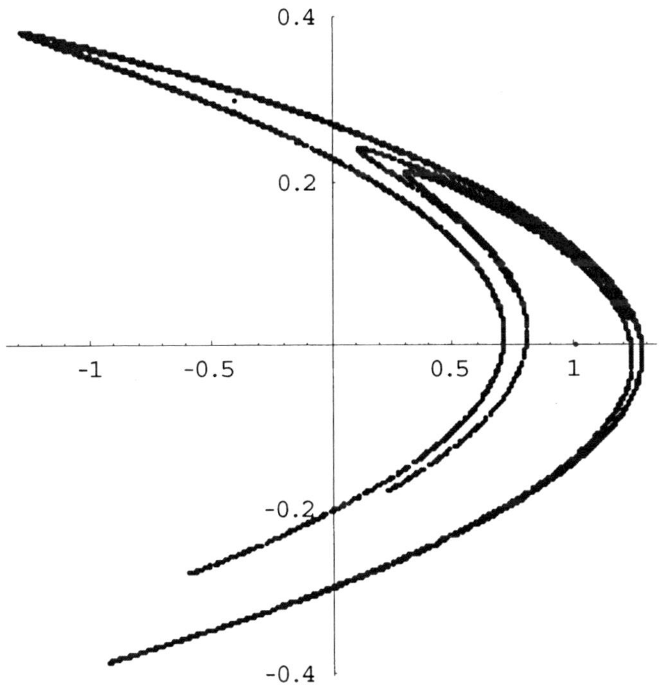

Figure 6.8.3: The Hénon attractor for $a = 1.4$ and $b = 0.3$

6.9 The Neimark-Sacker (Hopf) bifurcations

For illustration of the Hopf bifurcation, let us consider an example.

Consider the family of maps

$$\begin{bmatrix} x_1(t+1) \\ x_2(t+1) \end{bmatrix} = \left(1+\lambda - x_1(t) - x_2^2(t)\right)\begin{bmatrix} \cos\beta & -\sin\beta \\ \sin\beta & \cos\beta \end{bmatrix}\begin{bmatrix} x_1(t) \\ x_2(t) \end{bmatrix} = f_\lambda \begin{bmatrix} x_1(t) \\ x_2(t) \end{bmatrix}, \tag{6.9.1}$$

where $\beta = \beta(\lambda)$ is a smooth function of the parameter λ and $0 < \beta(0) < \pi$. The origin is a fixed point of the map F_λ for all λ with the Jacobian matrix

$$J = (1+\lambda)\begin{bmatrix} \cos\beta & -\sin\beta \\ \sin\beta & \cos\beta \end{bmatrix}.$$

The matrix has eigenvalues

$$\rho_{1,2} = (1+\lambda)e^{\pm i\beta}$$

with

$$|\rho_{1,2}| = |1+\lambda|.$$

Hence, at $\lambda = 0$, the eigenvalues cross the unit circle. Clearly, the origin is asymptotically stable for $\lambda > 0$. To analyze the bifurcation when $\lambda = 0$, we may write the map f_λ in polar coordinates (r, θ) by introducing

$$x_1(t) = r(t)\cos\theta(t), \quad x_2(t) = r(t)\sin\theta(t).$$

Under this transformation, equations (6.9.1) become

$$\begin{aligned} r(t+1) &= (1+\lambda)r(t) - r^3(t), \\ \theta(t+1) &= \theta(t) + \beta. \end{aligned} \tag{6.9.2}$$

The form of equations (6.9.2) enables us to detect the presence of an invariant circle by solving

$$r = (1 + \lambda)r - r^3.$$

The invariant circle is of radius

$$r^* = \sqrt{\lambda}.$$

The circle appears when λ crosses the value 0 as shown in figure 7.9.1. For $\lambda < 0$ the origin is asymptotically stable. The instance that λ becomes positive, the origin loses its stability to give rise to an attracting (asymptotically stable) circle with radius $r = \sqrt{\lambda}$. The dynamics on this circle are determined by the map $\theta \mapsto \theta + \beta$, which is a rotation by an angle β in the counterclockwise direction. This phenomenon is called a *Neimark-Sacker bifurcation* (or a *Hopf bifurcation*).

The above analysis is actually valid for a certain class of two-dimensional maps with one parameter. Consider a two-dimensional map

$$f : R^2 \times R \to R^2, \quad f(x_1, x_2, \lambda) \in C^r, \quad r \geq 5.$$

Suppose that

(i) the origin $x^* = (0, 0)$ be a fixed point; and

(ii) the Jacobian matrix

$$J = D_x f(x^*, \lambda^*)$$

have two complex conjugate eigenvalues

$$\rho(\lambda) = r(\lambda)e^{i\theta(\lambda)}$$

and $\bar{\rho}(\lambda)$, where

$$r(0) = 1, \quad r'(0) \neq 0, \quad \theta(0) = \theta_0,$$

(which imply $|\rho(0)| = 1$); and

(iii) $e^{ik\theta_0} \neq 0$ for $k = 1, 2, 3, 4, 5,$ i.e., $\rho(0)$ is not a low root of unity.

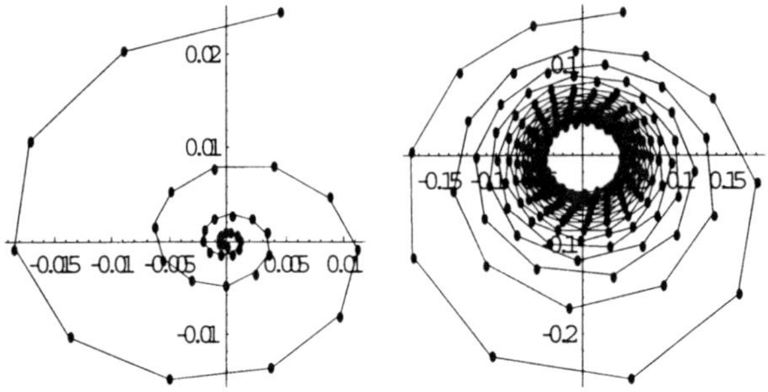

a) $\lambda < 0$: approaching the origin b) $\lambda = 0$

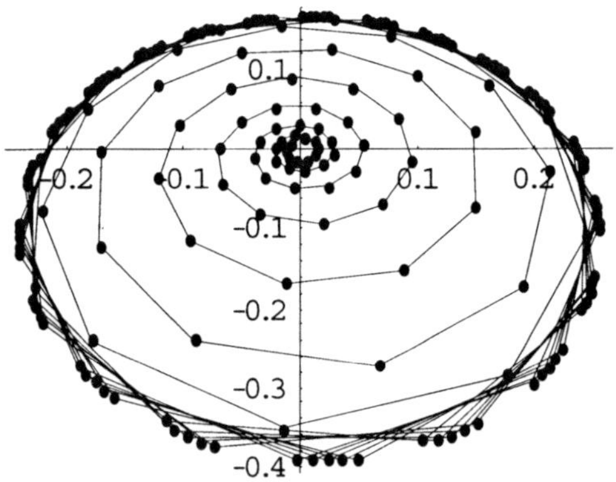

c) $\lambda > 0$: an attracting circle

Figure 6.9.1: Supercritical Neimark-Sacker bifurcation[21]

If the system satisfies these three conditions, then (1) by a change of basis in R^2, we may assume, without loss of generality, that

$$J = Df(0, 0, 0) = (1 + \lambda)\begin{bmatrix} \cos\beta & -\sin\beta \\ \sin\beta & \cos\beta \end{bmatrix}.$$

(2) from (iii), by a change of coordinates we may assume that the map f takes the form

[21] The simulation is for $\beta = 0.3$, a) $\lambda = -0.1$, b) $\lambda = 0$, and c) $\lambda = 0.06$.

$$f\begin{bmatrix} x_1 \\ x_2 \end{bmatrix} = N_\lambda \begin{bmatrix} r \\ \theta \end{bmatrix} + O\left(\left\|\begin{bmatrix} x_1 \\ x_2 \end{bmatrix}\right\|\right), \tag{6.9.3}$$

where

$$N_\lambda \begin{bmatrix} r \\ \theta \end{bmatrix} = \begin{bmatrix} (1+\lambda)r - F(\lambda)r^3 \\ \theta + \beta(\lambda) + G(\lambda) \end{bmatrix}, \tag{6.9.4}$$

with $F(0) \neq 0$. Moreover, the radius of the invariant circle is given by $\sqrt{\lambda / F(\lambda)}$.

Theorem 6.9.1. (Neimark-Sacker[22]) Suppose that f satisfies assumptions (i)-(iii). Then, for sufficiently small λ, f has an invariant closed curve enclosing the origin if $\lambda / F(\lambda) > 0$. Moreover, this curve is attracting if $F(0) > 0$; it is repelling if $F(0) < 0$.

Example (Hopf bifurcation in the Kaldor model[23]) The Kaldor model is describe by

$$Y(t+1) - Y(t) = \alpha\{I(Y(t), K(t)) - S(Y(t), K(t))\} \equiv \alpha F(Y(t), K(t)),$$
$$K(t+1) - K(t) = I(Y(t), K(t)) - \delta K(t),$$

in which variables and parameters are define as

$Y(t)$ and $K(t)$ = output level and capital stock in period t, respectively;
$I(Y, K)$ = investment function ($I_Y > 0$, $I_K < 0$);
$S(Y, K)$ = savings function ($0 < S_Y < 1$, $S_K > 0$);[24]

---

[22] See Rasband (1990).
[23] The original model was proposed by Kaldor (1940). Kaldor's contribution was in conjunction with the work of Kalecki (1937, 1939), who investigated similar models but concentrated on different aspects of stability. The analysis below is based on section 3.3 in Lorenz (1993). For the analysis and behavioral interpretations of the model for continuous case, see also Chang and Smyth (1971) and Gabisch and Lorenz (1989).
[24] The assumption of $S_K > 0$ is not convincing. In Chang and Smyth (1971), it is assumed $S_K < 0$. As we require $I_K - S_K < 0$, the different signs do not affect our analytical conclusion.

α and δ = a positive adjustment parameter and capital depreciation rate, respectively.

Suppose that the system has at least one equilibrium point. The determinant and trace of the Jacobian at a fixed point (Y^*, K^*) are

$$\det J = (\alpha F_Y + 1)(I_K + 1 - \delta) - \alpha F_K I_Y,$$
$$trJ = \alpha F_Y + I_K - \delta + 2.$$

The eigenvalues are complex conjugate if

$$\det J > \left(\frac{trJ}{2}\right)^2.$$

Assume that the inequality holds. A Hopf bifurcation occurs at a value $\alpha = \alpha_0$ if

$$\det J\big|_{\alpha=\alpha_0} = 1,$$

where

$$\alpha_0 = \frac{\delta - I_K}{F_Y(I_K + 1 - \delta) - F_K I_Y}.$$

The modulus crosses the unit circle with nonzero speed when the parameter α is changed[25]

$$\frac{d|\rho(a)|}{da} = \frac{d\sqrt{\det J}}{da} = \frac{\delta - I_K}{2a_0} > 0.$$

---

[25] The modulus of the complex eigenvalues, if they exist, of the characteristic equation,

$$\rho^2 + a\rho + b = 0$$

is given by $\sqrt{b}$.

We may apply the preceding procedure to describe the Hopf bifurcation in the discrete Kaldor model.

Exercise 6.9

1 Analyze the bifurcation structure of the map

$$\begin{bmatrix} x_1(t+1) \\ x_2(t+1) \end{bmatrix} = \left(1 + \lambda + x_1^2(t) + x_2^2(t)\right)\begin{bmatrix} \cos\beta & -\sin\beta \\ \sin\beta & \cos\beta \end{bmatrix}\begin{bmatrix} x_1(t) \\ x_2(t) \end{bmatrix}.$$

2 Consider the discrete predator-prey system

$$\begin{bmatrix} x_1(t+1) \\ x_2(t+1) \end{bmatrix} = \begin{bmatrix} \alpha x_1(t)(1 - x_1(t)) - x_1(t)x_2(t) \\ \dfrac{1}{\beta}x_1(t)x_2(t) \end{bmatrix},$$

where $x_1(t)$ denotes the prey population at generation t and $x_2(t)$ denotes the predator population at generation t. Show that a nontrivial fixed point of the map undergoes a Neimark-Sacker bifurcation.

6.10 The Liapunov numbers and chaos

It is well known that chaotic dynamics is characterized by an exponential divergence of initially chose points. In the case of one-dimensional maps, the Liapunov exponent is a measure of the divergence of two orbits starting with slightly different initial conditions. For a map on R^2, each orbit has two Liapunov numbers that measure the rates of separation from the current orbit point along two orthogonal directions. These directions are determined by the dynamics of the map. The first direction is the direction along which the separation between nearby points is the greatest. The second is the direction of greatest separation, chosen perpendicular to the first. The stretching factors in each of these two directions are the *Liapunov numbers of the orbit*. To formally define Liapunov numbers, we introduce

$$D_{(t)} = D_{f^t}(x_0),$$

to denote the Jacobian matrix of the t-th iterate of f. Let U be the unit circle centered on the first point of the orbit. Then $D_{(t)}(U)$ is an ellipse. This is a consequence of the fact that the linear transformation of a circle is an ellipse. The axis is longer than 1 in the expanding direction and shorter than 1 in the contracting direction.

Definition 6.10.1. Let f be a smooth map on R^2, and let $D_t = D_{f^t}(x_0)$. For $k = 1, 2, \cdots$, let $r^{k(t)}$ be the length of the k-th longest orthogonal axis of the ellipse $D_t(U)$ for an orbit with initial point x_0. Then $r^{k(t)}$ measures the contraction or expansion near the orbit that at x_0 during the first t iterations. The *k-th Liapunov number* of this trajectory is

$$L(k)x_0 = \lim_{t\to\infty} \sqrt[t]{r^{k(t)}},$$

if the limit exists. The *k-th Liapunov exponent* of the trajectory that starts at x_0 is

$$h(k)(x_0) = \ln(L(k)(x_0)), \quad k = 1, 2.$$

Clearly, $L(1) \geq L(2)$ and $h(1) \geq h(2)$.

Definition 6.10.2. Let f be a smooth map on R^2. An orbit

$$\{x_0, x_1, x_2, \cdots\},$$

is *asymptotically periodic* if it converges to a periodic orbit as $t \to \infty$; that is, there exists a periodic orbit $\{y_0, y_1, y_2, \cdots\}$ such that

$$\lim_{t\to\infty} \|x(t) - y(t)\| = 0, \text{ as } t \to \infty.$$

Definition 6.10.3. Let f be a map of R^2, and let $\{x_0, x_1, x_2, \cdots\}$ be a bounded orbit of f. The orbit is said to be *chaotic* if: (i)

$$\{x_0, x_1, x_2, \cdots\}$$

is not asymptotically periodic; (ii) no Liapunov number is exact, and (iii) the Liapunov exponent satisfies $h(1) > 0$.

Definitions 6.10.1 and 6.10.3 extend to m-dimensional case R^m, with the word "circle" replaced by the word "sphere", and the word "ellipse" replaced by the word "ellipsoid".

Chapter 7

Higher dimensional economic dynamics

This chapter applies the concepts and theorems of the previous chapter to examining behavior of different economic systems. Section 7.1 studies Dornbush's exchange rate model. We show how a monetary expansion will result in an immediate depreciation of the currency and sustain the inflation as the price level gradually adjusts upward. Section 7.2 studies a two-sector OLG model with the Leontief production functions. The economy produces two, consumption and investment, goods; it has two, consumption and investment, sectors. We provide conditions when the system is determinate or indeterminate. Section 7.3 introduces a one-sector real business cycle model with mild increasing returns-to-scale with government spending. Section 7.4 introduces endogenous fertility and old age support into the OLG model. Section 7.5 examines a model to capture the historical evolution of population, technology, and output. The economy evolves three regimes that have characterized economic development: from a Malthusian regime (where technological progress is slow and population growth prevents any sustained rise in income per capita) into a post-Malthusian regime (where technological progress rises and population growth absorbs only part of output growth) to a modern growth regime (where population growth is reduced and income growth is sustained). The model is defined within the OLG framework with a single good and it exhibits the structural patterns observed over history. Section 7.6 examines a model of unemployment and inflation. We demonstrate that the model which is built on the well-accepted assumptions behaves chaotically. Section 7.7 provides a model of long-run competitive two-periodic OLG model with money and capital. Section 7.8 introduces heterogeneous groups to the OSG model. Section 7.9 examines interdependence between economic growth and human capital accumulation within the OSG modeling framework.

7.1 An exchange rate model

We now consider the following Dornbush exchange-rate model[1]

$$y^d(t) = \delta(e(t) + \tilde{p} - p(t)) - \sigma(r(t) - p(t+1) + p(t)), \tag{7.1.1}$$
$$p(t+1) - p(t) = \alpha(y^d(t) - y), \tag{7.1.2}$$
$$m - p(t) = \phi y - \lambda r(t), \tag{7.1.3}$$
$$r(t) = \tilde{r} + e(t+1) - e(t), \tag{7.1.4}$$

where $\delta, \tilde{p}, \sigma, \alpha, y, m, \phi, \lambda, \tilde{r}$ are positive parameters, $y^d(t)$ and $p(t)$ are logarithms of domestic aggregate demand and the domestic price level at time t; r is the domestic nominal interest rate; and $e(t)$ is the logarithm of the exchange rate, that is, the price in home currency of one unit of foreign money. A rise in e implies that domestic money depreciates. Here, y and $\tilde{p}$ are the logarithms of the domestic commodity supply and foreign price level; and $\tilde{r}$ is the foreign nominal rate of interest and m is the logarithm of the nominal stock of money. Equation (7.1.1) implies that home aggregate demand is a decreasing function of the home-to-foreign price ratio and of the expected real rate of interest. Equation (7.1.2) tells that excess demand for goods and services drives price inflation. Equation (7.1.3) is a standard Keynesian LM schedule that relates the demand for money to real money and the nominal interest rate. Equation (7.1.4) is an arbitrage condition in asset markets which says that the home interest rate will exceed the foreign interest rate by an amount exactly equal to the expected rate of depreciation in the home country. The model consists of four equations with four endogenous variables, describing the evolution of a small open economy with perfect foresight. Here, by perfect foresight it means that the rate of inflation individuals expect equals the actual rate

$$p(t+1) - p(t) = \log\left(\frac{P(t+1)}{P(t)}\right) \approx \frac{P(t+1)}{P(t)} - 1,$$

for small changes in the actual level of prices $P(t)$.

To analyze the behavior of the model, we first eliminate two variables and reduce the dynamics to a two-dimensional problem. We can eliminate $y^d(t)$ and

[1] The model is proposed by Dornbusch (1976), for an open economy adaptation of the IS-LM structure. The example is from Azariadis (1993: 46-51).

$r(t)$ by substituting respectively equation (7.1.1) into equation (7.1.2) and equation (7.1.4) into equation (7.1.3). We thus have

$$(1-\alpha\sigma)p(t+1)-(1-\alpha\sigma+\alpha\delta)p(t)=$$
$$(\delta+\sigma)\alpha e(t)+\alpha\delta\widetilde{p}-\alpha\sigma r^{*}-\alpha\sigma e(t+1),$$
$$\lambda e(t+1)-\lambda e(t) = p(t)+ \phi y-\lambda\widetilde{r}-m. \tag{7.1.5}$$

The system has a unique equilibrium point

$$e^{*} = p^{*} - \widetilde{p} + \frac{y+\sigma\widetilde{r}}{\delta},$$
$$p^{*} = \lambda\widetilde{r} + m - \phi y. \tag{7.1.6}$$

At equilibrium, we also have

$$r(t)=\widetilde{r},\quad y^{d}(t)=\widetilde{y}.$$

At equilibrium the domestic price level is an increasing function of the exogenous variables m and $\widetilde{r}$, a decreasing function of y; the exchange rate is an increasing function of m and $\widetilde{r}$, a decreasing function of $\widetilde{p}$. We may transform dynamical system (7.1.5) into a homogeneous one by introducing

$$x_1(t)\equiv e(t)-e^{*},\quad x_2(t)\equiv p(t)-p^{*}.$$

Under the transformation, system (7.1.5) becomes

$$\begin{bmatrix} x_1(t+1) \\ x_2(t+1) \end{bmatrix} = \begin{bmatrix} 1 & \dfrac{1}{\lambda} \\ \dfrac{\alpha\delta}{1-\alpha\sigma} & 1-\dfrac{\alpha\delta+\alpha\sigma/\lambda}{1-\alpha\sigma} \end{bmatrix} \begin{bmatrix} x_1(t) \\ x_2(t) \end{bmatrix}, \tag{7.1.7}$$

or in matrix form

$$x(t+1)=Ax(t).$$

The characteristic equation is

$$z(\rho) = (\rho - \rho_1)(\rho - \rho_2) = \rho^2 - (TraecA)\rho + DetA,$$

where

$$TraecA = \rho_1 + \rho_2 = 2 - \frac{\alpha\delta + \alpha\sigma / \lambda}{1 - \alpha\sigma},$$

$$DetA = = \rho_1\rho_2 = 1 - \frac{\alpha\delta}{1 - \alpha\sigma}\left(\delta + \frac{\alpha}{\lambda}\right) - \frac{\alpha\delta}{\lambda(1 - \alpha\delta)}.$$

We calculate

$$\Delta = (TraecA)^2 - 4DetA = \left[\frac{\alpha}{1 - \alpha\sigma}\left(\delta + \frac{\alpha}{\lambda}\right)\right]^2 + 4\frac{\alpha\delta}{\lambda(1 - \alpha\delta)},$$

$$z(1) = (1 - \rho_1)(1 - \rho_2) = -\frac{\alpha\delta}{\lambda(1 - \alpha\delta)},$$

$$z(-1) = (1 + \rho_1)(1 + \rho_2) = 4 - \frac{\alpha}{1 - \alpha\sigma}\left[2\left(\delta + \frac{\alpha}{\lambda}\right) + \frac{\alpha}{\lambda}\right].$$

If $1 - \alpha\sigma > 0$, i.e.

$$\alpha < \frac{1}{\sigma},$$

the eigenvalues are real (because of $\Delta > 0$), and lie on opposites of 1 (because $z(1) < 0$). The sum of the two eigenvalues is less than 2 (because of $TraceA < 2$). If the equilibrium point is a saddle, we need the eigenvalues to lie on different sides of -1, that is

$$z(-1) > 0 \Leftrightarrow \alpha > \frac{4}{(2 + \lambda^{-1})(\delta + 2\sigma)}.$$

We conclude that if the coefficient α is small enough (price adjustment is sufficiently low), the steady state of the system is a saddle point. In the remaining of this section, we require α to be sufficiently small so that the two eigenvalues satisfy: $\rho_1 > 1$ and $|\rho_2| < 1$.

The general solution can be written as

$$e(t) = e^* + c_1 h_{11} \rho_1^t + c_2 h_{21} \rho_2^t,$$
$$p(t) = p^* + c_1 \rho_1^t + c_2 \rho_2^t, \tag{7.1.8}$$

where $(h_{ii}, 1)^T$ is the eigenvector associated with the eigenvalues, ρ_i, for $i = 1, 2$. We are interested in a special path which is not explosive and a situation in which for the given level of price the exchange rate adjusts as needed to keep the system on the unique convergent path. To exclude explosive paths, we take $c_1 = 0$. Then c_2 can be determined by

$$c_2 = p(0) - p^*.$$

It is straightforward to show that equations (7.1.7) can be written as

$$e(t) = e^* + h_{21}\left(p(t) - p^*\right).$$

It is direct to check $h_{21} < 0$ by

$$Ah_2 = \rho_2 h_2,$$

and $|\rho_2| < 1$. The equation thus describes the negatively sloped straight line through the equilibrium point $S = \left(e^*, p^*\right)$.

We may apply this equation to examine impact of changes in some parameters, say, m on the dynamics of the system. An increase in m will shift the equilibrium point. Equations (7.1.5) show that the equilibrium values of the two variables will rise by an equal amount, from $S = \left(e^*, p^*\right)$ to $\hat{S} = (\hat{e}, \hat{p})$ in the long term. As price is assumed to be sticky, the adjustment process is depicted as in figure 7.1.1. A monetary expansion results in an immediate depreciation of the currency. In initial stage, the exchange rate overshoots its long-run equilibrium value, so that inflation is actually accompanied by a gradual depreciation of the currency during the transition. This happens as the presumed stickiness of output prices puts the full burden of the immediate adjustment on the exchange rate. The instantaneous depreciation produces a disequilibrium in the goods market which adjusts over time as the output price adjusts. Following the sudden depreciation, and with output prices being constant, domestic goods become cheaper relative to foreign goods. This leads to an excess demand for domestic output and to a gradual increase in domestic prices.

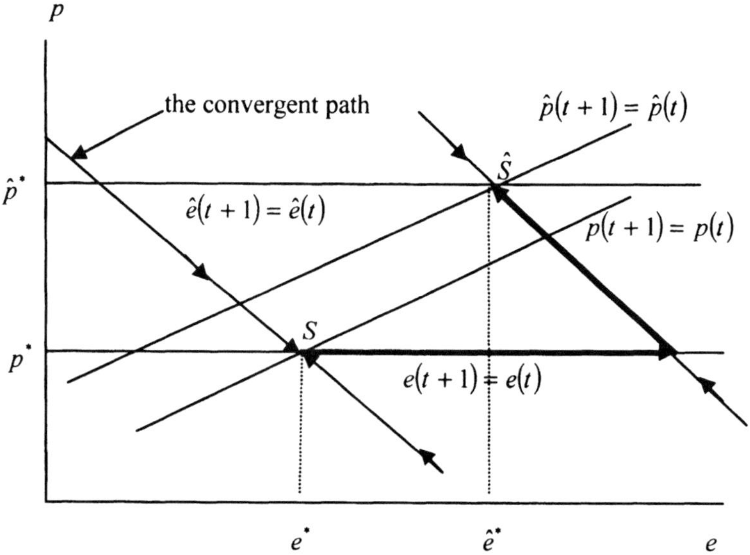

Figure 7.1.1: The overshooting adjustment process

7.2 A two-sector OLG model

This section deals with a two-sector overlapping generations model with fixed coefficients proposed by Ralf.[2] The economy produces two, consumption and investment, goods; it has two, consumption and investment sectors. The investment goods is taken as the numeraire. Individuals live two periods – in each period, there are old and young people. People work only in the first period of their lives supplying one unit of labor inelastically. There are $N(t)$ young individuals at t; the representative young individual splits his income $w(t)$ between consumption, $c(t)$, and saving, $s(t)$. The young agent's problem is defined as

$$Max\left(c^{\gamma}(t)+c^{\gamma}(t+1)\right)^{1/\gamma},\quad \gamma\in(-\infty,0)$$

$$\text{s.t.:}\quad s(t)=w(t)-p(t)c(t), \qquad c(t+1)=\frac{r(t+1)s(t)}{p(t+1)}$$

[2] See Ralf (2001). For two sector growth models, see also Galor (1992) and section 15.5 in Azariadis (1993).

where $r(t+1)$ are interest payments on saving, and $p(t)$ is the price of the consumption goods in terms of the investment goods. Suppose the elasticity of substitution between consumption of different periods

$$\sigma = \frac{1}{\gamma - 1},$$

is restricted to be less than 1 in absolute value. The optimal saving function is given by

$$s(t) = \frac{w(t)}{1 + (r(t+1)p(t)/p(t+1))^{\gamma_0}}, \tag{7.2.1}$$

where

$$\gamma_0 \equiv \frac{\gamma}{\gamma - 1}.$$

Once $p(t)$, $p(t+1)$, $w(t)$, and $r(t+1)$ are determined, we uniquely determine $s(t)$, as well as $c(t)$ and $c(t+1)$.

Using subscripts c and i respectively to denote the consumption and investment sectors, we have the production functions of the two sectors

$$C(t) = F_c(K_c(t), N_c(t)) = \min\{K_c(t), \alpha_c N_c(t)\},$$
$$I(t) = F_i(K_i(t), N_i(t)) = \mu \min\{K_i(t), \alpha_i N_i(t)\}, \quad \mu > 1,$$

where K_c, K_i, N_c and N_c denote capital and labor inputs of sector j, $j = i, c$, $C(t)$ is the aggregated consumption, and $I(t)$ is the total investment.

Introduce

$$k_j(t) = \frac{K_j(t)}{N_j(t)}, \quad n_j(t) = \frac{N_j(t)}{N(t)}.$$

We have

$$n_c(t) + n_i(t) = 1,$$
$$n_c(t)k_c(t) + n_i(t)k_i(t) = k(t). \tag{7.2.2}$$

Efficient production in the two sectors requires

$$K_j(t) = \alpha_j N_j(t).$$

That is

$$k_j(t) = \alpha_j.$$

Production in the consumption goods sector takes place with the capital intensity higher than in the investment goods sector, if $\alpha_c > \alpha_i$. Perfect competition in the markets yields

$$p(t) = \frac{w(t)}{\alpha_c} + r(t),$$
$$w(t) = \alpha_i \mu - \alpha_i r(t). \tag{7.2.3}$$

Investment is given as output of the investment goods sector, i.e.

$$i(t) = \alpha_i \mu n_i(t),$$

where $i \equiv I / N$. Capital stock is assumed to depreciate fully after one period. The output of the investment goods sector can be used in the following period. Hence

$$k(t+1) = i(t) = \alpha_i \mu n_i(t).$$

On the other hand, the saving in period t equals the investment in period t, i.e.

$$s(t) = i(t).$$

We have thus built the two-sector model. From

$$k(t+1) = \alpha_i \mu n_i(t),$$
$$k_j(t) = \alpha_j,$$

and equations (7.2.2), we have

$$k(t+1) = \mu \alpha \alpha_i (k(t) - \alpha_c), \tag{7.2.4}$$

where

$$\alpha \equiv \frac{1}{\alpha_i - \alpha_c}.$$

By equation (7.2.1), we solve

$$\frac{p(t+1)}{r(t+1)} = p(t)\left[\frac{w(t)}{s(t)} - 1\right]^{-1/\gamma_0}. \tag{7.2.5}$$

On the other hand from equations (7.2.3), we solve

$$\begin{aligned} w(t) &= \alpha \alpha_i \alpha_c (p(t) - \mu), \\ r(t) &= \alpha(\alpha_i \mu - \alpha_c p(t)). \end{aligned} \tag{7.2.6}$$

From the equation for w in equations (7.2.6) and

$$s(t) = \alpha_i \mu n_i(t), \quad n_i(t) = \alpha(k(t) - \alpha_c),^3$$

we obtain

$$\frac{w(t)}{s(t)} = \frac{\alpha_c}{\mu} \frac{p(t) - \mu}{k(t) - \alpha_c}. \tag{3.2.7}$$

From the equation for r in equations (7.2.6)

$$\frac{p(t+1)}{r(t+1)} = \frac{p(t+1)}{\alpha(\alpha_i \mu - \alpha_c p(t+1))}. \tag{7.2.8}$$

Substituting equations (7.2.7) and (7.2.8) into equation (7.2.5) yields

[3] This relation is obtained from equation (7.2.2) and $k_j = \alpha_j$, $j = c, i$.

$$p(t+1) = \frac{\mu\alpha_c}{\alpha_i + g(p(t), k(t))},$$
$$g(p(t), k(t)) \equiv \frac{1}{\alpha p(t)}\left[\frac{\alpha_c p(t) - \mu k(t)}{\mu k(t) - \alpha_c \mu}\right]^{1/\gamma_0}. \tag{7.2.9}$$

The dynamics of the economy is now described by the motion of equations (7.2.4) and (7.2.9). Suppose $\alpha_c > \alpha_i$. Then, for the variables to be positive, we should add the following constraints on the solution

$$\alpha_c > k(t) > \alpha_i,$$
$$\frac{\mu}{\alpha_c} k(t) < p(t) \le \frac{\alpha_i \mu}{\alpha_c}. \tag{7.2.10}$$

If $\alpha_c < \alpha_i$, we should require

$$\alpha_c < k(t) < \alpha_i,$$
$$\frac{\alpha_c \mu}{\alpha_i} \le p(t) < \frac{\mu}{\alpha_c} k(t). \tag{7.2.11}$$

From equation (7.2.4), we see that $k(t)$ is independent of $p(t)$. The equation has a unique steady state

$$k^* = \frac{\alpha_c}{1 - 1/\mu\alpha\alpha_i}.$$

From equation (7.2.9), a steady state of price is determined by

$$\Omega(p) \equiv \alpha_i p + \frac{1}{\alpha}\left[\frac{\alpha_c p - \mu k^*}{\mu k^* - \alpha_c \mu}\right]^{1/\gamma_0} - \mu\alpha_c = 0. \tag{7.2.12}$$

Under constraints (7.2.10) or (7.2.11), the equation

$$\Omega(p) = 0$$

has a unique solution.[4] It is straightforward to calculate the two eigenvalues, ρ_j, at the steady state

$$\rho_1 = \alpha \mu \alpha_i > 0,$$
$$\rho_2 = \left(\frac{\alpha_c p^* - \mu \alpha_i}{\alpha_c p^* - \mu k^*} \right) \frac{\mu k^* - \alpha_c p^* / \gamma}{\mu \alpha_i} < 0.$$

In economics, a concept called indeterminancy is currently often used. This concept is related to stability of steady state. If a steady state is asymptotically stable, for any given initial capital stock the economy will eventually approach the steady state. Hence, the agent is not able to distinguish rationally between the different perfect foresight paths. This property is called *indeterminancy*. *Determinacy* of the perfect foresight dynamics requires that the steady state is a saddle or a source. If $\alpha > 0$, i.e., $\alpha_i > \alpha_c$, the system is always determinate, since ρ_1 is greater than 1; the steady state is a source if $\rho_2 < -1$ and it is a saddle point if $\rho_2 > -1$. If $\alpha < 0$, i.e.

$$\alpha_i < \alpha_c,$$

the steady state may be either indeterminate or determinate, depending on the parameters.[5]

7.3 Growth with government spending

This section introduces a one-sector real business cycle model with mild increasing returns-to-scale with government spending.[6] The economy has a continuum of identical competitive firms in the economy, with the total number normalized to

---

[4] This can be proved by calculating the value of $\Omega(p)$ at the boundaries of the range of definition and then applying the mean value theorem.

[5] When the parameters are specified, we may have $\rho_1 = -1$. Period-2 solutions appear. It should be noted that in a two-sector model of optimal growth with infinite horizon by Nishimura and Yano (1995), the optimal time path of capital accumulation may be chaotic for any rate of time preference, if the production function in the consumption goods sector takes place with a higher capital intensity than in the investment sector.

[6] This section is based on a model developed by Guo (2004), which is influenced by Devereux *et al.* (1996, 2000).

one. Each firm produces output $y(t)$ according to a constant return-to-scale technology

$$y(t) = x(t)k^{\alpha}(t)h^{1-\alpha}(t), \quad 0 < \alpha < 1,$$

where $k(t)$ and $h(t)$ are capital and labor inputs, and $x(t)$ represents productive externalities. For each firm $x(t)$ is given. Factor markets are perfectly competitive; we thus have

$$r(t) = \alpha \frac{y(t)}{k(t)},$$
$$w(t) = (1 - \alpha)\frac{y(t)}{h(t)}. \tag{7.3.1}$$

Assume that $x(t)$ changes according to

$$x(t) = k^{\alpha\chi}(t)h^{(1-\alpha)\chi}(t),$$

where χ is the degree of productive externalities. Under the externalities, the production function becomes

$$y(t) = k^{\chi_k}(t)h^{\chi_h}(t), \tag{7.3.2}$$

where

$$\chi_k \equiv \alpha(1 + \chi), \quad \chi_h \equiv (1 - \alpha)(1 + \chi).$$

The economy is populated by a unit measure of identical infinitely lived household, each endowed with one unit of time. The representative household maximizes a discounted stream of expected utilities[7]

---

[7] It should be noted that we are only concerned with the case of perfect foresight dynamics with full capital utilization. Guo (2004) examines other cases, for instance, when capital may not be fully utilized.

$$\max\, E_0 \sum_{t=0}^{\infty} \theta^t \left[\log c(t) - \eta \frac{h^{1+\sigma}}{1+\sigma} \right], \quad 0 < \theta < 1, \ \ \sigma \geq 0, \ \ \eta > 0,$$

where θ is the discount factor, $c(t)$ is consumption, and σ denotes the inverse of the intertemporal elasticity of substitution for labor supply. The budget constraint is

$$c_i + i(t) + \tau(t) = w(t)h(t) + r(t)k(t),$$

where $i(t)$ is investment and $\tau(t)$ is a lump-sum tax. We may rewrite the budget constraint as

$$c_i + i(t) + \tau(t) = y(t),$$

where we use

$$y(t) = w(t)h(t) + r(t)k(t).$$

The motion of capital is given by

$$k(t+1) = (1-\delta)k(t) + i(t),$$

k_0 being given, where δ ($0 < \delta < 1$) is the capital depreciation rate. The first-order conditions for the optimization problem are given by[8]

$$\eta c(t) h^{\sigma}(t) = w(t),$$
$$\frac{1}{c(t)} = \theta \left[\frac{1}{c(t+1)} (1 - \delta + r(t+1)) \right], \tag{7.3.3}$$
$$\lim_{t\to\infty} \frac{k(t+1)}{c(t)} = 0.$$

Assume that government spending, $g(t)$, is constant. The period government budget constraint is

$$g = \tau(t).$$

---

[8] See appendix A.9 for dynamical optimization.

The aggregate resource constraint for the economy is given by

$$c(t)+k(t+1)-(1-\delta)k(t)+g=y(t).$$

It can be shown that this equation is redundant in the sense that we can obtain it through the household budget, the motion of capital and marginal conditions (7.3.1).

We now show that the dynamics is expressed by a two-dimensional dynamical system. First, from equations (7.3.1) and (7.3.2), and

$$\eta c(t)h^{\sigma}(t)=w(t)$$

in equations (7.3.3), we solve $h(t)$ as a function of $k(t)$ and $c(t)$ as follows

$$h(t)=\left[\frac{\eta}{1-\alpha}\frac{c(t)}{k^{\chi_k}(t)}\right]^{1/(\chi_h-1-\sigma)}. \tag{7.3.4}$$

From equations (7.3.3), we directly have

$$c(t+1)=\left[\left(1-\delta+\alpha k^{\chi_k-1}(t+1)h^{\chi_h}(t+1)\right)\right]\theta c(t), \tag{7.3.5}$$

where we use

$$r(t)=\frac{\alpha y(t+1)}{k(t+1)},$$

and equation (7.3.2). From the household budget and $g=\tau(t)$, we solve

$$i(t)=y(t)-c(t)-g.$$

Substituting this equation into the equation of motion of capital yields

$$k(t+1)=(1-\delta)k(t)+k^{\chi_k}(t)h^{\chi_h}(t)-c(t)-g, \tag{7.3.6}$$

where we use equation (7.3.2). Substituting equation (7.3.4) into equations (7.3.5) and (7.3.6), we obtain the two equations which contain only k and c

$$k(t+1) = (1-\delta)k(t) + \eta_0 c^{\chi_0}(t) k^{\chi_k(1-\chi_0)-1}(t) - c(t) - g,$$
$$c(t+1) = \left[\left(1-\delta+\alpha\eta_0 c^{\chi_0}(t+1) k^{\chi_k(1-\chi_0)-1}(t+1)\right)\right]\theta c(t), \tag{7.3.7}$$

where

$$\chi_0 \equiv \frac{\chi_h}{\chi_h - 1 - \sigma}, \quad \eta_0 \equiv \left(\frac{\eta}{1-\alpha}\right)^{\chi_0}.$$

It should be noted that the right-hand side of the first equation in equations (7.3.7) contains $c(t+1)$ and $k(t+1)$. It is straightforward to show that the system has a unique non-trivial steady state (k^*, c^*). The Jacobian matrix at (k^*, c^*) is given by

$$J = \begin{bmatrix} \rho_1 & \rho_2 \\ \dfrac{\rho_1\rho_3}{\rho_4} & \dfrac{1+\rho_2\rho_3}{\rho_4} \end{bmatrix},$$

where

$$\rho_1 = 1 - \delta - \frac{(1+\sigma)(\theta_0+\delta)(1+\chi)}{\chi_h - 1 - \sigma},$$
$$\rho_2 = \delta + \frac{(\theta_0+\delta)\left[\chi_h g/y^* + (1-g/y^*)(1+\sigma)\right]}{\alpha(\chi_h - 1 - \sigma)},$$
$$\rho_3 = -\frac{(\theta_0+\delta)\left[\chi_k(1+\sigma) + \chi_h - 1 - \sigma\right]}{(\chi_h - 1 - \sigma)(1+\theta_0)},$$
$$\rho_4 = 1 - \frac{(\theta_0+\delta)\chi_h}{(\chi_h - 1 - \sigma)(1+\theta_0)},$$

where $\theta_0 \equiv 1/\theta - 1$. Local indeterminacy requires that both eigenvalues of J are less than 1 in modulus. It can be shown that a necessary condition for the equilibrium point to display indeterminacy is

$$\chi_h > 1 + \sigma.$$

7.4 Growth with fertility and old Age support

This section introduces endogenous fertility and old age support into the OLG model introduced in section 3.3.[9] In the literature, two assumptions about population growth and economic conditions are made; the first is to assume that children are consumption goods and appear in the utility function of the parents; the second is to assume that children are valued as a source of old age support.[10] This assumption is made based on the observation that countries with more developed financial markets or better social security programs tend to have lower population growth rates than those that do not. For instance, in developing countries, children often start contributing to family income while living at home and prior to adulthood. Here, we assume that individuals give a constant fraction of their income to their parents in the form of old age support, and that parents incur a time cost of raising children which is increasing in the number of children.

The number of young agents at time t is denoted by $N(t)$ and there are $N(-1)$ agents in the initial old generation. The initial old generation is endowed with an aggregate initial capital stock of $K(0) > 0$. All young agents are identical and they are endowed with one unit of labor when young, and have no other endowments of goods or assets at any date. There is a single good in the economy, which is produced using capital and labor as inputs. For simplicity, it is assumed that individuals care only about old age consumption, denoted by c; their lifetime consumption utility is given by $u(c)$, which is assumed to be increasing in c, and concave. Agents work and raise children when young, and are retired when old. Young agents remit a fraction, μ, of their income to their parents as old age support, and save the rest which is equal to $(1 - \mu)$ of the income. Each young agent decides to have $n(t)$ children. A time cost of raising children, $\gamma(n(t))$, satisfies

$$\gamma(0) = 0, \quad \gamma' > 0, \quad \gamma'' \geq 0.$$

All agents take the wage rate, $w(t)$, at period t as given. The saving $s(t)$ is given by

$$s(t) = (1 - a)w(t)(1 - \gamma(n(t))). \tag{7.4.1}$$

---

[9] The model is proposed by Chakrabarti (1999).

[10] The first approach is accepted by, for instance, Barro and Becker (1989), Becker *et al.* (1990), Galor and Weil (1996); the second approach by, for instance, Raut and Srinivasan (1994). Chakrabarti tried to combine the two ideas within the same framework.

Old age consumption is thus given by

$$c(t+1) = r(t+1)s(t) + an(t)w(t+1)(1-\gamma(n(t+1))),$$

where r is the rate of interest. The production is carried out according to the constant returns to scale production function

$$F(t) = F(K(t), (1-\gamma(n(t)))N(t)),$$

where K is the capital stock. The intensive production function is given by $f(k)$, where

$$k \equiv \frac{K}{N(1-\gamma)}.$$

Assume that the production function is of the CES form

$$f(k) = (\alpha k^{\rho} + \beta)^{1/\rho}, \quad \alpha, \beta > 0, \quad -\infty < \rho < 1.$$

Each unit of the consumption goods saved at time t becomes one unit of capital at time $t+1$. The capital is used in production and it depreciates completely in the production process. Hence

$$K(t+1) = s(t)N(t).$$

The labor and capital markets are perfectly competitive. The marginal conditions are

$$\begin{aligned} r(t) &= f'(k(t)) = \alpha k^{\rho-1}(t)\left[\alpha k^{\rho}(t) + \beta\right]^{1/\rho-1}, \\ w(t) &= f(k(t)) - k(t)f'(k(t)) = \beta\left[\alpha k^{\rho}(t) + \beta\right]^{1/\rho-1}, \quad t \geq 0. \end{aligned} \tag{7.4.2}$$

The young agent's problem is to maximize $u(c(t+1))$ by choosing $n(t)$ subject to

$$c(t+1) = (1-a)r(t+1)w(t)(1-\gamma(n(t))) + an(t)w(t+1)(1-\gamma(n(t+1))).$$

With w and r as given, the problem of optimization has an interior optimum determined by

$$-(1-a)r(t+1)w(t)\gamma'(n(t)) + aw(t+1)(1-\gamma(n(t+1))) = 0. \tag{7.4.3}$$

Substituting equations (7.4.2) into equation (7.4.3) yields

$$f'(k(t+1)) = \frac{aw(k(t+1))[1-\gamma(n(t+1))]}{(1-a)w(k(t))\gamma'(n(t))}, \tag{7.4.4}$$

which states that the rate of return on capital is equal to the rate of return on children.[11] By the definition, we have

$$K(t+1) = N(t+1)[1-\gamma(n(t+1))]k(t+1).$$

With this equation and equation (7.4.1), we can rewrite

$$K(t+1) = s(t)N(t)$$

as

$$k(t+1) = \frac{(1-a)w(t)(1-\gamma(n(t)))}{[1-\gamma(n(t+1))]n(t)}. \tag{7.4.5}$$

In the remainder of this section, we specify the cost function as follows

$$\gamma(n(t)) = cn^2(t).$$

We also introduce a new variable

$$z(t) \equiv \frac{\alpha}{\beta}k^{\rho}(t).$$

[11] The rate of return on children $\gamma'(n(t))$ is given by equation (7.4.3), which equals the right-hand side of equation (7.4.4).

Multiplying both sides of equations (7.4.4) and (7.4.5) by $k(t+1)$ and utilizing equation (7.4.5), we obtain

$$z(t+1) = a\frac{1 - cn^2(t)}{2cn^2(t)}. \tag{7.4.6}$$

Applying equation (7.4.6) to equation (7.4.5) yields

$$1 - cn^2(t+1) = a_0 \frac{(z(t)+1)^{1/\rho - 1}}{n^{1-2/\rho}(t)}\left(1 - cn^2(t)\right)^{1-1/\rho}, \tag{7.4.7}$$

where

$$a_0 \equiv (1-a)\left(\frac{a}{2\alpha c}\right)^{-1/\rho}.$$

Difference equations (7.4.6) and (7.4.7) describe the sequence of $\{z(t), n(t)\}$. A fixed point of the system is determined by

$$p(z) \equiv z - a\frac{1 - cn^2}{2cn} = 0, \tag{7.4.8}$$

$$q(z) \equiv 1 - cn^2 - a_0 \frac{(z+1)^{1/\rho - 1}}{n^{1-2/\rho}}\left(1 - cn^2\right)^{1-1/\rho}. \tag{7.4.9}$$

Proposition 7.4.1. If $\rho > 0$, then equations (7.4.8) and (7.4.9) have a unique solution.

The proof is left to the reader. When $\rho > 0$, both equations (7.4.8) and (7.4.9) define downward sloping loci, as depicted in figure 7.4.1.

The linearized system of equations (7.4.8) and (7.4.9) at a fixed point is given by

$$J = \begin{bmatrix} 0 & -\dfrac{a}{cn^3} \\ -\dfrac{nz}{a\rho}\dfrac{1-\rho}{1+z} & -\dfrac{z}{a} - \dfrac{1-\rho}{\rho cn^2} \end{bmatrix}.$$

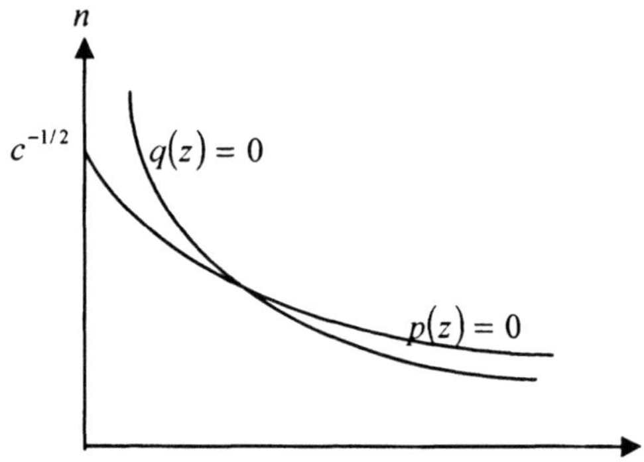

Figure 7.4.1: The unique fixed point for $\rho > 0$

We have

$$TraceJ = -\frac{z}{a} - \frac{1-\rho}{\rho c n^2},$$

$$DetJ = \frac{z}{\rho c n^2}\frac{1-\rho}{1+z}.$$

In the case of $\rho > 0$, $TraceJ < 0$ and $DetJ < 0$, J has one positive and one negative eigenvalues. In addition, we have

$$T < D < 1 + D.$$

We conclude that the positive eigenvalue is strictly less than one. It can be proved that if $0 < \rho \le 0.5$, then

$$TraceJ + DetJ < -1.$$

The steady state is a saddle. When $\rho > 0.5$, the fixed point can be either a sink or a saddle point.

Proposition 7.4.2.[12] If $\rho > 0$ and sufficiently large (with fixed a, α), then, (i) there is a critical value of c, denoted by c^*, such that for $c > (<)\ c^*$, the fixed point is a saddle (sink); (ii) At $c = c^*$, a flip bifurcation occurs. Thus, for c in a neighborhood of c^*, periodic solutions of period 2 can be observed.

In the case of $\rho < 0$, the configuration of equation (7.4.8) is the same as in the case of $\rho > 0$; the locus defined by equation (7.4.9) has the bell-shape, and attains a unique (local) maximum at the value

$$z = -\frac{1}{\rho}.$$

As demonstrated in figure 7.4.2, there are four possible cases.

These cases can be identified as follows.

For figure 7.4.2a

a	c	α	ρ	(z_1^*, n_1^*)	(z_2^*, n_2^*)
0.1	0.25	0.5	-0.5	$(0.965, 0.443)$	$(58.592, 0.058)^*$

For figure 7.4.2b

a	c	α	ρ	(z_1^*, n_1^*)	(z_2^*, n_2^*)
0.1	0.25	0.5	-2	$(0.884, 0.463)$	$(4.469, 0.210)^*$

For figure 7.4.2c

a	c	α	ρ	(z_1^*, n_1^*)
0.1	0.25	0.6062	-4	$(1.418, 0.369)^*$

For figure 7.4.2d

a	c	α	ρ
0.1	0.1	0.4	-4^*

---

[12] The proof of this proposition is referred to Chakrabarti (1999).

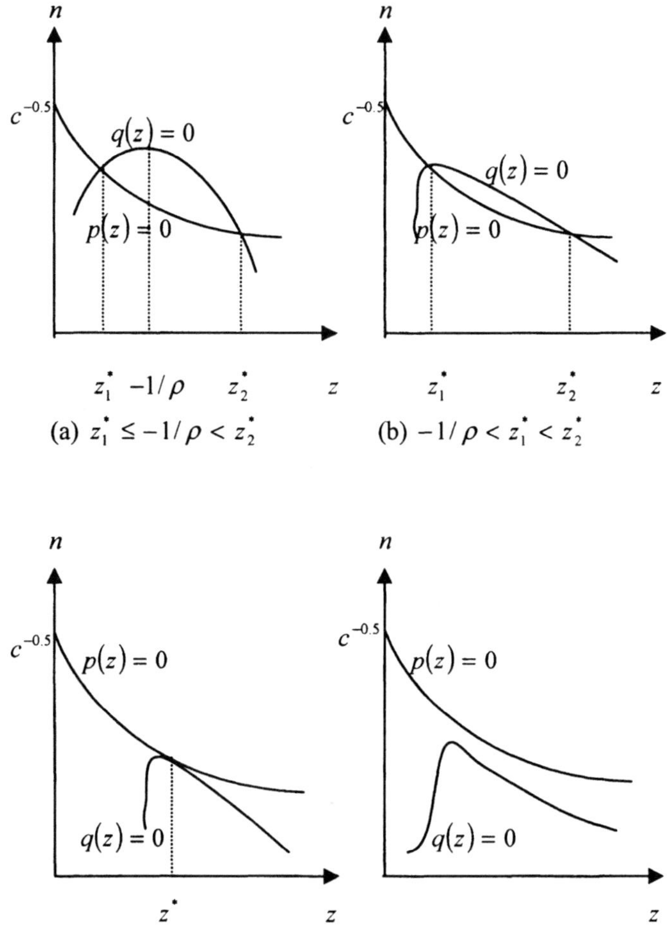

(a) $z_1^* \le -1/\rho < z_2^*$ (b) $-1/\rho < z_1^* < z_2^*$

(c) $z^* > -1/\rho$ (d) no fixed point

Figure 7.4.2: Fixed points for $\rho < 0$

It is straightforward to check the eigenvalues for the above examples. Further examination of the model is referred to Chakrabarti.

Exercise 7.4

1 Prove proposition 7.4.1.

7.5 Growth with different types of economies

This section introduces a growth model, proposed by Galor and Weil, that captures the historical evolution of population, technology, and output.[13] The economy evolves three regimes that have characterized economic development: from a Malthusian regime (where technological progress is slow and population growth prevents any sustained rise in income per capita) into a post-Malthusian regime (where technological progress rises and population growth absorbs only part of output growth) to a modern growth regime (where population growth is reduced and income growth is sustained). The model is defined within the OLG framework with a single good. The production uses land and efficiency unit of labor as inputs. The supply of land is exogenously fixed. The number of efficiency units is endogenous.

The output produced at time t is

$$Y(t) = H^{\alpha}(t)(A(t)X)^{1-\alpha},$$

where X and $H(t)$ are the quantities of land and efficiency units of labor employed in production at t, $0 < \alpha < 1$, and $A(t) > 0$ is endogenously determined technological level at t. The output per worker at t is

$$y(t) = h^{\alpha}(t)x^{1-\alpha}(t) \equiv y(h(t), x(t)),$$

where $y_h > 0$, $y_x > 0$ for any $(h, x) >> 0$

$$h \equiv \frac{H}{N}, \quad x \equiv \frac{AX}{N},$$

where $N(t)$ is the total labor force at t. Suppose that there are no property rights over land and the return to land is thus zero. The wage per efficiency unit of labor is therefore equal to its average product

$$w(t) = \left(\frac{x(t)}{h(t)}\right)^{1-\alpha} \equiv w(x(t), h(t)). \tag{7.5.1}$$

---

[13] See Galor and Weil (2000). Rather than analyzing behavior of the difference equations, this section shows how to construct a discrete dynamical system which explains very complicated economic phenomena.

Each individual born at period $t - 1$ lives two periods. In the first period, they consume a fraction of their parents' time. In the second period, they allocate their endowed one unit of time between child-rearing and labor force participation. In each period t a generation that consists of $N(t)$ identical individuals joins the labor force. The utility is represented by

$$u(t) = c^{1-\gamma}(t)[w(t+1)n(t)h(t+1)]^{\gamma}, \quad c(t) \geq \tilde{c} > 0,$$

where $\tilde{c}$ is a subsistence level, $n(t)$ is the number of children of individual t, $h(t+1)$ is the level of human capital of each child, and $w(t+1)$ is the wage per efficiency unit of labor at time $t+1$. The utility function is monotonically increasing and strictly quasi-concave. Let

$$\tau_0 + \tau e(t+1)$$

be the total time for a member of generation t of raising a child with a level of education quality $e(t+1)$. Define potential income as the amount that generation t would earn if they devoted their entire time endowment to labor force participation. That is, potential income is given by $w(t)h(t)$. This income is divided between child-rearing and working. Hence, in the second period of life, the individual daces the budget constraint

$$w(t)h(t)n(t)\{\tau_0 + \tau e(t+1)\} + c(t) \leq w(t)h(t).$$

It is assumed that the level of human capital of members of generation t, $h(t+1)$, is an increasing function of their education $e(t+1)$ and a decreasing function of the rate of progress in the state of technology from period t to $t+1$

$$g(t+1) \equiv \frac{A(t+1) - A(t)}{A(t)}.$$

That is

$$h(t+1) = h(e(t+1), g(t+1)),$$
$$h, h_e, h_{gg}, h_{eg} > 0, \quad h_g, h_{ee} < 0, \quad \forall (e, g) \geq 0.$$

The erosion effect is assumed to become higher as a result of technological progress, i.e.

$$\frac{\partial y(h(t), x(t))}{\partial g(t)} > 0.$$

It is further assumed that

$$\tau_0 h_e(0, 0) - \tau h(0, 0) < 0.$$

It is straightforward to show that under this assumption, there exists a value of $\widetilde{g}$ such that

$$\tau_0 h_e(0, \widetilde{g}) - \tau h(0, \widetilde{g}) = 0.$$

Denote $z(t)$ and $\widetilde{z}$ the level of potential income and the level of potential income at which the subsistence constraint is just binding; that is

$$z(t) = h(t)w(t), \quad \widetilde{z} = \frac{\widetilde{c}}{1 - \gamma}.$$

By equation (7.5.1), we have

$$z(t) = h^{\alpha}(t)x^{1-\alpha}(t) = h^{\alpha}(e(t), g(t))x^{1-\alpha}(t) \equiv z(e(t), g(t), x(t)). \tag{7.5.2}$$

Members of generation t choose $n(t)$ and $h(t+1)$ to maximize their intertemporal utility function subject to the budget constraints. It can be proved that the optimal solution is characterized by the following solution

$$n(t) = \begin{cases} \dfrac{\gamma}{\tau_0 + \tau e(g(t+1))}, & \text{if } z(t) \geq \widetilde{z}, \\ \dfrac{1 - \widetilde{c}/z(t)}{\tau_0 + \tau e(g(t+1))}, & \text{if } z(t) < \widetilde{z}, \end{cases} \tag{7.5.3}$$

$$\begin{cases} e(t+1) = 0, & \text{if } g(t+1) \leq \widetilde{g}, \\ e(t+1) = e(g(t+1)), \ e' > 0 & \text{if } g(t+1) > \widetilde{g}, \end{cases} \tag{7.5.4}$$

where $e(g)$ is an implicit function between e and g, i.e.

$$(\tau_0 + \tau e)h_e(e, g) = \tau h(e, g).$$

It is assumed that $e'' < 0$ for any $g(t) > \tilde{g}$.

We have described the behavior of the producers and consumers. We now describe technological change by the following equation[14]

$$g(t+1) \equiv \frac{A(t+1) - A(t)}{A(t)} = g(e(t)), \ \ g(0), \ \ g' > 0, \ \ g'' < 0. \tag{7.5.5}$$

The size of working population at time $t + 1$ is determined by

$$N(t+1) = n(t)N(t), \tag{7.5.6}$$

where N_0 is historically given. Utilizing

$$x(t) = \frac{A(t)X}{N(t)},$$

and equations (7.5.5) and (7.5.6), we have

$$x(t+1) = \frac{1 + g(t+1)}{n(t)} x(t).$$

Substituting equations (7.5.3) and (7.5.5) into the above equation yields

$$x(t+1) = \begin{cases} \dfrac{[\tau_0 + \tau e(g(e(t)))][1 + g(e(t))]}{\gamma} x(t), & \text{if } z(t) \geq \tilde{z}, \\ \dfrac{[\tau_0 + \tau e(g(e(t)))][1 + g(e(t))]}{1 - \tilde{c}/z(t)} x(t), & \text{if } z(t) < \tilde{z} \end{cases} \tag{7.5.7}$$

The construction of the model is thus completed. The system consists of equations (5.5.2)-(7.5.7). In the dynamical analysis, the economy is divided into two regimes:

---

[14] It should be noted that Galor and Weil also discussed implications of introducing $N(t)$ into the technological progress function g.

the subsistence regime characterized by $z(t) \le \tilde{z}$ and modern regime characterized by $z(t) > \tilde{z}$. Although the analysis is not complicated, it will take a long space to examine. The reader is encouraged to analyze the behavior of model, and then to read the analysis by Galor and Weil.

7.6 Unemployment, inflation and chaos

This section introduces a worker flow model with a nonlinear outflow rate from unemployment. The model is proposed by Neugart.[15] The unemployment $U(t)$ at period t follows the following identity

$$U(t+1) - U(t) = i(L - U(t)) - o(t)U(t), \tag{7.6.1}$$

where $i > 0$ denotes the (exogenous) inflow rate, $L = 1$ the labor force, and $o(t)$ is the outflow rate from unemployment. Here, the outflow rate is defined as the fraction of jobs that come to the market at time t to jobs searchers, i.e.

$$o(t) = \frac{J(t)}{U(t) + d(1 - U(t))},$$

where the parameter $0 < d < 1$ gives the on-the-job searchers as a constant fraction of employed workers and $J(t)$ denotes job creations.[16] We specify

$$J(t) = J_s + \gamma(m - \pi(t)),$$

where J_s is the job creation due to the structural characteristics of the economy and $\gamma(m - \pi(t))$ describes the cyclical component of job creations, where γ is a positive parameter, m the exogenous money growth rate, and $\pi(t)$ the inflation rate at t. Hence, the outflow rate from the unemployment can be expressed

[15] See Neugart (2004).

[16] It should be remarked that in Neugart (2000), the parameter is treated as an endogenous variable; in the appendix to Neugart (2004), the parameter i is treated as an endogenous variable.

$$o(t) = \frac{J_s + \gamma(m - \pi(t))}{U(t) + d(1 - U(t))}.$$

Substituting the above equation into equation (7.6.1) yields

$$U(t + 1) = i - \left[(1 - i) + \frac{J_s + \gamma(m - \pi(t))}{U(t) + d(1 - U(t))}\right]U(t) \equiv \Lambda(U(t), \pi(t)). \tag{7.6.2}$$

Assume that firms cannot raise prices to the same extent as nominal wages increase at t. This assumption is reflected in the requirement of $\delta > 1$ in the following formation of inflation rate dynamics

$$\pi(t) = \frac{1}{\delta}\left(\pi^e(t) + \frac{w_b(t) - w_p}{w_p}\right), \tag{7.6.3}$$

where $\pi(t)$ is the inflation rate at t, $\pi^e(t)$ is the expected inflation rate at t, $w_b(t)$ is the bargained wage rate at t, and w_p is the price determined real wage (which equals $(1 - \mu)y$, where $0 < \mu < 1$ is the fixed mark-up and $y = 1$ is the constant marginal productivity).

The bargained real wage $w_b(t)$ is given by

$$w_b(t) = y - (1 - b)U(t),$$

where $0 < b < 1$ is the "reservation wage". The inflation expectation is formed by

$$\pi^e(t) = a\pi(t - 1) + (1 - a)\pi^e(t - 1),$$

where $0 \le a \le 1$ is the weighted average rate. Under these assumptions, we may rewrite equation (7.6.3) as

$$\pi(t + 1) = \frac{1}{\delta}\left(\pi^e(t + 1) + \frac{\mu - (1 - b)U(t + 1)}{1 - \mu}\right). \tag{7.6.4}$$

From this equation and

$$\pi^e(t+1) = a\pi(t) + (1-a)\pi^e(t),$$

we have

$$\delta\pi(t+1) = a\pi(t) + (1-a)\pi^e(t) + \frac{\mu - (1-b)U(t+1)}{1-\mu}.$$

Substituting this equation into equation (7.6.4) for t, we solve

$$\pi(t+1) = \frac{a\mu}{\mu_0\delta} + \left(\frac{a}{\delta} + (1-a)\right)\pi(t) - \frac{ab_0}{\mu_0\delta}\Lambda(U(t), \pi(t)) \equiv \Phi(U(t), \pi(t)), \tag{7.6.5}$$

where

$$b_0 \equiv 1 - b, \quad \mu_0 \equiv 1 - \mu.$$

The dynamical system consists of equations (7.6.2) and (7.6.5), that is

$$U(t+1) = \Lambda(U(t), \pi(t)),$$
$$\pi(t+1) = \Phi(U(t), \pi(t)). \tag{7.6.6}$$

We may analyze its behavior in the same way as we did for the Hénon map in the previous chapter.

A fixed point is given by

$$U = \Lambda(U, \pi), \quad \pi = \Phi(U, \pi).$$

It is imposed that at a steady state, $\pi^* = m$, the inflation equals the money growth rate. It can be shown that the system has a unique steady state

$$(U^*, \pi^*) = \left(\frac{\mu + m\mu_0(1-\delta)}{b_0}, m\right),$$

provided that

$$0 \le \mu + m\mu_0(1-\delta) \le b_0.$$

The Jacobian matrix J is given by

$$J = \begin{bmatrix} a_{11} & a_{12} \\ a_{21} & a_{22} \end{bmatrix},$$

where

$$a_{11} \equiv \frac{\partial \Lambda(U,\pi)}{\partial U}\Big|_{(U^*,\pi^*)} = 1 - i - \frac{(1-U^*)id}{(U^* + d(1-U^*))U^*},$$

$$a_{12} \equiv \frac{\partial \Lambda(U,\pi)}{\partial \pi}\Big|_{(U^*,\pi^*)} = \frac{\gamma U^*}{U^* + d(1-U^*)},$$

$$a_{21} \equiv \frac{\partial \Phi(U,\pi)}{\partial U}\Big|_{(U^*,\pi^*)} = \frac{b_0}{\delta\mu_0}(1 - a - a_{11}),$$

$$a_{22} \equiv \frac{\partial \Phi(U,\pi)}{\partial U}\Big|_{(U^*,\pi^*)} = \frac{1}{\delta}\left(a + (1-a)\delta - \frac{b_0 a_{12}}{\mu_0}\right).$$

It is straightforward to calculate the two eigenvalues. We now provide some numerical examples carried out by Neugart. As discussed by Neugart, it is reasonable to choose the following parameters

$$a = 0.5, \quad \delta = 2, \quad b = 0.5, \quad \gamma = 0.5. \tag{7.6.7}$$

Under (7.6.7), if we further specify

$$d = 0.01, \quad m = 0.03, \quad i \approx 0.13199,$$

then one eigenvalue is equal to -1; a period-doubling bifurcation occurs. It can be shown that under (7.6.7), the eigenvalue with a positive root never crosses 1. Figure 7.6.1 depicts bifurcation diagram of U over i under (7.6.7) and

$$d = 0.01, m = 0.03.$$[17]

[17] The plot and the following one are from Neugart (2004). Neugart provides more charts for different values of the parameters.

Increases from $i \approx 0.13199$ will lead to further bifurcations. Mathematically, the dynamic mechanism is similar to bifurcations observed for the Hénon map.

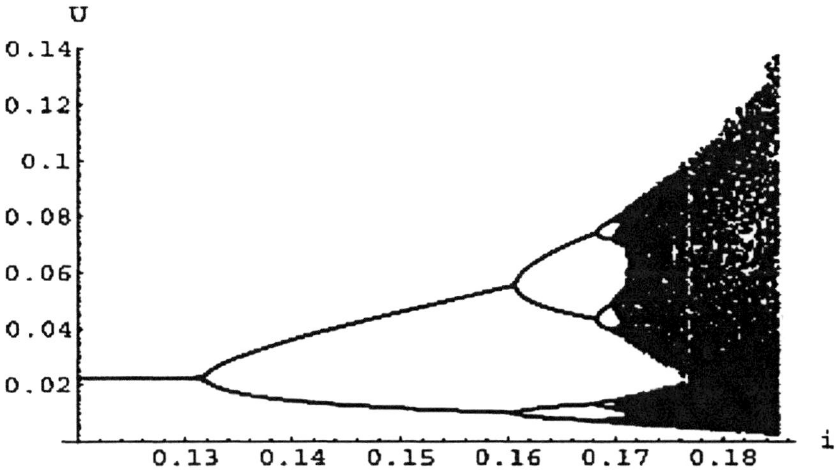

Figure 7.6.1: Bifurcation diagram of unemployment over the inflow rate

Figure 7.6.2 demonstrates the existence of an attractor with the inflation rate over the unemployment rate when the parameters are specified as in (7.6.7) and

$$d = 0.01,\ m = 0.03,\ i = 0.18.$$

The attractor looks like a Phillips curve. Nevertheless, in this model, there is not a stable trade-off between the inflation rate and unemployment rate. The system shifts up and down a negatively sloped "curve" erratically. A pair of inflation-unemployment rates today cannot tell where the economy will be in the long term.

7.7 Business cycles with money and capital

This section provides a model of the OLG model with money and capital.[18]

[18] The model and analysis in this section are based on Jullien (1988). A model in the same spirit is proposed by Grandmont (1985), even though Grandmont's model does not take account of capital accumulation. See also Tirole (1985), Farmer (1986), and von Thadden (1999). As far as technical issues related to existence of chaos are concerned, the reader is also referred to Yokoo (2000).

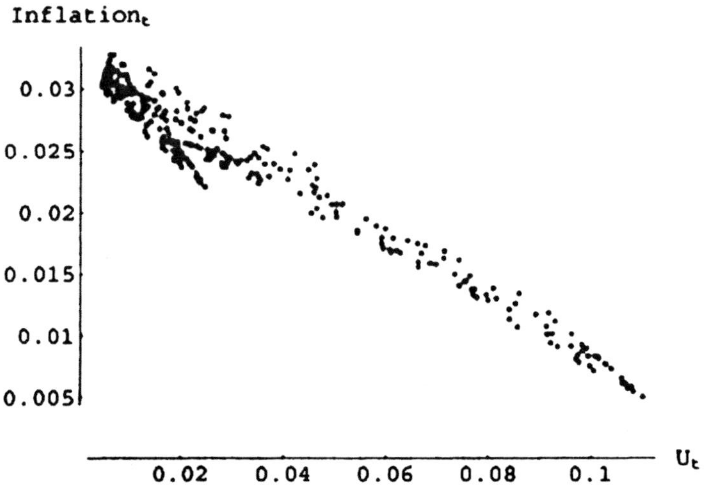

Figure 7.6.2: An attractor

A single good is produced and each agent lives two periods. The young generation sells one unit of labor at a real wage, $w(t)$, consumes quantity, $c_1(t)$, saves the real quantity, $s(t)$, by holding money and capital. Let $r(t+1)$ denote the real rate of interest between t and $t+1$. The old generation spends all its savings from the previous period on consumption, $c_2(t)$. A typical consumer maximizes his utility subject to the budget constraint as follows

$$\begin{aligned} &Max\ U(c_1(t), c_2(t)) \\ &\text{s.t.:}\ \begin{array}{l} c_1(t) + s(t) \le w(t), \\ c_2(t) \le r(t+1)s(t), \quad c_i(t) \ge 0, \quad i = 1, 2. \end{array} \end{aligned}$$

Under standard assumptions, the consumer's decision problem has a unique solution characterized by the saving function $s(w(t), r(t+1))$, which satisfies $0 < s(w, r) < w$.

Production is made through a neoclassical constant return to scale technology. Output per capita, $y(t)$, is a function of capital intensity, $k(t)$, i.e.

$$y(t) = f(k(t)),$$

where f is increasing, strictly concave on R_+ and C^2 on R_{++}, and satisfies

$$\lim_{k\to\infty} f'(k) \in [0, 1[, \quad \lim_{k\to 0} f'(k) = \infty,$$

$kf'(k)$ is non-decreasing, and

$$\lim_{k\to\infty} f(k) - kf'(k) = +\infty,$$
$$\lim_{k\to 0} f(k) - kf'(k) = 0.$$

Obviously, if

$$f = Ak^{\alpha}, 0 < \alpha < 1,$$

the above properties of f are satisfied. The CES production functions with an elasticity of substitution larger than 1 also have these properties. We also have

$$r(t) = f'(k(t)),$$
$$w(t) = f(k(t)) - k(t)f'(k(t)) \equiv W((k(t))).$$

It is known that W is C^1 on R_{++}, increasing, and maps R_+ into itself.

At date 0, the economy is endowed with a fixed quantity of capital k_0 and a quantity of money M. Let $p(t)$ and $m(t)$ ($M / p(t)$) denote respectively the price of goods and the real money. *A perfect foresight equilibrium* is a sequence of

$$\{p(t)\}_{t\geq\infty}, \ \{k(t)\}_{t\geq\infty}, \ \{r(t)\}_{t\geq\infty}, \ \{w(t)\}_{t\geq\infty},$$

which achieves a competitive temporary equilibrium with perfect foresight at each date. A perfect foresight equilibrium is described by a sequence of $\{k(t)\}_{t\geq\infty}$ and $\{m(t)\}_{t\geq\infty}$ such that

$$m(t) + k(t+1) = S(W(k(t)), f'(k(k+1))), \tag{7.7.1}$$
$$m(t+1) = f'(k(t+1))m(t), \ k(t) > 0, \ m(t) \geq 0, \ k_{00}, \text{ given.} \tag{7.7.2}$$

Equation (7.7.1) equalizes the demand and supply of assets; equation (7.7.2) equalizes the interest rate on money $p(t)/ p(t+1)$ and interest rate on capital $f'(k(t+1))$. We call an equilibrium non-monetary or monetary according to

$m_0 = 0$ or $m_0 > 0$. By equations (7.7.1) and (7.7.2), we can replace equation (7.7.1) with

$$m(t+1) + k(t+1)f'(k(k+1)) = S(W(k(t)), f'(k(k+1)))f'(k(k+1)).$$

It can be shown that the above equation determines a backward dynamics as

$$k(t) = \Lambda(k(t+1), m(t+1)). \tag{7.7.3}$$

According to the properties of the saving and production functions, it can be proved that Λ is C^1, increasing in each of its arguments, and Λ tends to 0 (resp. infinity) when k goes to 0 (resp. infinity) with the fixed m. A perfect foresight equilibrium can be described by a two-dimensional (backward) mapping

$$(k(t), m(t)) = \left(\Lambda(k(t+1), m(t+1)), \frac{m(t+1)}{f'(k(t+1))} \right) \equiv \Omega(k(t+1), m(t+1)). \tag{7.7.4}$$

We may express the system in vector form as

$$x(t) = \Omega(x(t+1)),$$

where $x \equiv (k, m)$. Here, we are interested in a monetary economy. A steady state (k^*, m^*) is determined by $f'(k^*) = 1$ and

$$m^* + k^* = s(W(k^*), 1).$$

We assume $s(W(k^*), 1) > k^*$. Then the conditions for the existence of a unique monetary steady state are satisfied. Under this assumption, there exists at least one inefficient non monetary steady state ($k > k^*$)[19] and we denote the lowest capital intensive one by k_s, i.e.

---

[19] A non-monetary steady state k verifies $s(W(k), f'(k)) = k$. It is inefficient if $k > k^*$, where $s(W(k^*), 1) > k^*$. For sufficiently large k, $W(k) < k$. Hence, for sufficiently large k, we have $s(W(k), f'(k)) < W(k) < k$. As s is continuous and it maps itself, it has at least one fixed point.

$$k_s = \inf\{k > k^* \mid \Lambda(k, 0) = k\}.$$

Here, we are interested in periodic perfect foresight equilibrium. To analyze global behavior of the system, it can be proved that all the periodic orbits must belong to a C^1 invariant curve. This will enable us to focus the dynamical analysis on the curve in one-dimension.

Theorem 7.7.1. There exists a compact set $K \in R_{++}^2$ and a function h, decreasing and C^1, from R_{++} to R_{++} such that if we define the sets

$$\Gamma = \{(k, m) \in R_+^2 \mid k = h(m)\},$$
$$\Gamma_+ = \{(k, m) \in R_+^2 \mid k > h(m)\},$$
$$\Gamma_- = \{(k, m) \in R_+^2 \mid k < h(m)\},$$

then $\{\Gamma, \Gamma_+, \Gamma_-\}$ is an invariant partition of R_{++}^2 and

$$x \in \Gamma_+ \Leftrightarrow \lim_{t \to \infty} m(t, x) = \infty,$$
$$x \in \Gamma_- \Leftrightarrow \lim_{t \to \infty} m(t, x) = 0,$$
$$x \in \Gamma \Leftrightarrow m(t, x) \in K \quad \text{for } t \text{ being sufficiently large.}$$

The proof is referred to Jullien. We depict the contents of the theorem as in figure 7.7.1. The real quantity of money increases infinitely along the orbit of a point greater than x^* and decreases to 0 along the orbit of a smaller point. Between these two behaviors it is possible to exhibit a set of points with an orbit bounded away from 0 and infinity. It can be shown that all cycles must belong to the curve Γ.

Introduce a one-dimensional map

$$\phi: R_{++} \to R_{++}, \quad m \to \frac{m}{f'(h(m))}.$$

The map, ϕ, is C^1. The cycles of ϕ are equivalent to the cycles of Ω in the senses of

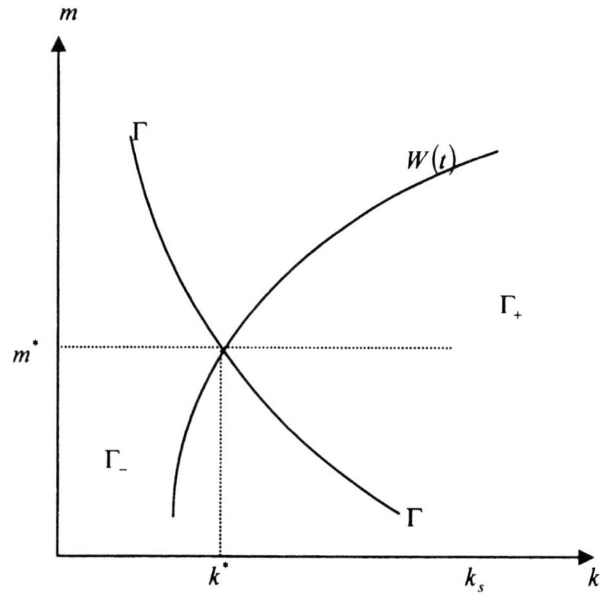

Figure 7.7.1: The partition of R_{++}

$$\Omega^{p}(x)=x \Leftrightarrow \phi^{m}(m)=m.$$

The following theorem is referred to Jullien.

Theorem 7.7.2. A sufficient condition for the existence of a cycle of order 2 is that

$$S(W(k^{*}),1)-k^{*}+2S_{R}(W(k^{*}),1)+2k^{*}S_{W}(W(k^{*}),1)-\frac{2}{f''(k^{*})}<0.$$

The proof is to verify the existence of solutions of

$$\phi^{2}(m)=m.$$

It should be noted that Jullien also discusses the conditions for the existence of periodic points of period 3, which guarantees economic chaos.

7.8 The OSG model with heterogeneous households

This section explores the interdependence between income distribution, wealth distribution, and economic growth within the OSG modeling framework.[20] The key feature of this theory is that it groups the population into different groups, whose consumption and saving behavior are homogenous within each group and are different among the groups.

The production side of our model is the same as in the OSG model in section 3.2. The population is classified into two groups, indexed by $j = 1, 2$. Each type of consumers has a fixed number, denoted by N_j. The aggregated labor force, N, is given by

$$N = h_1 N_1 + h_2 N_2,$$

where h_j are the level of human capital of group j, $j = 1, 2$. Production in period t uses inputs amount of capital, $K(t)$, and amount of labor services, N. It supplies amount of goods, $F(t)$. The production process is described by a neoclassical production function

$$F(t) = F(K(t), N).$$

The real interest rate and the wage of labor are given as before by

$$\begin{aligned} r(t) + \delta_k &= f'(k(t)), \\ w_j(t) &= h_j [f(k(t)) - k(t) f'(k(t))], \end{aligned} \tag{7.8.1}$$

where $k(t) \equiv K(t)/N$. Let $k_j(t)$ denote per capita wealth of group j in period t. According to the definitions, we have

$$K(t) = k_1(t) N_1 + k_2(t) N_2.$$

Dividing the two sides of the above equation by N yields

---

[20] The post-Keynesian approach is represented, for instance, by Kaldor (1940), Pasinetti (1960, 1974), Sato (1966), Samuelson and Modigliani (1966), Salvadori (1991), and Panico and Salvadori (1993). The recent literature on growth and distribution is referred to, for instance, Becker (1980), Sorger (2002), Nishimura and Shimomura (2002), Ghiglino and Sorger (2002), and Sorger (2000), and Zhang (2005c).

$$k(t) = k_1(t)n_1 + k_2(t)n_2,$$

where $n_j \equiv N_j / N$. As n_j are constant, we see that $k(t)$ is a function of $k_1(t)$ and $k_2(t)$. With equations (7.8.1), we see that $r(t)$ and $w_j(t)$ are functions of $k_1(t)$ and $k_2(t)$.

Group j 's per capita *current income,* $y_j(t)$, from the interest payment, $r(t)k_j(t)$, and the wage payment, $w_j(t)$, is defined by

$$y_j(t) = r(t)k_j(t) + w_j(t).$$

The per capita *disposable income* of consumer j is defined as the sum of the current income and the wealth available for purchasing consumption goods and saving

$$\hat{y}_j(t) = y_j(t) + k_j(t) = (1 + r(t))k_j(t) + w_j(t), \quad j = 1, 2.$$

The disposable income is used for saving and consumption. At each point of time, a consumer would distribute the total available budget among saving, $s_j(t)$, and consumption of goods, $c_j(t)$. The budget constraint is given by

$$c_j(t) + s_j(t) = \hat{y}_j(t). \tag{7.8.2}$$

In each period, consumers decide the two variables subject to the disposable income. We assume that utility level, $U_j(t)$, is dependent on the consumption level of commodity, $c_j(t)$, and the level of saving, $s_j(t)$, as follows

$$U_j(t) = c_j^{\xi_j}(t)s_j^{\lambda_j}(t), \quad \xi_j, \lambda_j > 0, \quad \xi_j + \lambda_j = 1, \quad j = 1, 2, \tag{7.8.3}$$

where ξ_j and λ_j are respectively group j 's propensities to consume and to hold wealth. Maximizing U_j subject to budget constraints (7.8.2) yields

$$c_j(t) = \xi_j\hat{y}_j(t), \quad s_j(t) = \lambda_j\hat{y}_j(t). \tag{7.8.4}$$

Per capita wealth of group j in period $t+1$ is equal to the saving made in period t. That is

$$k_j(t+1) = s_j(\hat{y}_j(t)), \quad j = 1, 2. \tag{7.8.5}$$

These mappings control the motion of the system. By equations (7.8.2) and (7.8.1), we have

$$\hat{y}_j(t) = h_j f(k(t)) + [k_j(t) - h_j k(t)] f'(k(t)) + \delta k_j(t), \quad j = 1, 2, \tag{7.8.6}$$

where $\delta \equiv 1 - \delta_k$. We see that $\hat{y}_j(t)$ are functions of $k_1(t)$ and $k_2(t)$.

As output is either consumed or saved, the sum of net saving and consumption equals output. That is

$$C(t) + S(t) - K(t) + \delta_k K(t) = F(t), \tag{7.8.7}$$

where $C(t)$ is the sum of consumption and

$$S(t) - K(t) + \delta_k K(t)$$

is the sum of net saving of the two groups

$$C(t) = \sum_j c_j(t) N_j,$$
$$S(t) - \delta K(t) = \sum_j (s_j(t) - \delta k_j(t)) N_j.$$

It can be shown that equation (7.8.7) is redundant in the sense that it can be derived from the other equations in the system. We thus omit equation (7.8.7) in later discussions.

As $\hat{y}_j(t)$ are functions of $k_1(t)$ and $k_2(t)$, we see that difference equations (7.8.5) determine the evolution of $k_1(t)$ and $k_2(t)$. It is straightforward to show that any other variable in the system is uniquely determined as a function of $k_1(t)$ and $k_2(t)$. It is sufficient for us to examine behavior of difference equations (7.8.5).

We are now concerned with dynamic properties of equations (7.8.5). Substituting $s_j(t) = \lambda_j \hat{y}_j(t)$ and equations (7.8.6) into equations (7.8.5) yields

$$k_j(t+1) = \lambda_j h_j f(k(t)) + [k_j(t) - h_j k(t)]\lambda_j f'(k(t)) + \lambda_j \delta k_j(t), \quad j = 1, 2. \tag{7.8.8}$$

A steady state is defined by

$$k_j = \lambda_j h_j f(k) + (k_j - h_j k)\lambda_j f'(k) + \lambda_j \delta k_j.$$

We rewrite the above equations as

$$k_j f'(k) + h_j(f(k) - kf'(k)) = \delta_j k_j,$$

where

$$\delta_j \equiv \frac{1}{\lambda_j} - \delta > 0, \quad j = 1, 2.$$

From the above equations, we exclude the possibility of $k_j = 0$ for any j for $k > 0$. Rearrange the above equations as follows

$$\left(\frac{f(k) - kf'(k)}{\delta_j - f'(k)}\right)h_j = k_j. \tag{7.8.9}$$

As f is strictly concave, $f(k) - kf'(k) > 0$ for $k > 0$. To guarantee $k_j > 0$, we should require

$$\delta_j - f'(k) > 0, \quad j = 1, 2.$$

Define

$$\bar{k} \equiv \max_j \{k : f'(k) = \delta_j\}.$$

Because

$$f''(k) < 0, \quad \lim_{k \to 0} f'(0) = \infty, \quad \lim_{k \to +\infty} f'(0) = 0,$$

we see that $\bar{k}$ is uniquely defined. Obviously

$$\delta_j - f'(k) > 0, \quad \text{for } k > \bar{k}, \quad j = 1, 2,$$

and $\delta_j - f'(k) < 0$ for some j if $k < \bar{k}$. We should require $k > \bar{k}$.

Multiplying the two sides of each equation by n_j and then adding the two equations in equations (7.8.9) yields

$$\left(\frac{n_1 h_1}{\delta_1 - f'(k)} + \frac{n_2 h_2}{\delta_2 - f'(k)}\right)\left(\frac{f(k)}{k} - f'(k)\right) = 1, \ k > \bar{k}, \tag{7.8.10}$$

where we use $n_1 k_1 + n_2 k_2 = k$ and neglect the trivial solution $k = 0$. We now show that this equation has a unique solution for $k > \bar{k}$. Introduce

$$\Phi(k) \equiv \left(\frac{n_1 h_1}{\delta_1 - f'(k)} + \frac{n_2 h_2}{\delta_2 - f'(k)}\right)\left(\frac{f(k)}{k} - f'(k)\right) - 1. \tag{7.8.11}$$

As $\Phi(k) \to +\infty$ as $k \to \bar{k}$ from the right and $\Phi(+\infty) \to -1$, we conclude that $\Phi(k) = 0$ has at least one solution for $k \in (\bar{k}, +\infty)$. Here, we use

$$\frac{f(k)}{k} \to 0 \text{ as } k \to \infty,$$

$$\lim_{k\to\infty} \frac{f(k)}{k} = \frac{\lim_{k\to\infty} f'(k)}{\lim_{k\to\infty} 1} = 0.$$

We calculate

$$\Phi' = \left[\frac{n_1 h_1}{(\delta_1 - f'(k))^2} + \frac{n_2 h_2}{(\delta_2 - f'(k))^2}\right]\left(\frac{f(k)}{k} - f'(k)\right) f''(k)$$
$$-\left(\frac{n_1 h_1}{\delta_1 - f'(k)} + \frac{n_2 h_2}{\delta_2 - f'(k)}\right)\frac{\Delta(k)}{k^2}. \tag{7.8.12}$$

We see that if

$$\Delta(k) \equiv f - kf' + k^2 f'' \geq 0,$$

then $\Phi'(k) \leq 0$ for $k \in (\bar{k}, +\infty)$. It should be noted that even if the term

$$f - kf' + k^2 f''$$

is negative, $\Phi'(k) \leq 0$ may hold. If

$$\Delta(k) \geq 0, \quad k \in (\bar{k}, +\infty), \tag{7.8.13}$$

the equation $\Phi(k) = 0$ has a unique solution. It is straightforward to check that if F takes on the Cobb-Douglas form, then (7.8.13) holds. If the production function takes on the CES form

$$f = A(ak^\sigma + 1)^{1/\rho},$$

where $\sigma < 1$, and a and A are positive, we calculate

$$\Delta(k) = \left(\frac{1 + (\sigma - 1)^2 ak^\sigma}{ak^\sigma + 1}\right) f > 0.$$

We see that when the production function takes on the Cobb-Douglass form or the CES form, the system has a unique equilibrium. For any positive solution $k^* \ (> \bar{k})$, we solve k_1 and k_2 by equations (7.8.9). We get r and w_j by equations (7.8.1), $\hat{y}_j$ by equations (7.8.6), and c_j and s_j by equations (7.8.4).

Proposition 7.8.1. The dynamic system has at least one equilibrium point. If

$$\Delta(k) \geq 0, \quad k \in (\bar{k}, +\infty),$$

the system has a unique equilibrium point, denoted by k^*. If

$$\left(\left(1 + \frac{1}{h_2 n_2}\right)\lambda_1 \lambda_2 + \frac{h_1 n_1}{h_2 n_2} \lambda_2^2\right)\left(f'(k^*) + \delta\right) > \lambda_1 - \lambda_2, \tag{7.8.14}$$

then the unique equilibrium point is stable.

The proof of the stability condition is provided in Appendix A.7.1.

If the difference between $\lambda_1 - \lambda_2$ is small, then the system is stable. For instance, as $f' \geq 0$, we see that if

$$\lambda_1 \lambda_2 \delta > \lambda_1 - \lambda_2,$$

condition (7.8.14) always holds. It should be noted that even if condition (7.8.14) is not held, it is possible that the unique equilibrium point is stable.

From equations (7.8.9), we get

$$\frac{k_1}{h_1} - \frac{k_2}{h_2} = \frac{(\lambda_1 - \lambda_2)(f - kf')}{\lambda_1 \lambda_2 (\delta_1 - f')(\delta_2 - f')}.$$

As $f - kf'$ and $\delta_j - f'$ are positive at the equilibrium point, we see that the sign of $k_1 / h_1 - k_2 / h_2$ is the same as that of $\lambda_1 - \lambda_2$. This relation tells that if one group's propensity to save is higher than the other group, then wealth per qualified capita of the former is greater than that of the latter. Hence, in the long term, differences in wealth per qualified capita are determined by differences in propensities to save among different groups. It should be noted that the sign of $k_1 - k_2$ is not determined solely by $\lambda_1 - \lambda_2$. For instance, if $\lambda_1 - \lambda_2 > 0$ and $h_2 - h_1 > 0$, $k_1 - k_2 < 0$ is possible. In fact, from equations (7.8.9), we have

$$k_1 - k_2 = \left(\frac{\lambda_1 h_1 - \lambda_2 h_2}{\lambda_1 \lambda_2} - (h_1 - h_2)(1 + f' - \delta_k) \right) \frac{(f - kf')}{(\delta_1 - f')(\delta_2 - f')}.$$

At equilibrium, we have $\hat{y}_j = k_j / \lambda_j$. From equations (7.8.9), we have

$$\hat{y}_1 - \hat{y}_2 =$$
$$((h_1 - h_2)(1 + \delta_k) - (h_1 \lambda_2 - h_2 \lambda_1)(1 + f')) \frac{(f - kf')}{(\delta_k + \xi_2 - f')(\delta_k + \xi_1 - \lambda_1 f')}.$$

We can obtain explicit conclusions about the signs of differences in the key variables only when one group does not have lower level of the propensity to save and lower level of human capital than the other group. That is $\lambda_1 \geq \lambda_2$ and

$h_1 \geq h_2$. Otherwise, the comparative economic conditions of the two groups are ambiguous.

We now examine impact of changes in one group's propensity to save and level of human capital upon the income and wealth distribution and national economic product. Taking derivatives of equation (7.8.12) with respect h_1 yields

$$-\Phi' \frac{dk}{dh_1} = \frac{(\lambda_1 - \lambda_2) n_1 n_2 h_2}{(\delta_1 - f')(\delta_2 - f')\lambda_1 \lambda_2} \left(\frac{f}{k} - f' \right), \tag{7.8.15}$$

where $\Phi' < 0$. It can be shown that if $\lambda_1 > (<) \lambda_2$, we have

$$\frac{f}{k} - \delta_2 < (>) \, 0, \quad \frac{f}{k} - \delta_1 > (<) \, 0,$$

and if $\lambda_1 = \lambda_2$, we have $f / k - \delta_j = 0$. Since

$$\frac{f}{k} - f' > 0, \quad \delta_j - f' > 0,$$

from equation (7.8.15) we conclude that $dk / dh_1 > (=, <) \, 0$ if $\lambda_1 > (<) \lambda_2$. As group 1 's level of human capital is increased, if group 1 's propensity to save is higher (lower) than group 2 's, then k increases (decreases). We see that improvement in one group's human capital does not necessarily lead to increases in the wealth per qualified capita in the economy. The difference in the preferences affects change directions in k. From equations (7.8.9), we obtain

$$\frac{dk_1}{dh_1} = \frac{k_1}{h_1} + \left(\frac{f}{k} - \delta_1 \right) \frac{k^2 f'' h_1}{(\delta_1 - f')^2} \frac{dk}{dh_1},$$

$$\frac{dk_2}{dh_1} = \left(\frac{f}{k} - \delta_2 \right) \frac{k^2 f'' h_2}{(\delta_2 - f')^2} \frac{dk}{dh_1}.$$

The following lemma summarizes our analytical conclusions.

Lemma 7.8.1. As group 1 's human capital is increased, (i) if group 1 's propensity to save is higher than group 2 's (i.e., $\lambda_1 > \lambda_2$), group 1 's wealth per capita may increase or decrease and group 2 's wealth per capita increases; (ii) if $\lambda_1 = \lambda_2$, then group 1 's wealth per capita increases but group 2 's wealth per capita is not affected; and (iii) if $\lambda_1 < \lambda_2$, group 1 's wealth per capita may be increased or decreased and group 2 's wealth per capita increases.

We see that as group 1 becomes more effective, group 2 's wealth per capita increases, at least does not decrease, in the long term. This conclusion does not hold when the economy has more than 2 groups.

From

$$K = kN, \quad F = fN, \quad r + \delta_k = f', \quad w_j = h_j(f - kf'),$$
$$s_j = k_j, \quad \hat{y}_j = \frac{k_j}{\lambda_j}, \quad c_j = \frac{\xi_j k_j}{\lambda_j},$$

we get

$$\frac{1}{K}\frac{dK}{dh_1} = \frac{1}{k}\frac{dk}{dh_1} + \frac{N_1}{N},$$
$$\frac{1}{F}\frac{dF}{dh_1} = \frac{f'}{f}\frac{dk}{dh_1} + \frac{N_1}{N},$$
$$\frac{dr}{dh_1} = f''\frac{dk}{dh_1},$$
$$\frac{dw_1}{dh_1} = (f - kf') - h_1 kf''\frac{dk}{dh_1},$$
$$\frac{dw_2}{dh_1} = -\, h_2 kf''\frac{dk}{dh_1},$$
$$\frac{1}{\hat{y}_j}\frac{d\hat{y}_j}{dh_1} = \frac{1}{c_j}\frac{dc_j}{dh_1} = \frac{1}{k_j}\frac{dk_j}{dh_1}.$$

From lemma 7.8.1, we can judge the signs of these variables. We note that the rate of interest may be either increased or decreased when group 1 's human capital is improved. This will be confirmed by simulation results later on. We obtain the following results.

Lemma 7.8.2. As group 1 's propensity to save increases, (i) if group 1 's propensity to save is higher than group 2 's (i.e., $\lambda_1 > \lambda_2$), group 1 's wealth per capita may be either increased or decreased and group 2 's wealth per capita increases; (ii) if $\lambda_1 = \lambda_2$, group 1 's wealth per capita increases and group 2 's wealth per capita is not affected; and (iii) if $\lambda_1 < \lambda_2$, group 1 's wealth per capita increases and group 2 's wealth per capita decreases.

Lemma 7.8.3. As group 1 's population increases, (i) if $\lambda_1 \neq \lambda_2$, group 1 's wealth per capita decreases and group 2 's wealth per capita increases; (ii) if $\lambda_1 = \lambda_2$, wealth per capita is not affected.

We now extend the two-group model to multiple groups of consumers. We may generally assume that the population is N_0 and the population can be classified into M groups, indexed by j, according to their preferences, wealth, human capital, and social status. We have $N_0 \geq M$. Two extreme cases are $M = N_0$ and $M = 1$. Let the number of group j be N_j. The aggregated labor force N is given by

$$N_0 = \sum_{j=1}^{M} h_j N_j,$$

where h_j are the level of human capital. We may also interpret the parameter h_j as the work time of group j.

The neoclassical production function is $F(t) = F(K(t), N)$. Let K, r, w_j, and δ_k $(0 \leq \delta_k < 1)$ be defined as before. The marginal conditions are then given by

$$r(t) + \delta_k = f'(k(t)),$$
$$w_j(t) = h_j w(t),$$
$$w(t) \equiv f(k(t)) - k(t) f'(k(t)),$$

where $k \equiv K / N$. The per capita *disposable income* of group j is

$$\hat{y}_j(t) = y_j(t) + k_j(t), \quad j = 1, 2, \cdots, M,$$

where

$$y_j(t) = r(t) k_j(t) + w_j(t).$$

The budget constraint is given by

$$c_j(t) + s_j(t) = \hat{y}_j(t).$$

Consumer j maximizes the utility level

$$U_j(t) = c_j^{\xi_j}(t) s_j^{\lambda_j}(t), \quad \xi_j, \lambda_j > 0, \quad j = 1, 2, \cdots, M,$$

subject to the budget constraints. The optimal solutions are

$$c_j(t) = \xi_j \hat{y}_j(t), \quad s_j(t) = \lambda_j \hat{y}_j(t).$$

As equations (7.8.8), we describe the wealth accumulation by

$$k_j(t+1) = \lambda_j h_j f(k(t)) + [k_j(t) - h_j k(t)]\lambda_j f'(k(t)) + \lambda_j \delta k_j(t),$$
$$j = 1, 2, ... M.$$

A steady state is defined by

$$k_j = \lambda_j h_j f(k) + (k_j - h_j k)\lambda_j f'(k) + \lambda_j \delta k_j.$$

We rewrite the above equations as

$$k_j f'(k) + h_j (f(k) - k f'(k)) = \delta_j k_j,$$

where $\delta_j \equiv 1/\lambda_j - \delta > 0$. Rearrange the above equations as follows

$$\left(\frac{f(k) - k f'(k)}{\delta_j - f'(k)} \right) h_j = k_j. \qquad (7.8.16)$$

Define

$$\bar{k} \equiv \max_j \{k : f'(k) = \delta_j\}.$$

As before, we can show that $\bar{k}$ is uniquely defined. We require $k > \bar{k}$. Multiplying the two sides of each equation by n_j and then adding the two equations in equations (7.8.16) yields

$$\left(\frac{f(k)}{k} - f'(k)\right)\sum_j \frac{n_j h_j}{\delta_j - f'(k)} = 1, \quad k > \bar{k}, \tag{7.8.17}$$

where we use

$$\sum_j n_j k_j = k$$

and neglect the trivial solution $k = 0$. Like equation (7.8.11), we introduce

$$\Phi(k) \equiv \left(\frac{f(k)}{k} - f'(k)\right)\sum_j \frac{n_j h_j}{\delta_j - f'(k)} - 1.$$

If

$$\Delta(k) \geq 0, \quad k \in (\bar{k}, +\infty),$$

the equation $\Phi(k) = 0$ has a unique solution.

Proposition 7.8.2. The dynamic system has at least one equilibrium point. If $\Delta(k) \geq 0, \ k \in (\bar{k}, +\infty)$, the system has a unique equilibrium.

We omit analyzing stability properties of the equilibrium because it is difficult to find explicit conclusions. The model of three groups is simulated with a wide range of the parameters and the unique equilibrium is stable. Nevertheless, I did not succeed in constructing a Liapunov function to confirm the stability. For illustration, we simulate the model when $M = 3$. The aggregated labor force N is

$$N = \sum_{j=1}^{3} h_j N_j$$

where h_j is the level of human capital of group j, $j = 1, 2, 3$. Suppose that the production function takes the Cobb-Douglas form. The production function and marginal conditions are

$$F = AK^{\alpha}N^{\beta}, \quad r = \frac{\alpha F}{K}, \quad w_j = \frac{\beta h_j F}{N},$$

where A is the total productivity. Let $K_j(t)$ and $Y_j(t)$ be respectively the capital stocks owned by group j and the current income level of group j. The total capital $K(t)$ and total current income $Y(t)$ are

$$K(t) = \sum_1^3 K_j(t), \quad Y(t) = \sum_1^3 Y_j(t),$$

where

$$K_j(t) = k_j(t)N_j, \quad Y_j(t) = y_j(t)N_j.$$

The consumers' behavior are described by

$$c_j(t) = \xi_j \hat{y}_j(t), \quad s_j(t) = \lambda_j \hat{y}_j(t).$$

To simulate the model, we specify the groups' human capital, preference parameters and the other parameters as follows

$$\alpha = 0.34, \quad A = 1.3, \quad \delta_k = 0.05,$$
$$\begin{pmatrix} h_1 \\ h_2 \\ h_3 \end{pmatrix} = \begin{pmatrix} 3 \\ 1 \\ 0.3 \end{pmatrix}, \quad \begin{pmatrix} \lambda_1 \\ \lambda_2 \\ \lambda_3 \end{pmatrix} = \begin{pmatrix} 0.8 \\ 0.65 \\ 0.3 \end{pmatrix}, \quad \begin{pmatrix} N_1 \\ N_2 \\ N_3 \end{pmatrix} = \begin{pmatrix} 1 \\ 5 \\ 10 \end{pmatrix}. \tag{7.8.18}$$

Group 1 is the rich class – with the highest level of human capital and highest propensity to own wealth. The population share of the rich in the total population is only $1/16$ percent. Group 2 is the middle class - with the middle level of human capital and middle-level propensity to own wealth. The population of this group is $5/16$ percent. Group 3 has the lowest level of human capital and the lowest propensity to own wealth. Under (7.8.18), we calculate the equilibrium values of the variables

$$F = 18.424, \ k = 4.558, \ r = 0.1249,$$

$$\begin{pmatrix} k_1 \\ k_2 \\ k_3 \end{pmatrix} = \begin{pmatrix} 31.334 \\ 3.373 \\ 1.941 \end{pmatrix}, \ \begin{pmatrix} w_1 \\ w_2 \\ w_3 \end{pmatrix} = \begin{pmatrix} 3.317 \\ 1.105 \\ 0.333 \end{pmatrix}, \ \begin{pmatrix} c_1 \\ c_2 \\ c_3 \end{pmatrix} = \begin{pmatrix} 7.713 \\ 1.175 \\ 0.385 \end{pmatrix}.$$

We see that in the long term the differences in living conditions are great among the groups. Table 7.8.1 gives the distribution of wealth, the levels of current income and levels of consumption among the three groups. Group 1 's share of the population is only 6.25 percent of the total population, but its shares of wealth, current income and consumption are respectively 62.49, 9.25, and 38.30 percent; group 3 's share of population is 62.5 percent, its shares of wealth, current income and consumption are respectively 3.87, 19.11, and 19.12.

Table 7.8.1: Distribution of wealth, income, and consumption

	population	wealth	income	consumption
group 1	6.25 %	62.49 %	9.25 %	38.30 %
group 2	31.25 %	33.64 %	41.44 %	42.58 %
group 3	62.5 %	3.87 %	19.11 %	19.12 %

The three eigenvalues are given by

$$\rho_1 = 0.857, \ \rho_2 = 0.734, \ \rho_3 = 0.342.$$

The steady state is stable. Figure 7.8.1 illustrates the motion of per capita wealth of the three groups. We see that the growth process is characterized of convergence under perfect competition. We can illustrate dynamics of the other variables in the system. As the dynamic system has a unique equilibrium, we may examine impact of changes in the parameters. First, we examine impact of change in human capital. We fix the parameter values as in (7.8.18) except h_1. We increase h_1 from $h_1 = 3$ to $h_1 = 3.5$. We calculate the new equilibrium values as

$$\Delta F = 1.130, \ \Delta K = 23.458, \ \Delta r = -0.004,$$

$$h_1 : \ 3 \Rightarrow 3.5 \Rightarrow \begin{pmatrix} \Delta k_1 \\ \Delta k_2 \\ \Delta k_3 \end{pmatrix} = \begin{pmatrix} 4.557 \\ 0.014 \\ 0.003 \end{pmatrix}, \ \begin{pmatrix} \Delta w_1 \\ \Delta w_2 \\ \Delta w_3 \end{pmatrix} = \begin{pmatrix} 0.611 \\ 0.017 \\ 0.005 \end{pmatrix}, \ \begin{pmatrix} \Delta c_1 \\ \Delta c_2 \\ \Delta c_3 \end{pmatrix} = \begin{pmatrix} 1.122 \\ 0.007 \\ 0.005 \end{pmatrix}. \quad (7.8.19)$$

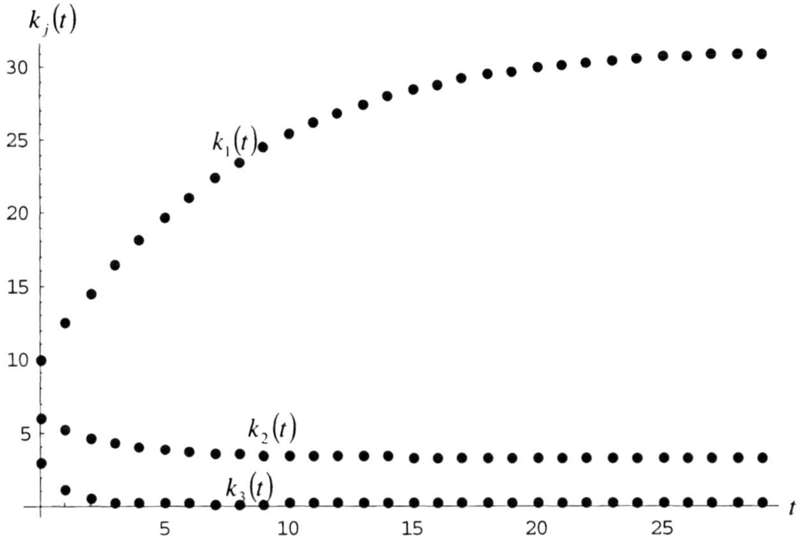

Figure 7.8.1: Dynamics of per capita wealth of the three groups

In (7.8.19), we denote the difference between the equilibrium values of the variables at the new equilibrium and old one state by Δ. From (7.8.19), we see that as group 1 's human capital is increased, the total output, the total wealth and per capita wealth of the three groups, the wage rates, and the levels of consumption are all increased. Hence, every group and the society as a whole benefit from human capital improvement of group 1. This property does not hold for the other two groups which have lower propensities to save and lower levels of human capital than group 1. As demonstrated in (7.8.20) which hold under (7.8.18) except for h_j, as group 2 's human capital is improved, group 3 which has the lowest level of human capital and lowest propensity to save will not benefit at all in terms of wage rate, per capita wealth and consumption level. As h_2 increases, group 1 and group 2 benefit but group 3 loses.

$$h_2 : \ 1 \Rightarrow 1.5 \Rightarrow$$

$$\Delta F = 3.874, \ \ \Delta K = 27.73, \ \ \Delta r = 0.004,$$

$$\begin{pmatrix} \Delta k_1 \\ \Delta k_2 \\ \Delta k_3 \end{pmatrix} = \begin{pmatrix} 0.600 \\ 1.167 \\ -0.002 \end{pmatrix}, \ \begin{pmatrix} \Delta w_1 \\ \Delta w_2 \\ \Delta w_3 \end{pmatrix} = \begin{pmatrix} -0.046 \\ 0.530 \\ -0.005 \end{pmatrix}, \ \begin{pmatrix} \Delta c_1 \\ \Delta c_2 \\ \Delta c_3 \end{pmatrix} = \begin{pmatrix} 0.148 \\ 0.848 \\ -0.005 \end{pmatrix},$$

$$h_3 : \ 0.3 \Rightarrow 0.8 \Rightarrow$$

$$\Delta F = 6.305, \ \ \Delta K = 26.250, \ \ \Delta r = 0.021,$$

$$\begin{pmatrix}\Delta k_1\\ \Delta k_2\\ \Delta k_3\end{pmatrix}=\begin{pmatrix}4.776\\ -0.046\\ 0.289\end{pmatrix},\ \begin{pmatrix}\Delta w_1\\ \Delta w_2\\ \Delta w_3\end{pmatrix}=\begin{pmatrix}-0.256\\ -0.085\\ 0.484\end{pmatrix},\ \begin{pmatrix}\Delta c_1\\ \Delta c_2\\ \Delta c_3\end{pmatrix}=\begin{pmatrix}1.176\\ -0.024\\ 0.574\end{pmatrix}. \tag{7.8.20}$$

This describes a situation that as the middle class gets more educated, the middle class becomes richer and the rich class becomes richer, the poor class becomes poorer. Similarly, as group 3 improves its human capital, group 2 's living conditions tend to deteriorate; but the other two groups' living conditions tend to be improved.

The effects of change in the technological parameter A are calculated as follows

$$\Delta F = 0.717,\ \Delta K = 24.765,\ \Delta r = -0.009,$$

$$A:\ 1.3 \Rightarrow 1.4 \Rightarrow \begin{pmatrix}\Delta k_1\\ \Delta k_2\\ \Delta k_3\end{pmatrix}=\begin{pmatrix}3.724\\ 0.401\\ 0.023\end{pmatrix},\ \begin{pmatrix}\Delta w_1\\ \Delta w_2\\ \Delta w_3\end{pmatrix}=\begin{pmatrix}0.129\\ 0.043\\ 0.013\end{pmatrix},\ \begin{pmatrix}\Delta c_1\\ \Delta c_2\\ \Delta c_3\end{pmatrix}=\begin{pmatrix}0.801\\ 0.161\\ 0.026\end{pmatrix}.$$

As A increases, the total output, the total wealth and per capita wealth of the three groups, the wage rates, and the levels of consumption are all increased. Every group and the society as a whole benefit from the technological change.

The effects of changes in the propensities to save of the three groups are illustrated as in (7.8.21):

$$\lambda_1:\ 0.8 \Rightarrow 0.82 \Rightarrow$$

$$\Delta F = 0.780,\ \Delta K = 25.308,\ \Delta r = -0.010,$$

$$\begin{pmatrix}\Delta k_1\\ \Delta k_2\\ \Delta k_3\end{pmatrix}=\begin{pmatrix}6.220\\ 0.042\\ 0.007\end{pmatrix},\ \begin{pmatrix}\Delta w_1\\ \Delta w_2\\ \Delta w_3\end{pmatrix}=\begin{pmatrix}0.140\\ 0.047\\ 0.014\end{pmatrix},\ \begin{pmatrix}\Delta c_1\\ \Delta c_2\\ \Delta c_3\end{pmatrix}=\begin{pmatrix}0.448\\ 0.021\\ 0.014\end{pmatrix},$$

$$\lambda_2:\ 0.65 \Rightarrow 0.68 \Rightarrow$$

$$\Delta F = 0.322,\ \Delta K = 21.430,\ \Delta r = -0.004,$$

$$\begin{pmatrix}\Delta k_1\\ \Delta k_2\\ \Delta k_3\end{pmatrix}=\begin{pmatrix}-0.651\\ 0.649\\ 0.003\end{pmatrix},\ \begin{pmatrix}\Delta w_1\\ \Delta w_2\\ \Delta w_3\end{pmatrix}=\begin{pmatrix}0.058\\ 0.019\\ 0.006\end{pmatrix},\ \begin{pmatrix}\Delta c_1\\ \Delta c_2\\ \Delta c_3\end{pmatrix}=\begin{pmatrix}-0.160\\ 0.088\\ 0.006\end{pmatrix},$$

$$\lambda_3:\ 0.3 \Rightarrow 0.4 \Rightarrow$$

$$\Delta F = 0.119, \ \Delta K = 19.764, \ \Delta r = -0.002,$$

$$\begin{pmatrix} \Delta k_1 \\ \Delta k_2 \\ \Delta k_3 \end{pmatrix} = \begin{pmatrix} -0.253 \\ 0.006 \\ 0.118 \end{pmatrix}, \begin{pmatrix} \Delta w_1 \\ \Delta w_2 \\ \Delta w_3 \end{pmatrix} = \begin{pmatrix} 0.021 \\ 0.007 \\ 0.002 \end{pmatrix}, \begin{pmatrix} \Delta c_1 \\ \Delta c_2 \\ \Delta c_3 \end{pmatrix} = \begin{pmatrix} -0.062 \\ 0.003 \\ 0.026 \end{pmatrix}. \tag{7.8.21}$$

As group 1 's propensity to save increases, the total output, the total wealth and per capita wealth of the three groups, the wage rates, and the levels of consumption are all increased. As group 2 's propensity to save increases, group 3 will benefit in terms of per capita wealth, wage rate, and consumption level but group 1 's consumption level and per capita wealth will be reduced. Similarly, an increase in group 3 's propensity will not benefit the rich group.

As each group's labor force is changed, the total output, wealth and distribution of wealth and income will be changed. We illustrates the effects of change in each group's labor force as in (7.8.22)

$$N_1: \ 1 \Rightarrow 1.2 \Rightarrow$$

$$\Delta F = 1.354, \ \Delta K = 24.386, \ \Delta r = -0.004,$$

$$\begin{pmatrix} \Delta k_1 \\ \Delta k_2 \\ \Delta k_3 \end{pmatrix} = \begin{pmatrix} -0.667 \\ 0.017 \\ 0.003 \end{pmatrix}, \begin{pmatrix} \Delta w_1 \\ \Delta w_2 \\ \Delta w_3 \end{pmatrix} = \begin{pmatrix} 0.060 \\ 0.020 \\ 0.006 \end{pmatrix}, \begin{pmatrix} \Delta c_1 \\ \Delta c_2 \\ \Delta c_3 \end{pmatrix} = \begin{pmatrix} -0.164 \\ 0.010 \\ 0.006 \end{pmatrix},$$

$$N_2: \ 5 \Rightarrow 5.5 \Rightarrow$$

$$\Delta F = 0.777, \ \Delta K = 20.602, \ \Delta r = 0.001,$$

$$\begin{pmatrix} \Delta k_1 \\ \Delta k_2 \\ \Delta k_3 \end{pmatrix} = \begin{pmatrix} 0.129 \\ -0.03 \\ -0.001 \end{pmatrix}, \begin{pmatrix} \Delta w_1 \\ \Delta w_2 \\ \Delta w_3 \end{pmatrix} = \begin{pmatrix} -0.010 \\ -0.004 \\ -0.001 \end{pmatrix}, \begin{pmatrix} \Delta c_1 \\ \Delta c_2 \\ \Delta c_3 \end{pmatrix} = \begin{pmatrix} 0.032 \\ -0.001 \\ -0.001 \end{pmatrix},$$

$$N_3: \ 10 \Rightarrow 11 \Rightarrow$$

$$\Delta F = 0.383, \ \Delta K = 19.225, \ \Delta r = 0.002,$$

$$\begin{pmatrix} \Delta k_1 \\ \Delta k_2 \\ \Delta k_3 \end{pmatrix} = \begin{pmatrix} 0.263 \\ -0.005 \\ -0.001 \end{pmatrix}, \begin{pmatrix} \Delta w_1 \\ \Delta w_2 \\ \Delta w_3 \end{pmatrix} = \begin{pmatrix} -0.021 \\ -0.007 \\ -0.002 \end{pmatrix}, \begin{pmatrix} \Delta c_1 \\ \Delta c_2 \\ \Delta c_3 \end{pmatrix} = \begin{pmatrix} 0.065 \\ -0.003 \\ -0.003 \end{pmatrix}. \tag{7.8.22}$$

We see that if the rich group has more labor force, the living conditions of all the groups will be improved. If group 2 or group 3 has more labor force, only group 1 benefits and both group 2 and group 3 lose in per capital terms. In our model, growth in the population reduces per capita output of the national economy; but this does not mean that every group's wealth and/or consumption level are reduced from the population growth. Some group(s) may benefit from the population growth.

7.9 Path-dependent evolution with education

This section builds a model of economic growth with endogenous physical capital and human capital accumulation.[21] Our model is built upon the three main models – Solow's one-sector growth model, Arrow's learning by doing model, and the Uzawa-Lucas's growth model with education - in the growth literature.[22] We will show that when the main mechanisms of economic growth in these three models are integrated into a single framework, they will produce certain economic phenomena that none of these models can produce.

Many aspects of the model are the same as the OSG model in section 3.3. The economy has an infinite future. We represent the passage of time in a sequence of periods, numbered from zero and indexed by $t = 0, 1, 2, \ldots$. There is one education sector in the system. We assume a homogenous and fixed national labor force N. The labor force is distributed between economic activities, teaching and study. We assume perfect competition in all markets and select commodity to serve as numeraire, with all the other prices being measured relative to its price. We introduce

$F(t)$ = output level of the production sector at time t;

$H(t)$ = level of the human capital of the population;

$N_i(t)$ and $K_i(t)$ = labor force and capital stocks employed by the production sector;

$N_e(t)$ and $K_e(t)$ = number of teachers and capital stocks employed by the education sector;

$w(t)$ and $r(t)$ = wage rate and rate of interest.

The production process is described by

$$F(t) = AK_i^{\alpha}(t)\left(H^m(t)TN_i(t)\right)^{\beta}, \quad A, \alpha, \beta > 0, \quad \alpha + \beta = 1,$$

where A, α and β are positive parameters and T is the work time. The marginal conditions are given by

[21] The model is a discrete version of the continuous model proposed by Zhang (2005d).

[22] See Arrow (1962) and Uzawa (1965). See also Zhang (2005a).

$$r(t) = (1 - \tau)\alpha A k_i^{-\beta}(t), \quad w(t) = (1 - \tau)\beta A H^m(t) k_i^\alpha, \tag{7.9.1}$$

where $k_i \equiv K_i / H^m T N_i$, δ_k is depreciation rate of capital, and τ is the tax rate on the producers.

We denote per capita wealth by $k(t)$, i.e., $k(t) \equiv K(t)/N$. Per capita current income from the interest payment, $r(t)k(t)$, and the wage payment, $w(t)$, is defined by

$$y(t) = r(t)k(t) + w(t).$$

We call $y(t)$ the *current income* in the sense that it comes from consumers' wages and current earnings from ownership of wealth. The per capita *disposable income* is defined as the sum of the current income and the wealth available for purchasing consumption goods and saving by

$$\hat{y}(t) = y(t) + k(t).$$

The disposable income is used for saving and consumption. In each period, a consumer would distribute the total available budget among saving, $s(t)$, consumption of goods, $c(t)$. The budget constraint is given by

$$c(t) + s(t) = \hat{y}(t).$$

At each point of time, households decide the two variables subject to the disposable income. We assume that households' utility, $U(t)$, is dependent on the consumption level of commodity, $c(t)$, and the saving, $s(t)$, as follows

$$U(t) = c^\xi(t) s^\lambda(t), \quad \xi, \lambda > 0, \quad \xi + \lambda = 1.$$

The consumer is to choose his most preferred bundle of consumption and saving, $(c(t), s(t))$, under his budget constraint. The optimal solution is given by

$$c(t) = \xi\hat{y}(t), \quad s(t) = \lambda\hat{y}(t). \tag{7.9.2}$$

According to the definitions, the household's wealth in period $t + 1$ is equal to the saving made in period t

$$k(t+1) = s(t) = \lambda\hat{y}(t). \tag{7.9.3}$$

As the disposable income is a function of the rate of interest, wage rate, and the wealth in period t, we see that the value of the right-hand side of equation (7.9.3) is determined in period t.

We now model the education sector and accumulation of human capital. It is assumed that there are two sources of improving human capital, through education and learning by producing. We will combine the two forces in a single equation. We propose that human capital dynamics is given by

$$H(t+1) - H(t) = \frac{\upsilon_e K_e^{\alpha_e}(t)\left(H^m(t)TN_e\right)^{\beta_v}\left(H^m(t)T_e N\right)^{\beta_e}}{N} + \frac{\upsilon_i F(t)}{NH^{\pi}(t)} - \delta_h H(t), \tag{7.9.4}$$

where T_e is the time used for education, δ_h (> 0) is the depreciation rate of human capital, υ_e, υ_i, α_e, β_v, and β_e are non-negative parameters. Equation (7.9.4) is a synthesis and generalization of Arrow's and Uzawa's ideas about human capital accumulation. The term $\upsilon_e K_e^{\alpha_e}\left(H^m TN_e\right)^{\beta_v}\left(H^m T_e N\right)^{\beta_e}$ describes the contribution to human capital improvement through education. We take account of learning by doing effects in human capital accumulation by the term $\upsilon_i F / H^{\pi}$.

We assume that education is free for students. The total tax income is used for paying the teachers and the capital stocks employed by the education sector. The education sector's budget is given by

$$w(t)TN_v(t) + \left(r(t) + \delta_k\right)K_e(t) = \tau F(t). \tag{7.9.5}$$

The education sector distributes its total resource $\tau F(t)$ to decide the number of teachers, $N_e(t)$, and the level of capital stocks, $K_e(t)$, to maximize the output of the education sector

$$\upsilon_e K_e^{\alpha_e}\left(H^m TN_e\right)^{\beta_v}\left(H^m T_e N\right)^{\beta_e}.$$

The education sector's problem is formed as follows

$$Max\ \upsilon_e K_e^{\alpha_e}(t)\left(H^m(t)TN_e\right)^{\beta_v}\left(H^m(t)T_e N\right)^{\beta_e},$$

subject to constraint (7.9.5). The optimal solution is given by

$$K_e(t) = \frac{\alpha_e \tau F(t)}{(\alpha_e + \beta_v)(r(t) + \delta_k)}, \quad N_e(t) = \frac{\beta_v \tau F(t)}{(\alpha_e + \beta_v) T w(t)}. \tag{7.9.6}$$

We see that K_e is negatively related to the rate of interest and positively to the education sector's total budget, and the number of teachers is negatively related to the wage rate and positively related to the total budget. Labor force and capital stocks are fully employed

$$N_i(t) + N_e(t) = N, \quad K_i(t) + K_e(t) = K(t).$$

We have thus built the model. We now examine properties of the dynamic system.

From equations (7.9.1) and (7.9.6), we obtain

$$K_e(t) = \frac{\alpha_e \tau A T N_i H^m k_i^\alpha}{(\alpha_e + \beta_v)(r(t) + \delta_k)}, \quad N_e = \frac{\beta_v \tau N_i}{\tau_0 \beta}, \tag{7.9.7}$$

where we use

$$F = T N_i H^m f,$$

in which

$$\tau_0 \equiv (\alpha_e + \beta_v)(1 - \tau).$$

From equations (7.9.7) and $N_i + N_e = N$, we solve

$$N_i = \frac{\tau_0 \beta N}{\tau_0 \beta + \beta_v \tau}, \quad N_e = \frac{\beta_v \tau N}{\tau_0 \beta + \beta_v \tau}. \tag{7.9.8}$$

From

$$r + \delta_k = (1 - \tau) \alpha A k_i^{-\beta}$$

and equations (7.9.7), we have

$$K_e(t) = \frac{\alpha_e \tau K_i(t)}{\alpha \tau_0}, \tag{7.9.9}$$

where we use the definition of k_i. From $K_i + K_e = K$ and equation (7.9.9), we solve

$$K_i(t) = a_i k(t), \quad K_e(t) = a_e k(t), \tag{7.9.10}$$

where

$$a_i \equiv \frac{\alpha \tau_0 N}{\alpha \tau_0 + \alpha_e \tau}, \quad a_e \equiv \frac{\alpha_e \tau N}{\alpha \tau_0 + \alpha_e \tau}.$$

The capital distribution between the two sectors is linear functions of the total capital. As $k_i = K_i / TH^m N_i$, we obtain

$$k_i(t) = n_i \frac{k(t)}{H^m(t)}, \tag{7.9.11}$$

where $n_i \equiv a_i / TN_i$. We see that $k_i(t)$ is a function of $k(t)$ and $H(t)$. Summarizing the above discussions, we have the following lemma.

Lemma 7.9.1. For any given levels of wealth and human capital, $k(t)$ and $H(t)$, in period t, all the other variables in the system are uniquely determined in this period. The values of the variables are given as functions of $k(t)$ and $H(t)$ by the following procedure: N_i and N_e by equations (7.9.8) $\rightarrow$ $K_i(t)$ and $K_e(t)$ by equations (7.9.10) $\rightarrow$ $k_i(t)$ by equation (7.9.11) $\rightarrow$ $f = Ak_i^\alpha$ and $F = TN_i H^m f$ $\rightarrow$ $r(t)$ and $w(t)$ by equations (7.9.1) $\rightarrow$ $y = rk + w$ and $\hat{y} = y + k$ $\rightarrow$ c and s by equations (7.9.2) $\rightarrow$ $U = c^\xi s^\lambda$.

We now show how to determine $k(t)$ and $H(t)$ in any period. From the definitions of $\hat{y}$, we have

$$\hat{y} = rk + w + k.$$

Substituting equations (7.9.1) and (7.9.11) into the above equation yields

$$\hat{y}(t) = \left(\alpha n_i^{-\beta} + \beta n_i^{\alpha}\right)(1 - \tau) A H^{\beta m}(t) k^{\alpha}(t) + (1 - \delta_k) k(t). \tag{7.9.12}$$

From equations (7.9.3), (7.9.4), (7.9.10), (7.9.12), and lemma 7.9.1, it is straightforward to show that the motion of $k(t)$ and $H(t)$ is explicitly given by the following two difference equations

$$k(t + 1) = \lambda^* H^{\beta m}(t) k^{\alpha}(t) + \delta k(t),$$
$$H(t + 1) = \upsilon_e^* k^{\alpha_e}(t) H^{(\beta_v + \beta_e)m}(t) + \upsilon_i^* H^{\beta m - \pi} k^{\alpha}(t) + \delta^* H(t), \tag{7.9.13}$$

where $\delta \equiv \lambda(1 - \delta_k)$, $\delta^* \equiv 1 - \delta_h$ and

$$\lambda^* \equiv \left(\alpha n_i^{-\beta} + \beta n_i^{\alpha}\right)(1 - \tau)\lambda A,$$
$$\upsilon_e^* \equiv \frac{\upsilon_e a_e^{\alpha_e} (TN_e)^{\beta_v} (T_e N)^{\beta_e}}{N},$$
$$\upsilon_i^* \equiv \frac{\upsilon_i ATN_i n_i^{\alpha}}{N}.$$

Lemma 7.9.1 guarantees that if we know values of $k(t)$ and $H(t)$, then we can explicitly solve all the other variables as functions of $k(t)$ and $H(t)$. Hence, to examine dynamic properties of the whole system, it is sufficient to examine the dynamic properties of equations (7.9.13).

A steady state of equations (7.9.13) is given by

$$k = \lambda^* H^{\beta m} k^{\alpha} + \delta k,$$
$$H = \upsilon_e^* k^{\alpha_e} H^{(\beta_v + \beta_e)m} + \upsilon_i^* H^{\beta m - \pi} k^{\alpha} + \delta^* H. \tag{7.9.14}$$

From the first equation in equations (7.9.14), we solve

$$k = \lambda_h H^m, \quad \lambda_h \equiv \left(\frac{\lambda^*}{1 - \delta}\right)^{1/\beta}. \tag{7.9.15}$$

Substituting equation (7.9.15) into the second equation in equations (7.9.14) yields

$$\Phi(H) \equiv \Phi_e(H) + \Phi_i(H) - \delta_h = 0, \tag{7.9.16}$$

where

$$\Phi_e(H) \equiv v_e^* \lambda_h^{\alpha_e} H^{x_e}, \quad \Phi_i(H) \equiv v_i^* \lambda_h^{\alpha} H^{x_i},$$
$$x_e \equiv (\alpha_e + \beta_v + \beta_e)m - 1, \quad x_i \equiv m - \pi - 1.$$

We see that the number of economic equilibrium points is equal to the number of solutions of the equation, $\Phi(H) = 0$ for $0 < H < \infty$. As shown in figure 7.9.1, the equation may have one or two equilibrium points. As shown in figure 7.9.1a, if $x_e < 0$ and $x_i < 0$, the equation monotonically decreases in H and it passes the horizontal axis only once. Figure 7.9.1b depicts the case of $x_e > 0$ and $x_i > 0$, the function monotonically increases in H and it passes the horizontal axis only once. Figure 7.9.1c depicts the case of $x_e > 0$ and $x_i < 0$ ($x_e < 0$ and $x_i > 0$).

In appendix 7.1, we check the conditions in figure 7.9.1. The following proposition shows that the properties of the dynamic system are determined by the two returns to scale parameters, x_e and x_i.

Proposition 7.9.1. (1) If $x_e < 0$ and $x_i < 0$, the system has a unique stable equilibrium; (2) If $x_e > 0$ and $x_i > 0$, the system has a unique unstable equilibrium; and (3) If $x_e > 0$ and $x_i < 0$ ($x_e < 0$ and $x_i > 0$), the system has either no equilibrium, one equilibrium point, or two equilibrium points. If it has two equilibrium points, the equilibrium point with low (high) level of H is stable (unstable).

We only interpret the stability condition $x_e < 0$ and $x_i < 0$. From the definitions of x_e and x_i, we may interpret x_e and x_i respectively as measurements of return to scales of the education sector and the industrial sector in the dynamic system. When x_e (< (>) 0, we say that the education sector displays decreasing (increasing) returns to scale in the dynamic economy. Proposition 7.9.1 shows that if the education and the production sectors display decreasing returns, then the dynamic system has a unique stable equilibrium.

If the two sectors exhibit decreasing returns to scale, the system will approach to its equilibrium in the long term. In a traditional society like the one constructed by Adam Smith where increases in human capital mainly come from division of labor and traditional education, the economic system tends to be dominated by stability. In a newly industrializing economy, education may exhibit increasing returns to scale and learning by doing may not be very effective in improving human capital. The economy may have multiple equilibrium points. We will demonstrate that if the society fails to explore increasing returns from education when the system is characterized by multiple equilibrium points, its development is not sustainable. When the system has a unique equilibrium, its behavior is easy to determine. When it has multiple equilibrium points, its behavior is path-dependent. We now demonstrate dynamics of the nonlinear system with multiple equilibrium points.

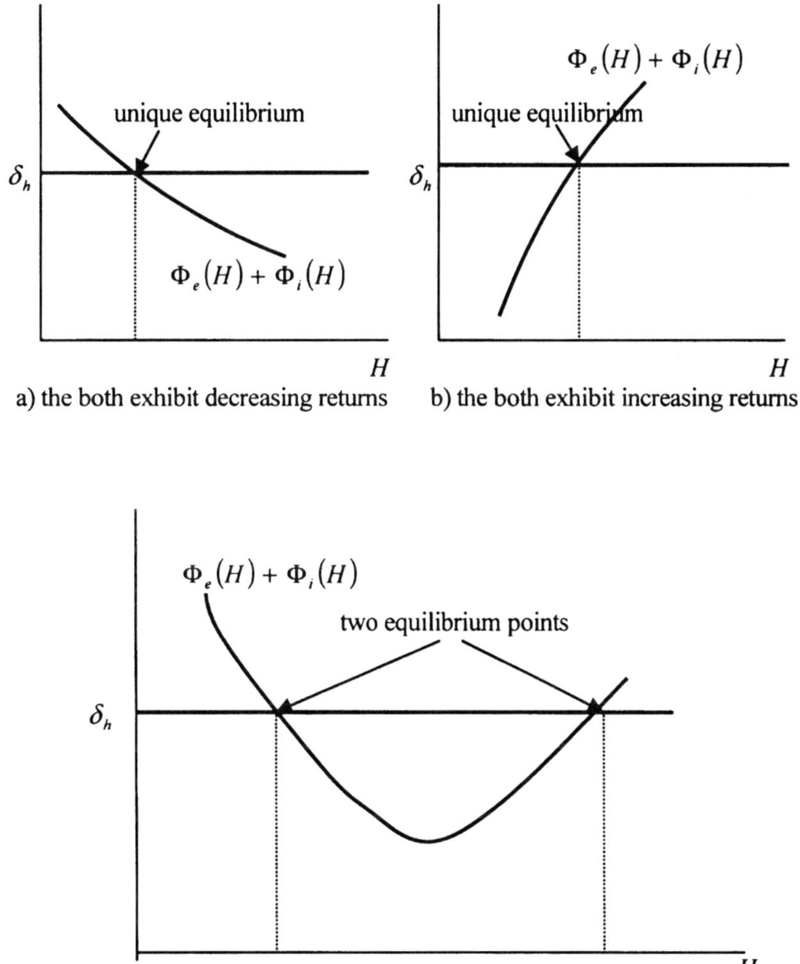

a) the both exhibit decreasing returns b) the both exhibit increasing returns

c) the one sector exhibits decreasing returns and the other increasing returns

Figure 7.9.1: The two sectors exhibit different returns to scale effects

We specify the parameters as follows

$$\alpha = 0.3,\ N = 1,\ A = 2,\ T_0 = 1,\ T_e = 0.15,\ \tau = 0.08,\ \lambda = 0.75,$$
$$\alpha_e = 0.35,\ \beta_e = 0.55,\ \beta_v = 0.55,\ v_e = 0.06,\ v_i = 0.01, \tag{7.9.17}$$
$$\delta_k = 0.05,\ \delta_h = 0.03,\ \pi = 0.3,\ m = 1.$$

The population devotes 15 percent of the total time to education and the tax rate is 8 percent. For illustration, we specify the parameter values related to productivity of the education sector by $\alpha_e = 0.35$, $\beta_e = 0.55$ and $\beta_v = 0.55$. An increase in qualified teachers or diligent students would improve the total output of the education sector. It should be remarked that the education sector exhibits diminishing returns in teaching, student's effort, or material input. Under (7.9.17), we have

$$x_e = 0.45,\ x_i = -0.3,\ N_e = 0.07,\ N_i = 0.93.$$

The economy exhibits increasing returns to scale in the education sector but decreasing returns in the production sector. About 92 percent of the labor force is engaged in production and 8 percent in education. The system has two equilibrium points

$$(k_1,\ H_1) = (31.782,\ 3.372),\ (k_2,\ H_2) = (181.469,\ 19.255).$$

At the equilibrium point $(k_2,\ H_2)$ the economy has higher levels of per capita capital and human capital than at the equilibrium point $(k_1,\ H_1)$. We can also calculate the equilibrium values of the other variables at the two equilibrium points. The simulation results are provided in Table 7.9.1.

Table 7.9.1: The values of the key variables at the two equilibrium points

	F	r	w	y	$\hat{y}$	c	U
$(k_1,\ H_1)$	10.85	0.11	23.44	27.05	58.84	14.71	33.53
$(k_2,\ H_2)$	61.98	0.11	764.03	784.72	966.19	241.54	550.61

We see that except the interest rate, the difference between the two values of any variable at the two equilibrium points is great. Our model demonstrates possibility of two equilibrium points for the same type of economy. Our analytical results show that two seemingly identical regions may follow radically different development paths, one leading to prosperity, the other to stagnation. Taiwan and Mainland China may provide a proper case for this result. Although they had similar backgrounds in terms of cultural heritages, values (except level of education), Taiwan and Mainland China had experienced totally different paths of industrialization during the period of 1950-1980 – the former had rapidly moved to the high equilibrium point, while the latter had remained near the low equilibrium. See Zhang for analyzing these dynamical processes.[23]

We now show how the system evolves when it starts from a state far away from equilibrium. We simulate the model with five different initial states as illustrated in figure 7.9.2. The five initial states are marked as $1, \cdots, 5$. The figure shows that the processes starting from states $1, \cdots, 4$, approach the equilibrium point (k_1, H_1) with low levels of consumption and human capital. The process starting from state 5 will go infinite in the long term. It passes above the equilibrium point (k_2, H_2). We may interpret this case as sustainable development as it will not end up at the "poverty trap" (k_1, H_1). It is important to note that the process started at state 1 will end up at the poverty trap, though the initial level of wealth is high. In contrast, the process started at state 5 is sustainable, even though the initial level of wealth is low. As human capital is the key factor for increasing returns, the economy with low level of human capital cannot escape poverty traps in the long term even if it has a plenty of material wealth.

An implication of the above results is that even big donations from international organizations to poor countries may matter little in helping them from avoiding poverty in the long term. For instance, let us consider a case that a poor country receives donation of a discrete quantity of capital from the World Bank. If the economy is near its low equilibrium and the donation raises k to a high level near to 175 without much improving the level of human capital, the economy would enjoy temporary high levels of income and consumption, but it will return back to its poverty state in the long term.

We now study the impact of change in the tax rate. In our system, the tax income is totally spent on education. We may thus interpret increases in the tax rate as encouraging education by the government. Intuitively, an increase in tax rate may either increase or decrease output because education costs resources and may have little impact on human capital improvement. We now show under what conditions an increase in the tax rate may promote the economy. Taking derivatives of equations (7.9.8) with respect to τ, we get the impact on the number of teachers

[23] Zhang (2003a).

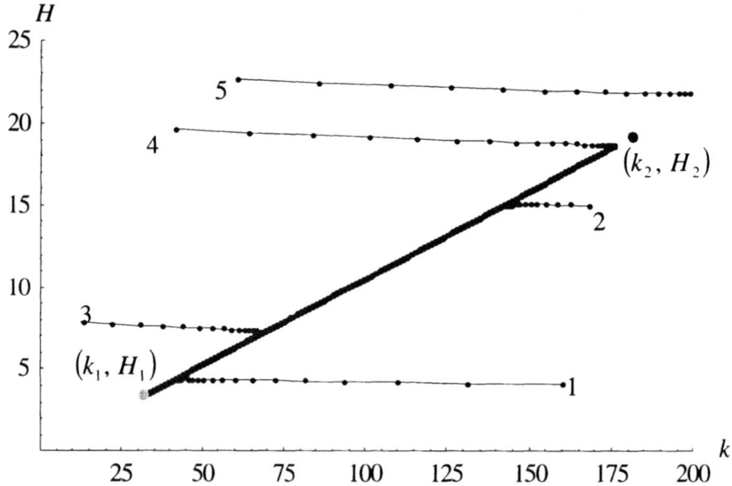

Figure 7.9.2: Path-dependent dynamic processes

$$\frac{dN_i}{d\tau} = -\frac{dN_e}{d\tau} = -\left[\frac{1}{1-\tau} + \frac{\alpha\beta_v - \alpha_e\beta}{\tau_0\beta + \tau\beta_v}\right]N_i < 0.$$

As τ is increased, the number of teachers is increased and the number of workers is reduced. An increase in the share of educational expenditure in the total output tends to make some people to shift from the production sector to the education sector. Taking derivatives of equation (7.9.16) with respect to τ yields

$$-\frac{d\Phi}{dH}\frac{dH}{d\tau} = \left(\frac{\alpha_e(\beta_v + \alpha_e)}{(\alpha\tau_0 + \alpha_e\tau)} + \frac{\beta_v}{N_e}\frac{dN_e}{d\tau} + \alpha_e b_1\right)\Phi_e + \left(\frac{1}{N_i}\frac{dN_i}{d\tau} + \alpha b_2 + \alpha b_1\right)\Phi_i, \tag{7.9.18}$$

where

$$b_1 \equiv \frac{1}{\lambda_h}\frac{d\lambda_h}{d\tau} = \frac{\alpha b_2(n_i - 1)}{\alpha + \beta n_i} - \frac{1}{(1-\tau)\beta},$$

$$b_2 \equiv \frac{1}{n_i}\frac{dn_i}{d\tau} = \frac{(\alpha_e + \beta_v)(\alpha\beta_v - \alpha_e\beta)}{(\alpha\tau_0 + \alpha_e\tau)(\tau_0\beta + \tau\beta_v)}.$$

We notice that the sign of the right-hand side of equation (7.9.18) is ambiguous. It should be remarked that (1) if $x_e < 0$ and $x_i < 0$, $d\Phi / dH < 0$; (2) If $x_e > 0$ and $x_i > 0$, $d\Phi / dH > 0$; and (3) If $x_e > 0$ and $x_i < 0$ ($x_e < 0$ and $x_i > 0$), $d\Phi / dH < (>) \, 0$ at the equilibrium point with low (high) level of H. By $k = \lambda_h H^m$, we get

$$\frac{dk}{d\tau} = \frac{mk}{H}\frac{dH}{d\tau} - b_1 k.$$

If $dH / d\tau$ is negative and b_1 positive, then per capita wealth will definitely be reduced as the society spends more resources on education. By equations (7.9.10), we obtain

$$\frac{dK_i}{d\tau} = \left(\frac{1}{k}\frac{dk}{d\tau} - \frac{1}{1-\tau} + \frac{\alpha\beta_v - \beta\alpha_e}{\alpha\tau_0 + \alpha_e\tau}\right)K_i,$$

$$\frac{dK_e}{d\tau} = \left(\frac{1}{k}\frac{dk}{d\tau} + \frac{\alpha(\beta_v + \alpha_e)}{\alpha\tau_0 + \alpha_e\tau}\right)K_e.$$

We see that if k is increased, the capital used by the education sector will be increased, even though the level of capital employed by the production sector may be reduced.

As the conclusions from the comparative statics analysis are ambiguous, we illustrate impact of changes in the tax rate by simulation. Here, we are interested in the path-dependent case. We still specify the parameter values as in equations (7.9.17) except the tax rate τ. Let us consider the case that the expenditure on education is increased from 8 percent of the GDP to 8.5 percent, that is, $\tau: 0.08 \Rightarrow 0.085$. We calculate

$$N_e = 0.075, \quad N_i = 0.925.$$

We see that the number of teachers is increased. The system has two equilibrium points given by

$$(k_1, H_1) = (34.172, 3.654), \quad (k_2, H_2) = (143.511, 15.346).$$

As the government further encourages education, the equilibrium point with low level of human capital is upgraded; while the other one is downgraded. The distance between the two equilibrium points is shortened. We can also calculate the equilibrium values of the other variables at the two equilibrium points. The simulation results are provided in table 7.9.2. The symbol + (–) in the table means the difference between the variable values in table 7.9.2 and table 7.9.1 is positive (negative). We see that the output at the equilibrium point with lower level of human capital is increased as the tax rate is increased; but the output at the other equilibrium point is reduced.

Table 7.9.2: The values of the key variables at the two equilibrium points

	F	r	w	y	$\hat{y}$	c
(k_1, H_1)	11.7(+)	0.12(+)	27.4(+)	31.4(+)	65.5(+)	16.4(+)
(k_2, H_2)	49.0(-)	0.12(+)	484.0(-)	500.4(-)	644.0(-)	160.9(-)

Figure 7.9.3 shows how the equilibrium points are shifted – the points with larger sizes is the new steady states and the other two points with smaller sizes are the old steady states. The initial states in figure 7.9.3 are correspondingly the same as in figure 7.9.2. It is important to note that the path started at initial state 4 will not end up poverty trap; instead the economic development is sustainable as a consequence of strengthened education policy.

The identification of poverty traps in theoretical models has important implications for policy making. When an economy is trapped into stable poverty, it cannot escape traps by small policy intervention or foreign aids. The economy needs either large short-term policy interventions or large amount of foreign aids to sustain economic progress. Moreover, even large interventions may fail if the society does not learn, for instance, about how to improve skills, how to live with inequality caused by rapid economic growth, and how to handle with class mobility due to differences in human capital. In an evolutional economy, as demonstrated in our example, it is possible that national growth is affected by government education policy. If the other parameters remain fixed and the government encourages education, our example indicates that it is relatively easier for the economy to escape poverty trap with the encouraging policy. We see that initial conditions matter.

We now study the impact of change in the propensity to save. From equations (7.9.8), we see that change in the preference has no impact on the labor distribution, i.e., $dN_v / d\lambda = 0$ and $dN_i / d\lambda = 0$. Taking derivatives of equation (7.9.16) with respect to λ yields

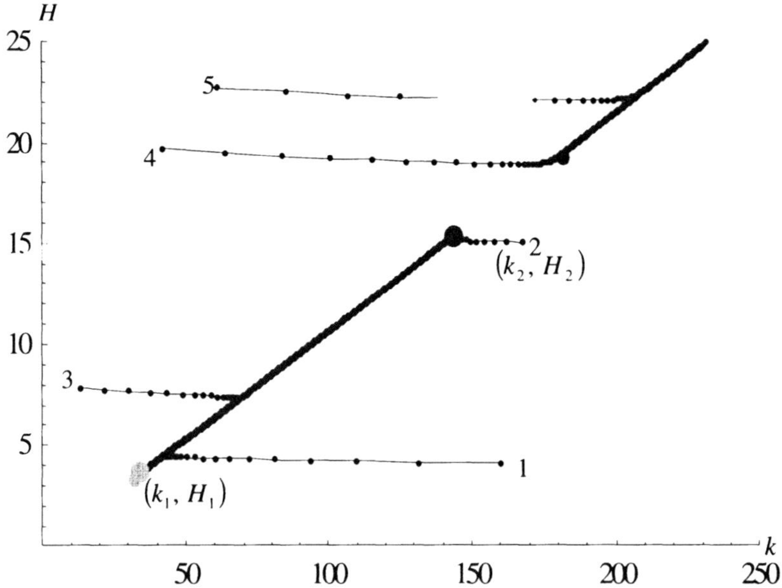

Figure 7.9.3: The impact of change in the education policy

$$-\frac{d\Phi}{dH}\frac{dH}{d\lambda} = \frac{\alpha_e \Phi_e + \alpha \Phi_i}{\lambda \beta (1 - \delta)}.$$

As the propensity to save increases, (1) if $x_e < 0$ and $x_i < 0$, the equilibrium level of human capital increase; (2) if $x_e > 0$ and $x_i > 0$, the equilibrium level of human capital will decrease; and (3) If $x_e > 0$ and $x_i < 0$ ($x_e < 0$ and $x_i > 0$), the equilibrium level of human capital may either increase or decrease. In a traditional society where no sector exhibits increasing returns to scale, an increase in the propensity to save will increase human capital. On the other hand, in an economy where its current equilibrium is located at the higher equilibrium level of H, an increase in the propensity to save may reduce the level of human capital.

By $k = \lambda_h H^m$, we get

$$\frac{dk}{d\lambda} = \frac{mk}{H}\frac{dH}{d\lambda} + \frac{k}{\lambda \beta (1 - \delta)}.$$

If $dH / d\lambda > 0,$ an increase in the propensity to save will increase the level of per capita wealth. It should be remarked that if no sector exhibits increasing sector, Adam Smith's argument that saving benefits the nation as well as the individuals hold. But if $dH / d\lambda < 0,$ we see that increases in the propensity to save may reduce the equilibrium wealth if $k / \lambda\xi$ is small. Malthus argued that higher propensity to save may not benefit the economic growth due to decreasing returns to scale in population growth; Keynes argued that a high propensity to save will not benefit economic growth because of loss of the total output due to unemployment. Our model shows that a high propensity to save may harm national economic growth even if we assume that the population is constant and labor force is fully employed.

By equations (7.9.10), we obtain

$$\frac{1}{K_i}\frac{dK_i}{d\tau} = \frac{1}{K_v}\frac{dK_v}{d\tau} = \frac{1}{k}\frac{dk}{d\tau}.$$

The capital stocks employed by the two sectors are changed in the same direction as that of k. By $k_i(t) = n_i k / H^m,$ we have

$$\frac{dk_i}{d\lambda} = \frac{k_i}{\lambda\beta(1 - \delta)} > 0.$$

From equations (7.9.1) and $F = AK^{\alpha}(H^m N_i)^{\beta},$ we get

$$\frac{dr}{d\lambda} = -\beta r \frac{dk_i}{d\lambda} < 0,$$

$$\frac{1}{w}\frac{dw}{d\lambda} = \frac{m}{H}\frac{dH}{d\lambda} + \alpha\frac{dk_i}{d\lambda},$$

$$\frac{1}{F}\frac{dF}{d\lambda} = \frac{\beta m}{H}\frac{dH}{d\lambda} + \frac{\alpha}{k}\frac{dk}{d\lambda} + \frac{\beta}{N_i}\frac{dN_i}{d\lambda}.$$

We see that the rate of interest will definitely decrease as the propensity to save is increased. If $dH / d\lambda > 0,$ the wage rate also increases. The impact on the output is ambiguous even when $dH / d\lambda > 0$ because the number of workers is reduced.

For illustration, we simulate the model. We still specify the parameter values as in (7.9.17) except the propensity to save λ. Let us consider the case that the propensity to save is increased from 0.75 to 0.76, that is, $\lambda : 0.75 \Rightarrow 0.76$. The two equilibrium points are given now by

$$(k_1, H_1) = (42.217, 4.189), \quad (k_2, H_2) = (157.315, 15.611).$$

We can also calculate the equilibrium values of the other variables at the two equilibrium points. The simulation results are provided in table 7.9.3.

Table 7.9.3: The values of the key variables at the two equilibrium points

	F	r	w	y	$\hat{y}$	c
(k_1, H_1)	13.7(+)	0.11(-)	36.9(+)	41.5(+)	83.7(+)	20.1(+)
(k_2, H_2)	51.3(-)	0.11(-)	512.4(-)	529.6(-)	686.7(-)	164.9(-)

Figure 7.9.4 shows the simulation results – the points with larger sizes is the new steady states and the other two points with smaller sizes are the old steady states.

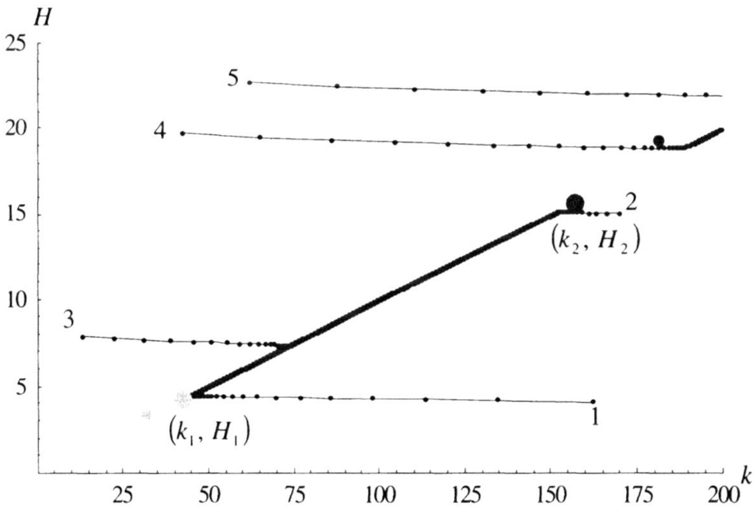

Figure 7.9.4: The impact of change in the propensity to save

We see that like the encouraging education policy in the previous discussion, a higher propensity to save makes it easier for a poor society to escape poverty trap. Our model suggests a way to escape poverty traps. The way is to strengthen the propensity to own wealth and encourage education. These policies would make economic growth sustainable. Foreign aids are important for rapid industrialization as the aids may help the economy to move to a necessary state for sustainable economic development. The economic miracles of Japan, Taiwan, Korea, and Singapore after the Second World War are all characterized by high saving rates, encouraging education policy, and proper international environment in the form of foreign aids or/and foreign investment. Mainland China is currently repeating similar processes by heavily investing in education, saving as much as possible, and attracting aids and foreign investment as much as possible.

Appendix

A.7.1: Proving proposition 7.8.1

The Jacobian matrix at equilibrium is

$$J = \begin{bmatrix} \lambda_1 a_{11} & \lambda_1 a_{12} \\ \lambda_2 a_{21} & \lambda_2 a_{22} \end{bmatrix},$$

where

$$\begin{aligned} a_{11} &= f' + \delta + \frac{n_1}{n_2} a_{12} = 1 - (f - kf')\frac{h_1}{k_1} + \frac{n_1}{n_2} a_{12}, \\ a_{12} &= (k_1 - h_1 k) n_2 f'', \\ a_{21} &= (k_2 - h_2 k) n_1 f'', \\ a_{22} &= f' + \delta + \frac{n_2}{n_1} a_{21} = 1 - (f - kf')\frac{h_2}{k_2} + \frac{n_2}{n_1} a_{21}. \end{aligned} \tag{A.7.1.1}$$

According to the definitions of k and n_j, we have

$$(k_1 - h_1 k) n_1 + (k_2 - h_2 k) n_2 = 0.$$

From this equation, we have: $a_{12}a_{21} < 0$. We see that $(k_1 - h_1 k)$ and $(k_2 - h_2 k)$ have the opposite signs if

$$k_1 - h_1 k \neq 0.$$

From equations (7.8.9) we have

$$k_j - h_j k = \left(\frac{f - k\delta_j}{\delta_j - f'} \right) h_j,$$

where $\delta_j - f' > 0$. We see that if $\delta_2 > \delta_1$, $k_2 - h_2 k$ (and $f - k\delta_2$) must be negative. If

$$\delta_2 = \delta_1, \quad k_2 - h_2 k$$

(and $f - k\delta_2$) will equal zero. As

$$\delta_j = \frac{1}{\lambda_j} - \delta > 0,$$

$\delta_2 > \delta_1$ means $\lambda_1 > \lambda_2$. In the remainder of the proof, we require $\lambda_1 > \lambda_2$. We can similarly prove the case of $\lambda_1 \leq \lambda_2$.

Under $\lambda_1 > \lambda_2$, we have

$$\frac{k_2}{k} - h_2 < 0, \quad \frac{f}{k} - \delta_2 < 0, \quad \frac{k_1}{k} - h_1 > 0, \quad \frac{f}{k} - \delta_1 > 0, \quad a_{12} < 0, \quad a_{21} > 0.$$

The characteristic equation is given by

$$M(\rho) = \rho^2 - Trace(J)\rho + Det(J) \qquad \text{(A.7.1.2)}$$

where

$$Trace(J) = \lambda_1 a_{11} + \lambda_2 a_{22},$$
$$Det(J) = \lambda_1 \lambda_2 (a_{11} a_{22} - a_{12} a_{21}).$$

From equations (A.7.1.1) and (A.7.1.2) we calculate

$$Det(J) = \lambda_1 \lambda_2 \left[\left(f' + \delta + \frac{n_1}{n_2} a_{12} \right) \left(f' + \delta + \frac{n_2}{n_1} a_{21} \right) - a_{12} a_{21} \right]$$

$$= \lambda_1 \lambda_2 (f' + \delta)^2 > 0, \tag{A.7.1.3}$$

where we also use

$$\frac{n_1 a_{12}}{n_2} + \frac{n_2 a_{21}}{n_1} = 0.$$

From $\delta_j - f' > 0$, we obtain

$$1 > \lambda_j (1 + f') > 0.$$

From equation (A.7.1.3), we have

$$0 < DetJ < 1.$$

It is straightforward to calculate

$$M(1) = 1 - Trace(J) + Det(J) =$$

$$1 - (\lambda_1 + \lambda_2)(f' + \delta) + \lambda_1 \lambda_2 (f' + \delta)^2 - \left(\frac{n_1 \lambda_1}{n_2} a_{12} + \frac{n_2 \lambda_2}{n_1} a_{21} \right)$$

$$= \lambda_1 \lambda_2 (\delta_1 - f')(\delta_2 - f') - \left(\frac{n_1 \lambda_1}{n_2} a_{12} + \frac{n_2 \lambda_2}{n_1} a_{21} \right) > 0, \tag{A.7.1.4}$$

where we use $\delta_j - f' > 0$ and

$$\frac{n_1 \lambda_1}{n_2} a_{12} + \frac{n_2 \lambda_2}{n_1} a_{21} = \frac{n_1 \lambda_1}{n_2} a_{12} + \frac{n_2 \lambda_2}{n_1} a_{21} - \lambda_1 \left(\frac{n_1}{n_2} a_{12} + \frac{n_2}{n_1} a_{21} \right)$$

$$= -(\lambda_1 - \lambda_2) \frac{n_2}{n_1} a_{21} < 0.$$

From

$$1 + Det(J) > Trace(J), \quad 0 < DetJ < 1, \tag{A.7.1.5}$$

we see $2 > Trace(J)$. We now try to find conditions for $Trace(J) > 0$. First we have

$$Trace(J) = (\lambda_1 + \lambda_2)(f' + \delta) - (\lambda_1 - \lambda_2)(k_2 - h_2 k) n_2 f'',$$

where we use equations (A.7.1.3) and (A.7.1.4). As $\Delta(k) \geq 0$ implies

$$k^2 f'' \geq -\frac{f}{k} + f',$$

from the above inequality we have

$$Trace(J) \geq (\lambda_1 + \lambda_2)(f' + \delta) - (\lambda_1 - \lambda_2) k_2 n_2 f'' - (\lambda_1 - \lambda_2) h_2 n_2 \left(\frac{f}{k} - f' \right).$$

As

$$-(\lambda_1 - \lambda_2) k_2 n_2 f'' > 0,$$

$Trace(J) > 0$ is guaranteed if

$$(\lambda_1 + \lambda_2)(f' + \delta) - (\lambda_1 - \lambda_2) h_2 n_2 \left(\frac{f}{k} - f' \right) \geq 0. \tag{A.7.1.6}$$

From

$$\frac{f}{k} < \delta_2 = \frac{1}{\lambda_2} - \delta,$$

we have

$$(\lambda_1 + \lambda_2)(f' + \delta) - (\lambda_1 - \lambda_2)h_2 n_2\left(\frac{f}{k} - f'\right) >$$
$$((1 + h_2 n_2)\lambda_1 + h_1 n_1 \lambda_2)(f' + \delta) - \left(\frac{\lambda_1}{\lambda_2} - 1\right)h_2 n_2,$$

where we use

$$h_1 n_1 + h_2 n_2 = 1.$$

If

$$\left(\left(1 + \frac{1}{h_2 n_2}\right)\lambda_1 \lambda_2 + \frac{h_1 n_1}{h_2 n_2}\lambda_2^2\right)(f' + \delta) > \lambda_1 - \lambda_2, \tag{A.7.1.7}$$

then (A.7.1.6) holds. We see that if the difference $\lambda_1 - \lambda_2$ is not large, (A.7.1.7) holds. In summary, under $\Delta(k) \geq 0$ and (A.7.1.7) we have

$$0 < DetJ \ (= \rho_1 \rho_2) < 1,$$
$$2 > Trace(J) \ (= \rho_1 + \rho_2) > 0. \tag{A.7.1.8}$$

We also have

$$M(-1) = 1 + Det(J) + Trace(J) > 0, \quad M(1) > 0.$$

We calculate

$$TraceJ^2 - 4DetJ = (\lambda_1 + \lambda_2)^2 (f' + \delta)^2 + \left(\frac{n_1 \lambda_1}{n_2} a_{12} + \frac{n_2 \lambda_2}{n_1} a_{21}\right)^2$$
$$+ 2(\lambda_1 + \lambda_2)(f' + \delta)\left(\frac{n_1 \lambda_1}{n_2} a_{12} + \frac{n_2 \lambda_2}{n_1} a_{21}\right) - 4\lambda_1 \lambda_2 (f' + \delta)^2$$
$$= (\lambda_1 - \lambda_2)^2 (f' + \delta)^2 + \left(\frac{n_1 \lambda_1}{n_2} a_{12} + \frac{n_2 \lambda_2}{n_1} a_{21}\right)^2$$
$$+ 2(\lambda_1 + \lambda_2)(f' + \delta)\left(\frac{n_1 \lambda_1}{n_2} a_{12} + \frac{n_2 \lambda_2}{n_1} a_{21}\right).$$

The sign of $TraceJ^2 - 4DetJ$ is ambiguous. In the case of

$$TraceJ^2 - 4DetJ > 0,$$

the two eigenvalues are real and distinct. The two eigenvalues have the same sign. From (A.7.1.8), we have

$$0 < \rho_1, \rho_2 < 1.$$

The equilibrium point is a sink and the adjustment is monotone. If

$$TraceJ^2 - 4DetJ < 0,$$

the two eigenvalues are imaginary and the eigenvalues have modulus less than 1. The equilibrium point is a steady focus. We thus conclude that under (A.7.1.7) the unique equilibrium point is stable.

A.7.2: Proving proposition 7.9.1

First, we find conditions such that

$$\Phi(H) = 0,$$

has positive solutions. We exclude the case of

$$x_e = x_i = 0.$$

It is easy to check that if $x_e > 0$ and $x_i > 0$ (which guarantees $\Phi(0) < 0$, $\Phi(\infty) > 0$ and $\Phi'(H) > 0$ for $H > 0$) or $x_e < 0$ and $x_i < 0$ (which guarantees $\Phi(0) > 0$, $\Phi(\infty) < 0$ and $\Phi'(H) < 0$ for $H > 0$), the system has a unique positive equilibrium point. We now show that in the cases of $x_e > 0$ and $x_i < 0$ (or $x_e < 0$ and $x_i > 0$), the system has two equilibrium points, or one equilibrium point, or no one.

We just prove one case, $x_e > 0$ and $x_i < 0$. The other case can be similarly checked. Since $x_e > 0$ and $x_i < 0$, $\Phi(0) > 0$ and $\Phi(\infty) > 0$. This implies that $\Phi(H) = 0$ has either no solution or multiple solutions. Since

$$H\Phi' = x_e \Phi_e + x_i \Phi_i,$$

where Φ_e and Φ_i are positive, we see that $\Phi'(H)$ may be either positive or negative, depending upon the values of x_e and x_i. If $\Phi(H) = 0$ has more than two positive solutions, there are at least two positive values of H such that $\Phi'(H) = 0$. Since

$$\frac{dH\Phi}{dH} > 0,$$

strictly holds for $H > 0$, it is impossible for

$$\Phi'(H) = 0,$$

to have more than one solution. Accordingly, $\Phi(H) = 0$ has either no solution or one solution, or two solutions. A necessary and sufficient condition for the existence of two equilibrium points is that there exists a value of H^* such that $\Phi(H^*) < 0$.

The Jacobian matrix is given by

$$J = \begin{bmatrix} \alpha + \beta\delta & (1-\delta)\dfrac{\beta mk}{H} \\ \dfrac{\alpha_e H \Phi_e}{k} + \dfrac{\alpha H \Phi_i}{k} & (\beta_v + \beta_e) m \Phi_e + (\beta m - \pi)\Phi_i + \delta^* \end{bmatrix}.$$

The characteristic equation is given by

$$M(\rho) = \rho^2 - Trace(J)\rho + Det(J),$$

where

$$Trace(J) = \alpha + \beta\delta + (\beta_v + \beta_e) m \Phi_e + (\beta m - \pi)\Phi_i + \delta^*,$$
$$Det(J) = \left((\beta_v + \beta_e) m \Phi_e + (\beta m - \pi)\Phi_i + \delta^*\right)(\alpha + \beta\delta)$$
$$- (1-\delta)\beta m(\alpha_e \Phi_e + \alpha \Phi_i).$$

We see that $Trace(J) > 0$. We calculate

$$Trace(J)^2 - 4Det(J) = ((\alpha + \beta\delta) - ((\beta_v + \beta_e)m\Phi_e + (\beta m - \pi)\Phi_i + \delta^*))^2$$
$$+ (1 - \delta)(\alpha_e\Phi_e + \alpha\Phi_i)4\beta m > 0.$$

As

$$Trace(J)^2 - 4Det(J) > 0,$$

we conclude that the eigenvalues are real and distinct. We calculate

$$M(1) = 1 - Trace(J) + Det(J) = -(x_e\Phi_e + x_i\Phi_i)(1 - \delta)\beta, \tag{A.7.2.1}$$

in which we use

$$H = \Phi_e H + \Phi_i H + \delta^* H, \quad 1 - \delta > 0.$$

We rewrite $Det(J)$ as follows

$$Det(J) = (\beta_v + \beta_e)m\Phi_e + (\beta m - \pi)\Phi_i - \beta(1 - \delta)(x_e\Phi_e + x_i\Phi_i)$$
$$+ \delta^* - \beta(1 - \delta), \tag{A.7.2.2}$$

where we use

$$\Phi_e + \Phi_i = \delta_h$$

at the steady state. As

$$\delta^* - \beta(1 - \delta) = \alpha - \delta_h + \beta\delta,$$

we can always assume this term to be positive if the depreciation rate of human capital is assumed less than α.

We now examine the case of $x_e < 0$ and $x_i < 0$. From equations (A.7.2.1) and (A.7.2.2), evidently

$$M(1) > 0, \quad M(0) = Det(J) > 0.$$

As the eigenvalues are real and distinct and

$$\rho_1 + \rho_2 = TraceJ > 0, \quad \rho_1\rho_2 = DetJ > 0,$$

we conclude that they are both positive. As $M(1) > 0$, the two eigenvalues are on the same side of $+1$. Since

$$\begin{aligned} Trace(J) &= 1 + \alpha + \beta\delta + (\beta_v + \beta_e)m\Phi_e + (\beta m - \pi)\Phi_i - \delta_h \\ &= 1 + \alpha + \beta\delta + x_e\Phi_e + x_i\Phi_i - m(\alpha_e\Phi_e + \alpha\Phi_i) < 2, \end{aligned} \tag{A.7.2.3}$$

where we use

$$\Phi_e + \Phi_i = \delta_h, \quad 1 + \alpha + \beta\delta < 2,$$

we conclude that $0 < \rho_1, \rho_2 < 1$. We thus proved that if $x_e < 0$ and $x_i < 0$, the unique equilibrium is stable.

We now consider the case of $x_e > 0$ and $x_i > 0$. From equations (A.7.2.1), $M(1) < 0$. This implies that the two distinct eigenvalues are on the two sides of $+1$, i.e., $\rho_1 < 1$ and $\rho_2 > 1$. The unique equilibrium is unstable. Furthermore, it is straightforward to confirm the following relation

$$M(-1) = 1 + Trace(J) + Det(J) = M(1) + 2Trace(J).$$

Substituting equations (A.7.2.1) and (A.7.2.3) into the above equation yields

$$\begin{aligned} M(-1) &= (x_e\Phi_e + x_i\Phi_i)(1 + \alpha + \delta\beta) + 2(1 + \alpha + \beta\delta) - 2m(\alpha_e\Phi_e + \alpha\Phi_i) \\ &= (2 - \delta_h - m\delta_h)(1 + \alpha + \beta\delta) \\ &+ [(1 + \alpha_e + \beta_v + \beta_e)(1 + \alpha + \delta\beta)m - 2\alpha_e]m\Phi_e \\ &+ [(1 + m - \pi)(1 + \alpha + \delta\beta) - 2m\alpha]\Phi_i, \end{aligned}$$

where we use $\Phi_e + \Phi_i = \delta_h$ and the definitions of x_e and x_i. As it is reasonable to assume the following three terms to be positive

$$\begin{aligned} &2 - \delta_h - m\delta_h > 0, \\ &(1 + \alpha_e + \beta_v + \beta_e)(1 + \alpha + \delta\beta)m - 2\alpha_e > 0, \\ &(1 + m - \pi)(1 + \alpha + \delta\beta) - 2m\alpha > 0, \end{aligned} \tag{A.7.2.4}$$

we see that $M(-1) > 0$. Since

$$M(1) < 0, \quad M(-1) > 0,$$

we have $\rho_1 \in (-1, 1)$. Hence, the equilibrium point under (A.7.2.4) is a saddle point.

We now check the case of $x_e > 0$ and $x_i < 0$, when there are two equilibrium points. The other case, $x_e > 0$ and $x_i < 0$, can be similarly proved. First, we note that

$$x_e \Phi_e + x_i \Phi_i < (>)\ 0$$

at the equilibrium point with the lower (higher) level of H^*. Let H_l and H_h ($H_l < H_h$) stand for the two equilibrium points of H. At H_l

$$x_e \Phi_e + x_i \Phi_i < 0.$$

It can be seen that the stability conditions are the same as in the case of $x_e < 0$ and $x_i < 0$. We thus have $0 < \rho_1, \rho_2 < 1$. At H_h

$$x_e \Phi_e + x_i \Phi_i > 0.$$

The stability conditions are the same as in the case of $x_e > 0$ and $x_i > 0$. We thus conclude that the equilibrium point with the higher value of H is unstable.

Chapter 8

Epilogue

To conclude this study, we mention two important issues, time scales and economic structures, for understanding economic evolution.

As time passes, economic issues with which economists are concerned have shifted. Even since the time of Adam Smith, the economic variables that economists have dealt with appear to have been invariant. But the ways in which these variables are combined and the speeds at which they change have constantly varied and the dominant economic doctrines have shifted over time and space. The complexity of economic reality is constantly increasing in modern time. This is partially because of the expanded capital and knowledge stocks of mankind. Knowledge, in fields of philosophy, arts, literature, music, technology and sciences, expands man's imagination and extends possibilities of human action, not to mention that the knowledge reservoir can directly satisfy the desires of an unlimited number of people at the same time. Knowledge is not only power and sources of money, but also the most durable capital goods for human mind. Increases in machines, housing and infrastructures has enriched human environment, increased accessibility to various locations, and enlarged variety of human behavior. The explosion of knowledge and capital in modern times has resulted in very complicated human action fields.

Time is at the center of the chief difficulty of almost every economic problem. The role of time in decision-makings and action is becoming increasingly complicated as variety of action and social networks are expanding. It is a difficult issue to decide the length of time which affects a special decision making since each kind of human decisions are made with different time scales and two persons may have different time scales with regard to the same kind of decision making. Because of the high variety of human behavior and time scales, in order to analyze a single person's economic behavior as a whole we have to conduct the analysis within a framework with varied time scales. Human behavior is connected in direct or indirect ways in human action fields; but we may miss interdependence between some elements if we do not properly recognize the role of time.

Another dimension in analysis is space. Man, action, capital, knowledge and time can become culturally and socially meaningful only if we locate them over space. Each human being is born into a unique existence and each piece of land has its unique attributes in affecting human action. Space means individual characteristics and accordingly requires refined classification. This is particularly important in analyzing modern economies. Fast technological changes, richness of material living conditions, complicated international interactions, and many other modern phenomena have increased complexity of spatial economies. The subsystems such as ecological, economical and social subsystems, which could be once decomposable as separate elements in analyzing the social system at least in short terms over a homogenous space, have to be treated as a part of the whole system. Some economic relations cannot be recognized if we don't explicitly introduce spatial and temporal dimensions. It will take some time for what is happening in a scientific lab to affect economic reality. Without spatial dimension, we can hardly analyze actual processes of, for instance, how Japanese economy may actually affect the world economy. In fact, the choice of spatiotemporal scale is a delicate and obligatory process and must be made before actual study of any special economic problem. The explicit awareness of this necessity is important for understanding both economic reality and structure of economics. For instance, for human life what is good to one's taste (assessment on a short timescale) may be harmful to one's health (assessment on a longer timescale). One can hardly explain differences between Keynes and Schumpeter's economic visions without differentiating their temporary scales. Temporal scales in the economist's vision have complicated interdependent relationships with actual analyses and abstraction of reality.

We are in an era of high economic complexity. This implies that economic decisions have to be made within a large context in which internal structures of each subsystem and connections of different subsystems have to be taken into account within a genuinely dynamic framework. The bringing-up of children, lower and higher education, family structure, and family values are all connected in a subtle and complicated way in economic networks. We have to consider reciprocal relations of different aspects of social and economic factors rather than considering these facts in isolation. Simple one-sector growth models without economic structures will hardly provide any useful information about the complexity of modern economies. We need to enlarge analytical frameworks to handle multiple hierarchical levels, multiple space degrees and multiple time scales.

An economic system is composed of many people, and the psychology of people and the relations (which are reflected in values, institutions and customs) among people are constantly changing. The difficult task is to find out whether or not there are durable (if not permanent) patterns or orders in human behavior and in human societies and to explicitly construct descriptions (usually, models) for these orders if they exist. In order to construct a comprehensive theory it is necessary to understand general patterns of people's behavior in a society over time and space. The difficult task is how to construct such a comprehensive economic theory.

It is significant to examine economic systems with a spatiotemporal structural vision. The key words are space, time, and structure. It is hard to give a precise concept of structure. Here, a structure means a sum of elements and relationships between those elements. In other words, structure stands for the way the elements and constituent parts of a whole are arranged with respect to each other. Structure represents a whole in which each element depends on the others by virtue of its relation with them. According to Thom,[1] structure is defined as a spatiotemporal morphology described by significant spatial discontinuities and by the syntax that determines how these sets of discontinuities form into relatively stable systems. In evolving structures relations depend on time. The structure includes properties, which are properties of the whole rather than only properties of its component parts. Any change in one element or one relationship will cause a modification in other elements or relationships. By means of the cooperation of the individual parts of different subsystems new properties may emerge that are not present in the subsystems. Economic evolution involves not only changes in variable levels and functions but also in organizational structures that concern the way elements are connected within subsystems, the way subsystems are embedded in large ones, and the way that organizational structures emerge or disappear. As mentioned in the introduction, advances in theory of complex systems provide promising ways for understanding the dynamics of structural changes in socio-economic systems. Theory of complex systems provides many deep insights into structural evolution. The modern study of economic chaos permits the discovery of chaotic economic structures disguised by very complicated fluctuations. The concept of structural stability in theory of complex systems is essentially significant in the study of structural evolution.

Hierarchy is a main character of economic structures. Economic systems consist of a hierarchical structuring among the component parts. Hierarchy here means, following Herbert Simon[2], a set of Chinese boxes of a particular kind. Opening any given box discloses a whole small set of boxes; and opening any one of these component boxes discloses a new set in turn. Power distribution is an important indicator of this structure. For instance, in common situations the state organizes the regions within the country, and the regional governments in turn organize the lower-levels within them. Each society is characterized by its own hierarchical structure. In social evolution, these structures may be either stable or unstable, depending on material, affective, cognitive, and spiritual, factors. In the traditional societies economic structures often remained quite stable over many generations; in modern societies structural changes may occur several times within a short period of time.

---

[1] Thom (1977).
[2] Simon (1973).

Hierarchy is not only the character of human societies; even sciences exhibit hierarchical structures. Dawkins sees scientific theories and areas as a hierarchical structure, on different levels, corresponding to levels of description of phenomena.[3] Philosophers and some scientists have sought ultimate reality in the structure of matter at increasingly finer scales in order to provide the most elementary explanation, while astronomers have sought the structure of the universe in increasingly wider domains. In natural science, the complexities of ecosystems are explained by examining those of organisms, organisms are explained by referring them back to the growth of spatially organized proteins and other macromolecule, the complex organization of organisms is explained back to the linear complexity of their DNA code, the complexity of DNA is referred back to combinations of simpler atoms, and so on. We should have national macroeconomics based on regional economics; regional economics should be referred back to urban and rural economics; spatial economics referred back to family-level and company-level economics. In a broader perspective, psychology and behavior sciences should be the starting point of microeconomics. Chemists will to explain psychological processes in terms of natural laws. The processes can be further going on. Darwin's remarks that it is not necessary to refer every phenomenon back down this chain of reductions in order to understand it. In natural sciences, chemistry can be considered as a 'fixed parameter' for the purpose of understanding DNA. In economics, macroeconomics can be considered as 'given' for labor economics and family economics. It is obviously important to construct a grand theory, which connects all the levels within a compact framework.[4]

Connections between levels in a social hierarchy are usually not simple. An economic hierarchical system may operate on different scales. Its variables and substructures may operate or change in different process rates. Since higher levels usually strongly and quickly affect low levels in the hierarchical structure, higher levels usually tend to be changed in lower frequencies. But this asymmetry in change speeds is not always held. To study the hierarchical nature of complex systems, we have to accept a different perspective - a different spatiotemporal scale used. There are gaps between any two levels of social hierarchy. For instance, we may have a reasonable understanding of single male or female behavior and we know how men and women get married and form families. But the functioning of families is far more complicated. Micro level phenomena such as family ties have significant implications for macro economies. An economic theory without endogenous family structure can hardly explain modern economic reality since on one hand family structures have been affected by economic development, on the

[3] Dawkins (1986).

[4] The foundation for a grand economic theory has been laid through my efforts over years. I have first applied nonlinear science to economics (Zhang, 1991) to 'modernize' the vision and methodology of economic analysis and then have developed the analytical framework with time and space for analyzing various economic issues with the vision of nonlinear science (Zhang, 1996, 2000, 2001, 2002, 2003b, 2005b).

other hand economic development is the consequences of cooperative (and competitive) behavior among family members.

All these intrinsic difficulties related to economic structures heavily affect the efficiency of modeling economic systems. Multiple levels have to be described in long-term studies. This requires economic theory to have internal structures to represent the complexity of subsystems and connections of the subsystems. Such structural models will eventually turn out to be complicated. Indeed, we may find out some special characteristics of the system under consideration and thus are able to simplify the analysis. For instance, some hierarchical systems are decomposable, at least in short timescales. This means that it is possible to effectively isolate and describe a part of the system for a given timescale. We may analyze behavior of the independent subsystem in isolation from the rest of the hierarchy to which it belongs. A study of dynamics of a particular process on a particular level can thus be conducted by taking behavior of higher levels as fixed and 'enslaving' behavior of the low levels as structurally determined flows. In other words, for the chosen time scale the behavior of higher levels are so slow that they can be effectively negligible and the behavior of lower levels are so fast that perturbations generated by the behavior of lower levels can also be effectively neglected. For instance, an economic analysis may be conducted in a time scale short enough to assume changes in ecological processes negligible and long enough to average out noise from processes occurring at individual levels. It should be remarked that this method might be invalid especially in 'revolutionary' periods. At such critical points, neither the dynamics of higher levels nor the perturbations generated by the behavior of lower levels are negligible. The model used to describe the dynamic interaction of the chosen subsystem is no longer able to provide reliable information about possible behavior of the subsystem.[5]

An important feature of economic structures is that they are intrinsically complicated at each level. Individuals, groups or clubs, regions and nations, even as they develop under practically similar conditions, are never exactly the same. Detailed studies of their evolution have provided many examples of an intrinsic complexity. For instance, random fluctuations in tastes may affect microeconomic evolutionary processes on a large scale. The economic structure represents the values and principles of the economic organization. The system may be analyzed by dividing the whole system into different levels, each representing a subsystem, which consists of relatively uniform elements that interact with each other either in simple or complicated ways. To find and describe these interactions are the key elements for analyzing order and disorder at any given level. Economists have sought structural invariants on macro, meso and micro levels. The construction of a theory with structure is not arbitrary and gratuitous. We first have to determine issues under examination, scales (both of variables, time and space) and domains, and analytical methods. Here, when assuming the habitual three-dimensional

[5] Synergetic economics by Zhang (1991) deals with nonlinear economic dynamics with different time scales and speeds of changes.

representation of space, with time as a fourth dimension, scale is defined as the smallest volume within the interior of which it is agreed not to try to distinguish the nonuniformity of a property being measured and as the shortest interval of time during which it is agreed not to try to distinguish variations of a given property. The domain is defined as the greatest volume and the longest time interval over which the study will be extended. For instance, the whole economy can be studied by employing several scales. The variables used at one scale may be treated as a coarser scale, macroscopic in comparison with the first by taking averages of larger volumes and longer intervals of time. In building a sophisticated economic theory, one has to construct, without making any mistakes, a long chains of assertions, has to be aware of what one is doing at each step of the construction process, and has to speculate about where one is going. The constructor has be to able to guess what is true and what is false at each level and be able to judge what is useful and what is not in the whole framework.

Appendix

The appendix is arranged as follows. A.1 introduces matrix theory. A.2 shows how to solve linear equations, based on matrix theory. A.3 introduces metric spaces and some basic concepts and theorems related to metric spaces. A.4 defines some basic concepts in study of functions and states the implicit function theorem. A.5 gives a general expression of the Taylor Expansion. A.6 is concerned with convexity of sets and functions and concavity of functions. A.7 shows how to solve unconstrained maximization problems. In A.8, we introduce conditions for constrained maximization. A.9 introduces theory of dynamic optimization.

A.1 Matrix theory

We present some important concepts and theorems from linear algebra and matrix theory. Some elementary concepts, such as identity matrices and null matrices, matrix operations, and proofs of theorems are omitted.[1]

Let vectors A be a nonempty set of vectors in R^n. A vector x in R^n is *linearly dependent* on the set A. if there exist vectors $y_1, y_2, ..., y_m$ and scalars $a_1, a_2, ..., a_m$ such that

$$x = \sum_{j=1}^{m} a_j y_j .$$

For any nonempty set A of vectors in R^n, $\langle A \rangle$ is the set of all vectors in R^n that are dependent on A. $\langle A \rangle$ is a subspace of R^n. A vector of the form $\sum_{j=1}^{m} a_j y_j$ is called a *linear combination*. A set A of vectors is a *basis* of the

[1] This part on matrix theory is based on Gilbert and Gilbert (1970). See also Chiang (1984), Berman and Plemons (1979), Zhang (1999fz), and Peterson and Sochachi (2002).

subspace U if (i) A "spans" U and (ii) A is linearly independent. If U is any subspace of R^n, the number of vectors in a basis of U is called the *dimension* of U and is abbreviated as $\dim(U)$. The dimension of R^n is n.

Let

$$U = \{U_1, U_2, \ldots, U_r\}$$

be a set of vectors in R^n and

$$V = \{V_1, V_2, \ldots, V_s\}$$

be a set of vectors in $\langle U \rangle$, a *matrix of transition* from U to V is a matrix

$$A = \lfloor a_{ij} \rfloor_{rxs}$$

such that

$$V_j = \sum_{i=1}^{r} a_{ij} U_i, \quad j = 1, 2, \ldots, s.$$

Definition A.1.1. A square matrix $A = \lfloor a_{ij} \rfloor_{nxn}$ is *nonsingular* if and only if A is a matrix of transition from one basis of R^n to another basis of R^n. A square matrix that is not nonsingular is called *singular*.

We denote the identity matrix by I_n ($\equiv \lfloor \delta_{ij} \rfloor_{n\times n}$ where δ_{ij} is the Kronecker delta).

For any $m \times n$ rectangular matrix, if the maximum of linearly independent rows that can be found in such a matrix is r, the matrix is said to be of *rank* r, denoted by $Rank(A)$ or $Rank\ A$. The rank also tells us the maximim number of linearly independent columns in the same matrix. As a square matrix has n linearly independent rows (or columns), it must be of rank of n. If A is $m \times n$ matrix over R and P is any invertible $n \times n$, then we have

$$Rank(A) = Rank(AP).$$

Definition A.1.2. An $n \times n$ matrix B is an *inverse* of the $n \times n$ matrix $A = \lfloor a_{ij} \rfloor_{nxn}$ if

$$AB = I_n = BA.$$

Also a square matrix is called *invertible* if it has an inverse.

Theorem A.1.1. An $n \times n$ matrix A is invertible if only if A is nonsingular. The inverse of an invertible matrix is unique.

If $A = \lfloor a_{ij} \rfloor_{nxn}$ is invertible, its unique inverse is denoted by A^{-1}. If

$$A_1, A_2, ..., A_m$$

are square invertible matrices of order n over R^n, then $A_1 A_2 \cdots A_m$ is invertible and

$$(A_1 A_2 \; A_m)^{-1} = A_m^{-1} \cdots A_2^{-1} A_1^{-1}.$$

For any $m \times n$ matrix A, the transpose of A is denoted by A^T. If a square matrix A is invertible, then A^T is also invertible and

$$(A^T)^{-1} = (A^{-1})^T.$$

The following concept is only referred to square matrices.

Definition A.1.3. The *determinant* of the square matrix $A = \lfloor a_{ij} \rfloor_{nxn}$ is the scalar $\det(A)$ defined by

$$\det(A) = \sum_j (-1)^t a_{1j_1} a_{1j_2} \cdots a_{1j_n},$$

where $\sum_j$ denotes the sum of all terms of the form

$$(-1)^t a_{1j_1} a_{1j_2} \cdots a_{1j_n}$$

as $j_1, j_2 \ldots, j_n$ assumes all possible permutations of the numbers of the numbers 1, 2, ... n, and the exponent t is the number of interchanges used to carry $j_1, j_2 \ldots, j_n$ into the natural order 1, 2, ... n.

The notations $\det A$ and $|A|$ are used interchangeably with $\det(A)$. When $n = 2$ and $n = 3$, we have

$$|A_{2\times 2}| = \begin{vmatrix} a_{11} & a_{12} \\ a_{21} & a_{22} \end{vmatrix} = a_{11}a_{22} - a_{21}a_{12},$$

$$|A_{3\times 3}| = \begin{vmatrix} a_{11} & a_{12} & a_{13} \\ a_{21} & a_{22} & a_{23} \\ a_{31} & a_{32} & a_{33} \end{vmatrix} = a_{11}a_{22}a_{22} + a_{12}a_{23}a_{31} + a_{13}a_{32}a_{21} \tag{A.1.1}$$

$$- a_{31}a_{22}a_{13} - a_{32}a_{23}a_{11} - a_{33}a_{12}a_{21}.$$

Definition A.1.4. The *minor* of the element a_{ij} in $A = \lfloor a_{ij} \rfloor_{nxn}$ is the determinant M_{ij} of the $(n-1)\times(n-1)$ submatrix of A obtained by deleting row i and column j of A. The *cofactor*, denoted by A_{ij}, of a_{ij} in $A = \lfloor a_{ij} \rfloor_{nxn}$ is the product of $(-1)^{(i+j)}$ and M_{ij}, that is,

$$A_{ij} = (-1)^{(i+j)} M_{ij}.$$

The *adjoint* of A, denoted by $adj(A)$, is given by

$$adj(A) = [A_{ij}]^T_{nxn}.$$

Theorem A.1.2. If $A = \lfloor a_{ij} \rfloor_{nxn}$, then

$$a_{i1}A_{k1} + a_{i2}A_{k2} \cdots + a_{in}A_{kn} = \delta_{ik}\det(A), \quad i, k = 1, 2, \ldots, n,$$
$$a_{1j}A_{1k} + a_{2j}A_{2k} \cdots + a_{nj}A_{nk} = \delta_{jk}\det(A), \quad j, k = 1, 2, \ldots, n.$$

By the above formula and equations (A.1.1) with $\delta_{ii} = 1$, we can calculate the value of $\det(A)$ of any dimensional matrix, in principle. For instance, when $n = 4$, we have

$$\det(A_{4\times4}) = a_{11}A_{11} + a_{21}A_{21} + a_{31}A_{31} + a_{41}A_{41},$$

where A_{i1} are calculated from the corresponding 3×3 matrices.

It can be shown that if $A = \lfloor a_{ij} \rfloor_{n\times n}$ is invertible, then

$$A^{-1} = \frac{1}{\det(A)} adj(A).$$

Definition A.1.5. If A is $n\times n$ matrix, an *eigenvector* of A is a nonzero column vector v in R^n such that

$$Av = \rho v$$

for some scalar ρ; the scalar ρ is called an *eigenvalue* of A.

Theorem A.1.3. If A is an $n\times n$ matrix, a number ρ is an eigenvalue of A if and only if

$$\det(\rho I_{n\times n} - A) = 0.$$

The equation

$$\det(\rho I_{n\times n} - A) = 0$$

is called the *characteristic equation* of the matrix A. Upon expanding the determinant, $\det(\rho I_{n\times n} - A)$, we will have a polynomial of degree n in ρ. The polynomial is called the *characteristic polynomial* of the matrix A.

An $n\times n$ matrix B is said to be *similar* to the $n\times n$ matrix A if there is an invertible $n\times n$ matrix P such that

$$B = P^{-1}AP.$$

A square matrix is said to be *diagonalizable* if it is similar to a diagonal matrix. It can be proved that a square matrix A is diagonalizable if only if there is a basis for R^n consisting of eigenvectors of A. Not square matrix is not diagonalizable. There is something close to diagonal form called the *Jordan canonical form* of a square matrix. A *basic Jordan block* associated with a value ρ is expressed

$$\begin{bmatrix} \rho & 1 & 0 & \cdots & 0 & 0 \\ 0 & \rho & 1 & \cdots & 0 & 0 \\ \vdots & \vdots & \vdots & & \vdots & \vdots \\ 0 & 0 & 0 & \cdots & \rho & 1 \\ 0 & 0 & 0 & \cdots & 0 & \rho \end{bmatrix}.$$

The Jordan canonical form of a square matrix is compromised of such Jordan blocks.

Theorem A.1.4. Suppose that A is an $n \times n$ matrix and suppose that

$$\det(\rho I - A) = (\rho - \rho_1)^{m_1} (\rho - \rho_2)^{m_2} \cdots (\rho - \rho_k)^{m_k},$$

where

$$\rho_1, \rho_2, \cdots, \rho_k$$

are distinct roots of the characteristic polynomial of A. Then A is similar to a matrix of the form

$$\begin{bmatrix} B_1 & 0 & \dots & 0 \\ 0 & B_2 & \cdots & 0 \\ \vdots & \vdots & & \vdots \\ 0 & 0 & \cdots & B_k \end{bmatrix},$$

where each B_i is an $m_i \times m_i$ matrix of the form

$$\begin{bmatrix} J_{i_1} & 0 & \dots & 0 \\ 0 & J_{i_2} & \cdots & 0 \\ \vdots & \vdots & & \vdots \\ 0 & 0 & \cdots & J_{i_l} \end{bmatrix},$$

and each J_{i_j} is a basic Jordan block associated with ρ_i.

A.2 Systems of linear equations

A system of linear equations is

$$a_{i1}x_1 + a_{i2}x_2 + \cdots + a_{in}x_n = b_j, \quad j = 1, 2, \ldots m,$$

or in the matrix form

$$A_{m\times n}x_{n\times 1} = b_{m\times 1}. \tag{A.2.1}$$

A solution of the system is a set of values of x that satisfies

$$Ax = b.$$

In this system, A is called the *coefficient matrix*, x the *matrix of unknowns*, and b the *matrix of constants*. The matrix $[A, b]$ is called the *augmented matrix* of the system.

Theorem A.2.1. The system $Ax = b$ has a solution if and only if

$$Rank([A,b]) = Rank(A).$$

Theorem A.2.2. If

$$Rank([A,b]) = Rank(A) = r,$$

then the solution to

$$A_{m\times n}x_{1\times n} = b_{1\times n}$$

can be expressed in terms of $n - r$ parameters.

Theorem A.2.3. Consider a system of linear equations $A_{n\times n}x_{n\times 1} = b_{n\times 1}$. If

$$\det(A) \neq 0,$$

then the unique solution is given by

$$x = A^{-1}b.$$

Theorem A.2.4. Consider a system of linear equations

$$A_{n\times n}x_{n\times 1} = b_{n\times 1}.$$

If $\det(A) \neq 0$, then the unique solution of the system is given by

$$x_j = \frac{\sum_{k=1}^{n} b_k A_{kj}}{\det(A)}, \quad j = 1, 2, \cdots n,$$

where A_{ij} are cofactors of A.

The above formula is called *Cramer's Rule*. We note that

$$\sum_{k=1}^{n} b_k A_{kj}$$

is the determinant of the matrix obtained by replacing the jth column of A by the column of constants b.

A.3 Metric spaces

Metric spaces are used to represent states of economic systems. A metric measures the "distance" between two states.

Definition A.3.1. A *metric space* is a pair (X, d) consisting of a set X and a function

$$d : X \times X \to [0, \infty)$$

called a metric such that
(i) $d(x, y) = 0$ if and only if $x = y$;
(ii) $d(x, y) = d(y, x)$;
(iii) $d(x, y) + d(y, z) \leq d(x, z)$.

A sequence s is a function from the set of nonnegative integers to X. The n th point of the sequence is the point s_n. A *subsequence* of a sequence s is any sequence p of the form $p_j = p_{\alpha_j}$ where α is a strictly increasing function from Z^+ to Z^+. A *Cauchy sequence* is a sequence $\{x_n\}$ such that given $\varepsilon > 0$ there exists a positive integer k such that

$$d(x_n, x_m) < \varepsilon$$

whenever n and m are greater than k.

Definition A.3.2. A metric space X is *complete* if every Cauchy sequence in X converges to a point in X.

A point p is a *limit point* of a subset Y of X provided that there is a sequence of points in Y which converges to p. A subset Y of X is *closed* if it contains all its limit points. The *closure* of Y denoted by $cl(Y)$ is the set of all limit points of Y. The *ball* of radius r centered at x is the set

$$B(x, r) = \{y, y \in X, d(x, y) < r\}.$$

A subset of X is *open* if for each point x belonging to U there exists $r > 0$ such that the ball $B(x, r)$ is contained in U. The radius r may depend on x. A subset A of a metric space X is said to be *dense* in X if $cl(A) = X$. Another way of saying this is that A is dense in X if every ball centered at a point of X contains a point of A, no matter how small the radius of the ball.

Theorem A.3.1.
(1) If Y is a closed set then $X - Y$ is an open set.
(2) If U is an open set then $X - U$ is a closed set.

Suppose that X and Y are metric spaces. A function f from X to Y is *continuous* if for every open subset V of Y the set $f^{-1}(V)$ is an open subset of X. f is a *homeomorphism* if f is continuous and invertible, and the inverse of f is continuous.

Theorem A.3.2. A function f from X to Y is continuous if and only if given any point x in X and any sequence $\{x_n\}$ converging to x then the sequence $\{f(x_n)\}$ converges to $f(x)$.

Theorem A.3.3. If f is continuous and A is a subset of X, then

$$f(cl(A)) \subset cl(f(A)).$$

The *interior* of a subset S of a metric space X is the set defined by

$$\text{int}(S) = X - cl(X - S).$$

A *neighborhood* of a point x in a metric space X is a subset N such that x belongs to the interior of X. The *exterior* of a subset S of a metric space X is the set

$$ext(S) = X - cl(S).$$

The *boundary* of a subset S of a metric space X is the set

$$\partial S = cl(S) \cap cl(X - S).$$

A subset S of a metric space X is *disconnected* if there exists disjoint open sets U and V each having nonempty intersection with S such that S is contained in the union of U and V. A subset S of a metric space X is *connected* if it is not disconnected.

Theorem A.3.4. If N is a subset of a metric space X and S is a connected subset of X whose intersection with both the interior and the exterior of N is nonempty, then the intersection of S with the boundary of N is nonempty.

A subset K of a metric space X is *compact* if every sequence of points in K has a convergent subsequence. A metric space X is *locally compact* if each point x in X has a neighborhood which is compact. A subset K of a metric space X is *bicompact* if given any collection of open sets whose union contains K there exists a finite number of these sets whose union also contains K.

Theorem A.3.5. A subset K of a metric space X is compact if and only if it is bicompact.

Theorem A.3.6. Suppose that f is a continuous function from a metric space X to a metric space Y. Suppose that C is a connected subset of X and K is a compact subset of X. Then (1) $f(C)$ is connected; and (2) $f(K)$ is compact.

Theorem A.3.7. Suppose that g is a continuous real-valued function defined on a compact subset K of a metric space. Then there exist points x and y in K such that $g(K)$ is contained in the interval $[g(x), g(y)]$. Thus g achieves its maximum and its minimum.

A function f from a metric space (X, d) to a metric space (X, D) is *uniformly continuous* if given $\varepsilon > 0$ there exists $\delta > 0$ such that

$$d(x_1, x_2) < \delta$$

implies that

$$D(f(x_1), f(x_2)) < \varepsilon.$$

Theorem A.3.8. Suppose that f is a continuous function from a compact metric space X to a metric space Y. Hence f is uniformly continuous.

Theorem A.3.9. Suppose that X and Y are compact metric spaces. Let

$$f : X \to Y$$

be continuous, one-to-one, and onto. Then the inverse of f is continuous and hence f is a homeomorphism.

A.4 The implicit function theorem

First we state a few theorems from analysis.

Definition A.4.1. Suppose that V_1 and V_2 are two normed linear spaces with respective norms $\| \|_1$ and $\| \|_2$. Then

$$F : V_1 \to V_2$$

is *continuous* at $x_0 \in V_1$ if for all $\varepsilon > 0$ there exits a $\delta > 0$ such that $x \in V_1$ and

$$\|x - x_0\|_1 \in \delta$$

implies that

$$\|F(x) - F(x_0)\|_2 \leq \varepsilon.$$

And F is said to be continuous on the set $U \in V_1$ if it is continuous at each point $x \in U$, and we write $F \in C(U)$.

Theorem A.4.1. (the intermediate-value theorem) If the function

$$f : [a, b] \to R$$

is continuous and

$$f(a) < 0 < f(b),$$

then there exists a point $c \in (a, b)$ so that

$$f(c) = 0.$$

Definition A.4.2. The function $f : R^n \to R^n$ is *differentiable* at $x_0 \in R^n$ if there is a linear transformation $Df(x_0)$ that satisfies

$$\lim_{|h| \to 0} = \frac{|f(x_0 + h) - f(x_0) - Df(x_0)h|}{|h|} = 0.$$

The linear transformation v is called the *derivative* of f at x_0.

The following theorem gives us a method for computing the derivative in coordinates.

Theorem A.4.2. If the function $f : R^n \to R^n$ is differentiable at $x_0 \in R^n$, then the partial derivatives $\partial f_i / \partial x_j$ all exist at x_0 and for all $x \in R^n$

$$Df(x_0)x = \sum_{j=1}^{n} \frac{\partial f}{\partial x_j}(x_0)x_j.$$

Thus if f is a differentiable function, the derivative Df is given by the $n \times n$ Jacobian matrix

$$Df = \left[\frac{\partial f_i}{\partial x_j}\right].$$

For U an open subset of R^n, the higher order derivatives $D^k f(x_0)$ are defined in a similar way.

Definition A.4.3. Suppose that

$$f : U \to R^n$$

is *differentiable* on U. Then $f \in C^1(U)$ if the derivative Df is continuous on U.

We can define

$$f \in C^k(U),\ k = 2, 3, \cdots$$

in a similar manner.

Theorem A.4.3. Suppose that U is an open subset of R^n and that $f : U \to R^n$. Then $f \in C^1(U)$ if and only if the partial derivatives $\partial f_i / \partial x_j$ all exist and are continuous on U.

It can be shown that $f \in C^k(U)$ if and only if the partial derivatives

$$\frac{\partial^k f_i}{\partial x_{j_1} \cdots \partial x_{j_k}},$$

with $i,\ j_1, \cdots, j_k = 1, \cdots, k$ exist and are continuous on U.

Theorem A.4.4. (the inverse function theorem) Let U be an open set in R^n and

$$f : U \to R^n$$

be a C^k function with $k \geq 1$. If a point $\bar{x} \in U$ is such that the $n \times n$ matrix $Df(\bar{x})$ is invertible, then there is an open neighborhood V of $\bar{x}$ in U such that

$$f : V \to f(V)$$

is invertible with a C^k inverse.

The inverse function theorem implies that if the matrix $f_x(0)$ is nonsingular, then there is a locally defined smooth function

$$x = g(y),\ \ g : R^n \to R^n$$

such that

$$f(g(y)) = y,$$

for all y in some neighborhood of the origin of R^n. The function g is called the inverse function for f and is denoted by

$$g = f^{-1}.$$

If

$$y = g(x), \ \ g : R^n \to R^m$$

and

$$z = f(y), \ f : R^m \to R^k$$

are two maps, then their *superposition*

$$h = f \circ g$$

is a map

$$z = h(x), \ R^n \to R^k,$$

defined by the formula

$$h(x) = f(g(x)).$$

Let $f_y(y)$ denote the Jacobian matrix f evaluated at a point $y \in R^m$

$$f_y(y) = \left(\frac{\partial f_i(y)}{\partial y_j} \right)_{k \times m}, \quad i = 1, 2, ..., k, \ \ j = 1, 2, ..., m.$$

We similarly define $h_x(x)$ as

$$h_x(x) = \left(\frac{\partial f_i(y)}{\partial y_j} \right) \left(\frac{\partial y_p}{\partial x_q} \right).$$

We consider a map

$$(x, \ y) \mapsto F(x, y),$$

where

$$F : R^n \times R^m \to R^n$$

is a smooth map defined in a neighborhood of $(x, y) = (0, 0)$ and $F(0,0) = 0$. Let $F_x(0,0)$ denote the matrix of first partial derivatives of F with respect to x evaluated at $(0, 0)$

$$F_x(0,0) = \left(\frac{\partial F_i(x, y)}{\partial x_j} \right)_{(x,y)=(0,0)} .$$

Theorem A.4.4. (the implicit function theorem) If the matrix $F_x(0,0)$ is nonsingular, then there is a smooth locally defined function $y = f(x)$, $f : R^n \to R^m$ such that

$$F(x, f(x)) = 0,$$

for all x in some neighborhood of the origin of R^n. Moreover

$$f_x(0) = -[F_x(0,0)]^{-1} F_y(0,0).$$

The degree of smoothness of the function f is the same as that of F.

Theorem A.4.5. (the submanifold theorem) Let U be an open set in R^n and let

$$f : U \to R^p$$

be a differentiable function such that $Df(x)$ has rank p whenever

$$f(x) = 0.$$

Then $f^{-1}(0)$ is an $(n - p)-$ dimensional manifold in R^n.

Morse Lemma. Let

$$f : R^n \to R$$

be a sufficiently differentiable function. If x^* is a *nondegenerate critical point* of f, that is

$$Df(x^*) = 0,$$

and the *Hessian matrix*

$$\left[\frac{\partial^2 f(x^*)}{\partial x_i \partial x_j}\right]$$

is nonsingular, then there is a local coordinate system $(y_1, \cdots, y_n)$ in a neighborhood U of x^* with

$$y_i(x^*) = 0,$$

for all i, such that

$$f(y) = f(x^*) - \sum_{i=1}^{k} y_i^2 + \sum_{i=1+k}^{n} y_i^2,$$

for all $y \in U$. The integer k is the number of negative eigenvalues of the Hessian matrix.

Sard's Theorem. Let U be an open set in R^n and let

$$f : U \to R^p$$

be a differentiable function. Let C be the set of critical points of f, that is, the set of all $x \in U$ with

$$rank Df(x) < p.$$

Then $f(C)$ has measure zero in R^p.

A.5 The Taylor expansion and linearization

Given a successively differentiable one-variable function $f(x)$, the *Taylor expansion* around a point x^* gives the series

$$f(x) = f(x^*) + f'(x^*)(x - x^*) + \frac{f''(x^*)}{2!}(x - x^*)^2 + \\ + \frac{f^{(n)}(x^*)}{n!}(x - x^*)^n + R(x),$$

where a polynomial involving higher powers (than n) of $(x - x_0)$ appears on the right. For a two-variables function, $f(x, y)$, the Taylor expansion around a point (x^*, y^*) is given by

$$f(x, y) = f(x^*, y^*) + f_x(x^*, y^*)(x - x^*) + f_y(x^*, y^*)(y - y^*) + \\ \frac{1}{2!}\left[f_{xx}(x^*, y^*)(x - x^*)^2 + 2f_{xy}(x^*, y^*)(x - x^*)(y - y^*) + f_{yy}(x^*, y^*)(y - y^*)^2\right] \\ + \ldots + R(x, y).$$

The linearization of a function is obtained by simply dropping all terms of order higher than one from the Taylor series of the function. For instance, the linear approximation of a one-variable function $f(x)$ gives

$$f(x) = f(x^*) + f'(x^*)(x - x^*).$$

In the case of two variables

$$f(x, y) = f(x^*, y^*) + f_x(x^*, y^*)(x - x^*) + f_y(x^*, y^*)(y - y^*).$$

We now give the Taylor expansion for any dimension around the origin. Let U be a region in R^n containing the origin $x = 0$. We denote the set of all continuous functions

$$f : U \to R^m$$

by $C^0(U, R^m)$ and the set of all differentiable functions with continuous first derivatives by $C^1(U, R^m)$. Analogously, we will use $C^k(U, R^m)$ to indicate the functions with continuous derivatives up through order k. If $f \in C^k(U, R^m)$ with a sufficiently large k, the function f is called *smooth*. A C^∞ function has continuous partial derivatives of any order. Any function $f \in C^k(U, R^m)$ can be represented near $x = 0$ in the Taylor expansion

$$f(x) = \sum_{|i|}^{k} \frac{1}{i_1! i_2! \ldots i_n!} \frac{\partial^{|i|} f(x)}{\partial x_1^{i_1} \partial x_2^{i_2} \ldots \partial x_n^{i_n}} \Bigg|_{x=0} x_1^{i_1} x_2^{i_2} \ldots x_n^{i_n} + R(x),$$

where

$$|i| \equiv i_1 + i_2 + \ldots + i_n, \quad R(x) = O\left(\|x\|^{k+1}\right) = o\left(\|x\|^k\right),$$

namely

$$\frac{\|R(x)\|}{\|x\|^k} \to 0 \quad as \ \|x\| \to 0,$$

in which $\|x\| = \sqrt{x^T x}$. Here, we give precise definitions of O and o. Let f and g be two given functions. We say that

$$f(x) = O(g(x)), \quad as \ x \to 0,$$

if there are constants $\alpha > 0$ and $A > 0$ such that

$$|f(x)| \le A|g(x)|$$

for $|x| < \alpha$. We say that

$$f(x) = o(g(x)), \quad as \ x \to 0,$$

if for any $\varepsilon > 0$ there is a $\delta > 0$ such that

$$|f(x)| \leq \varepsilon |g(x)|$$

for $|x| < \delta$.

A C^∞-function f is called *analytical* near the origin if the corresponding *Taylor series*

$$\sum_{|i|}^{\infty} \frac{1}{i_1! i_2! \ldots i_n!} \frac{\partial^{|i|} f(x)}{\partial x_1^{i_1} \partial x_2^{i_2} \ldots \partial x_n^{i_n}} \bigg|_{x=0} x_1^{i_1} x_2^{i_2} \ldots x_n^{i_n},$$

converges to $f(x)$ at any point x sufficiently close to $x = 0$.

A.6 Concave and quasiconcave functions

Consider $f : \Omega \to R$ where Ω is a domain that is a convex subset of R^n (such as $\Omega = R^n$ or

$$\Omega = R_+^n = \{x \in R^n, x \geq 0\}.$$

Definition A.6.1. The function

$$f : \Omega \to R,$$

defined on the convex set $\Omega \subset R^n$ is *concave* if

$$f(\alpha x' + (1-\alpha)x) \geq \alpha f(x') + (1-\alpha) f(x), \tag{A.6.1}$$

for all x and $x' \in \Omega$ and all $\alpha \in [0, 1]$. If the inequality is strict for all $x' \neq x$ and all $\alpha \in (0, 1)$, then we say that the function is *strictly concave*.

Figure A.6.1a illustrates a strictly concave function of variable. Figure A.6.1b depicts a concave but not strictly concave function.

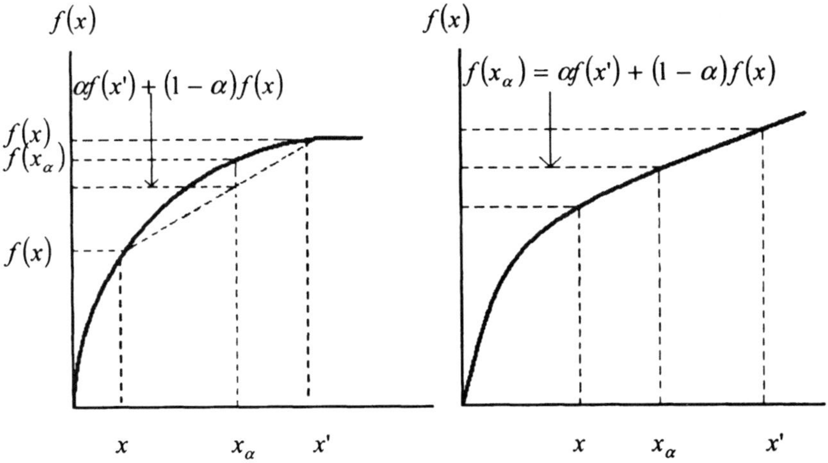

a) strictly concave function b) concave but not strictly concave function

Figure A.6.1: Concave functions

Condition (A.6.1) is equivalent to the following property

$$f\left(\sum_{i=1}^{M}\alpha_i x^i\right) \geq \sum_{i=1}^{M}\alpha_i f(x^i), \tag{A.6.2}$$

for any collection of vectors $x^i \in A$ and numbers

$$\alpha_i \geq 0, \quad i = 1, 2, \cdots, M$$

such that

$$\sum_{i=1}^{M}\alpha_i = 1.$$

The properties of *convexity* and *strict convexity* for a function f are defined analogously but with the inequality in equation (A.6.1) reversed. Note that f is concave if and only if $-f$ is convex. For a strictly convex function, a straight line connecting any two points in its graph should lie entirely above its graph, as shown in figure A.6.2.

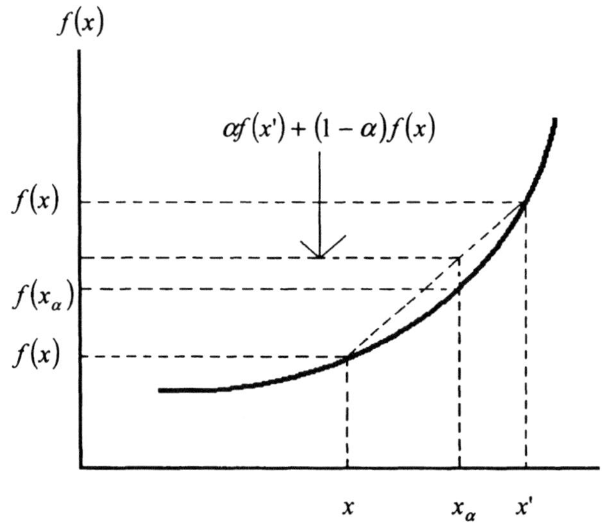

Figure A.6.2: Strictly convex functions

Theorem A.6.1. Assume $f \in C^1$. The function $f : \Omega \rightarrow R$ is concave if only if

$$f(x+z) \leq f(x) + \nabla f(x) \cdot z, \tag{A.6.3}$$

for all

$$x \in \Omega, \quad z \in R^n, \quad x + z \in \Omega.$$

The function f is strictly concave if inequality (A.6.3) holds strictly for all $x \in \Omega$ and all $z \neq 0$.

Condition (A.6.3) says that any tangent to the graph of a concave function f must lie weakly above the graph of f. The corresponding characterization of convex and strictly convex functions simply entails reversing the direction of the inequality in (A.6.3).

Definition A.6.2. The $N \times N$ matrix A is *negative semidefinite* if

$$z^T A z \leq 0 \tag{A.6.4}$$

for all $z \in R^n$. If the inequality is strict for all $z \neq 0$, then the matrix A is *negative definite*. Reversing the inequalities in condition (A.6.4), we get the concepts of *positive semidefinite* and *positive definite* matrices.

Theorem A.6.2. Assume $f \in C^2$. The function

$$f : \Omega \to R$$

is concave if and only if $D^2 f(x)$ is negative semidefinite for every $x \in \Omega$. If $D^2 f(x)$ is negative definite for every $x \in \Omega$, the function is strictly concave.

Definition A.6.3. The function $f : \Omega \to R$, defined on the convex set, $\Omega \subset R^n$,[2] is *quasiconcave* if its upper contour sets

$$\{x \in \Omega : f(x) \geq t\}$$

are convex sets; that is, if $f(x) \geq t$ and $f(x') \geq t$ imply that

$$f(\alpha x + (1 - \alpha)x') \geq t, \tag{A.6.5}$$

for any $t \in R$, x, $x' \in \Omega$, and $\alpha \in [0, 1]$. If the concluding inequality in (A.6.5) is strict whenever $x \neq x$ and $\alpha \in [0, 1]$, then we say that f is *strictly quasiconcave*.

From the above definition, we see that the function is quasiconcave iff

$$f(\alpha x + (1 - \alpha)x') \geq \text{Min}\{f(x), f(x')\}, \tag{A.6.6}$$

for all x, $x' \in \Omega$, and $\alpha \in [0, 1]$. We see that a concave function is quasiconcave. But the converse is not true. Concavity is a stronger property than quasiconcavity.

Note that f is *quasiconvex* if its lower contour sets are convex; that is if

---

[2] The set $\Omega \subset R^n$ is convex if

$$\alpha x + (1 - \alpha)x' \in \Omega$$

whenever x, $x' \in \Omega$, and $\alpha \in [0, 1]$.

$$f(x) \le t \text{ and } f(x') \le t \text{ implies that } f(\alpha x + (1-\alpha)x') \le t,$$

for any $t \in R$, $x, x' \in \Omega$, and $\alpha \in [0, 1]$. If the above concluding inequality is strict whenever $x \neq x$ and $\alpha \in [0, 1]$, then we say that f is *strictly quasiconvex*.

Theorem A.6.3. Assume $f \in C^2$. The function $f : \Omega \to R$ is quasiconcave iff

$$\nabla f(x)(x' - x) \ge 0, \tag{A.6.7}$$

whenever

$$f(x') \ge f(x),$$

for all $x, x' \in \Omega$. If

$$\nabla f(x)(x' - x) > 0,$$

whenever $f(x') \ge f(x)$ and $x \neq x$, then f is strictly quasiconcave. In the other direction, if f is strictly quasiconcave and if $\nabla f(x) \neq 0$ for all $x \in \Omega$, then

$$\nabla f(x)(x' - x) > 0$$

whenever $f(x') \ge f(x)$ and $x \neq x$.

Theorem A.6.4. Assume $f \in C^2$. The function $f : \Omega \to R$ is quasiconcave iff for every $x \in \Omega$, $D^2 f(x)$ is negative semedefinite in the subspace

$$\{z \in R^n : \nabla f(x)z = 0\},$$

that is, if and only if

$$z^T D^2 f(x) z \le 0 \tag{A.6.8}$$

whenever

$$\nabla f(x)z = 0,$$

for every $x \in \Omega$. If the Hessian matrix $D^2 f(x)$ is negative definite in the subspace

$$\{z \in R^n : \nabla f(x)z = 0\}$$

for every $x \in \Omega$, then f is strictly quasiconcave.

A.7 Unconstrained maximization

Definition A.7.1. Consider the function $f: R^n \to R$. The vector x^* is a *local maximizer* of f if there is an open neighborhood of x^*, $\Omega \subset R^n$, such that

$$f(x^*) \geq f(x)$$

for every $x \in \Omega$. If $f(x^*) \geq f(x)$ for every $x \in R^n$ (i.e., if we can take $\Omega = R^n$), then we say that x^* is a *global maximizer* (or simply a maximizer). The concepts of *local* and *global minimizers* are defined analogously.

Theorem A.7.1. Suppose that f is differentiable and that $x^* \in R^n$ is a local maximizer or local minimizer of f. Then

$$\frac{\partial f}{\partial x_i} = 0$$

at x^* for every i, $i = 1, 2, \cdots$.

The condition $\partial f / \partial x_i = 0$ for every i can be expressed in more concise notation

$$\nabla f(x^*) = 0.$$

A vector $x^* \in R^n$ such that $\nabla f(x^*) = 0$ is called a *critical point*. The above theorem tells that every maximizer or minimizer is a critical point. The converse does not hold. For instance, consider

$$f(x_1, x_2) = x_1^2 - x_2^2.$$

At the origin we have $\nabla f = 0$. Thus the origin is a critical point, but it is neither a local maximizer nor minimizer. The following second-order conditions provide the sufficient condition for maximization.

Theorem A.7.2. Suppose that the function

$$f : R^n \to R$$

is C^2 and that

$$\nabla f(x^*) = 0.$$

(i) If $x^* \in R^n$ is a local maximizer, then the $n \times n$ matrix, the *Hessian* $D^2 f(x^*)$, is negative semidefinite. (ii) If the Hessian $D^2 f(x^*)$ is negative definite, then $x^* \in R^n$ is a local maximizer.

It should be noted that in the borderline case in which $D^2 f(x^*)$ is negative semidefinite but not negative definite, we cannot assert that x^* is a local maximizer. Consider, for instance

$$f = x^3.$$

We have $D^2 f(0)$ is negative demidefinite; but $x^* = 0$ is neither a local maximizer nor a local minimizer of this function.

If we replace "negative" by "positive" in the above theorem, the same is true for local minimizers. A critical point x^* of f for which the Hessian $D^2 f(x^*)$ is indefinite is called a *saddle point* of f. A saddle point x^* is a minimizer of f in some directions and maximizer of f in other directions. For instance, the origin is a saddle point of

$$f(x_1, x_2) = x_1^2 - x_2^2$$

as the Hessian is

$$\begin{bmatrix} 2 & 0 \\ 0 & -2 \end{bmatrix}.$$

A saddle point is neither a local maximizer nor a local minimizer.

Theorem A.7.3. Let

$$f : R^n \to R$$

be a C^2 function whose domain is an open set Ω in R^n. Suppose that

$$\frac{\partial f}{\partial x_i}(x^*) = 0 \text{ at } x^* \text{ for every } i, \ i = 1, 2, \cdots .$$

(i) If the n leading principal minors of the Hessian $D^2 f(x^*)$ alternate in sign

$$|f_{11}| < 0, \begin{vmatrix} f_{11} & f_{12} \\ f_{21} & f_{22} \end{vmatrix} > 0, \begin{vmatrix} f_{11} & f_{12} & f_{13} \\ f_{21} & f_{22} & f_{23} \\ f_{31} & f_{32} & f_{33} \end{vmatrix} < 0, \cdots,$$

at x^*, then x^* is a strict maximizer of f.

(ii) If the n leading principal minors of the Hessian $D^2 f(x^*)$ are all positive

$$|f_{11}| > 0, \begin{vmatrix} f_{11} & f_{12} \\ f_{21} & f_{22} \end{vmatrix} > 0, \begin{vmatrix} f_{11} & f_{12} & f_{13} \\ f_{21} & f_{22} & f_{23} \\ f_{31} & f_{32} & f_{33} \end{vmatrix} > 0, \cdots,$$

at x^*, then x^* is a strict minimizer of f.

(iii) If some of the n leading principal minors of the Hessian $D^2 f(x^*)$ violate the sign patterns in (i) and (ii) at x^*, then x^* is a saddle point of f.

Theorem A.7.4. Any critical point $x^* \in R^n$ of a concave function f is a global maximizer.

Similarly, any critical point of a convex function f is a global minimizer of f.

A.8 Constrained maximization

Consider the problem of maximizing f under m equality constraints

$$\underset{x \in R^n}{Max}\ f(x)$$
$$\text{s.t.: } g_j(x) = \bar{b}_j, \quad j = 1, 2, \cdots, m, \tag{A.8.1}$$

where the functions

$$f, \ g_1, \cdots, g_m$$

are defined on R^n. We generally assume $n \geq m$. Let us introduce the *constraint set*

$$\Omega \equiv \{x \in R^n : g_j = \bar{b}_j, \quad j = 1, 2, \cdots, m\}.$$

The *feasible point* $x^* \in \Omega$ is a *local constrained maximizer* in problem (A.8.1) if there exists an open neighborhood of x^*, say $M \subset \Omega$, such that

$$f(x^*) \geq f(x) \text{ for all } x \in M \cap \Omega.$$

The point x^* is a *global constrained maximizer* if it solves problem (A.8.1). The first-order condition is stated as follows.

Theorem A.8.1. Suppose that the objective and constraint functions of problem (A.8.1) are differentiable and that $x^* \in \Omega$ is a local maximizer. Assume also that the $m \times n$ matrix

$$\left[\frac{\partial g_j}{\partial x_i}\right]_{m\times n},$$

has rank m.[3] Then there are numbers $\rho_j \in R$, one for each constraint, such that

$$\frac{\partial f(x^*)}{\partial x_i} = \sum_{j=1}^{m} \rho_j \frac{\partial g_j(x^*)}{\partial x_i}, \quad i = 1, 2, \cdots, n, \tag{A.8.2}$$

or, in more concise notation

$$\nabla f(x^*) = \sum_{j=1}^{m} \rho_j \nabla g_j(x^*).$$

The numbers ρ_j, are referred to as *Lagrange multipliers*.

Often the first-order conditions (A.8.2) are presented in a different way. Given variables $x \in R^n$ and

$$\rho = (\rho_1, \rho_2, \cdots, \rho_m),$$

we can define the *Lagrangian function*

$$L(x, \rho) = f(x) - \sum_j \rho_j g_j(x). \tag{A.8.3}$$

The above theorem says that if x^* is a local constrained maximizer (and if the constrain qualification is satisfied), then for some values

$$\rho_1, \rho_2, \cdots, \rho_m,$$

all of the partial derivatives of the Lagrangian function are null; that is

[3] This is called the *constraint qualification*. It says that the constraints are independent at x^*.

$$\frac{\partial L(x^*, \rho)}{\partial x_i} = 0, \quad i = 1, \cdots, n,$$

$$\frac{\partial L(x^*, \rho)}{\partial \rho_j} = 0, \quad j = 1, \cdots, m.$$

There is also a second-order theory for problem (A.8.3). Suppose that at x^* the constraint qualification is satisfied and that there are Langrange multiplies satisfying (A.8.2). If x^* is a local maximizer, then

$$D_x^2 L(x^*, \rho) = D^2 f(x^*) - \sum \rho_j \nabla g_j(x^*).$$

is negative semidefinite on the subspace

$$\{z \in R^n : \nabla g_j(x^*) z = 0, \ \text{for all } j\}.$$

In the other direction, if $x^* \in \Omega$ and satisfies the first-order condition, and if $D_x^2 L(x^*, \rho)$ is negative definite on the subspace

$$\{z \in R^n : \nabla g_j(x^*) z = 0, \ \text{for all } j\},$$

then x^* is a local maximizer.

As a local constrained minimizer of f is a local constrained maximizer of $-f$, theorem A.8.1 and the discussion of second-order conditions above are also applicable to the characterization of local constrained minimizers.

We now generalize our analysis to problems that may have inequality constraints. Consider the problem of maximizing f under m equality constraints and q inequalities

$$\begin{aligned} &\underset{x \in R^n}{Max}\ f(x) \\ \text{s.t.:}\ &g_j(x) = b_j, \quad j = 1, 2, \cdots, m, \\ &h_k(x) \le c_k, \quad k = 1, 2, \cdots, q, \end{aligned} \tag{A.8.4}$$

where the functions

$$f,\ g_1, \cdots, g_m,\ h_1, \cdots, h_q$$

are defined on R^n. We generally assume $n \geq m + q$. We say that the constraint qualification is satisfied at $x^* \in \Omega$ if the constraints that hold at x^* with equality are independent, that if, if the vectors in

$$\{\nabla g_j(x^*), \text{ for all } j\} \cup \{\nabla h_k(x^*), \text{ for all } k\},$$

are linearly independent.

Theorem A.8.2. (the Kuhn-Tucker conditions) Suppose that the objective and constraint functions of problem (A.8.4) are differentiable and that $x^* \in \Omega$ is a local maximizer of the problem. Assume also that the constraint qualification is satisfied. Then there are multipliers $\rho_j \in R$, one for each equality constraint, and $\rho_k \in R_+$, one for each inequality constraint, such that
(i) for every $i = 1, \cdots, m$

$$\frac{\partial f(x^*)}{\partial x_i} = \sum_{j=1}^{m} \rho_j \frac{\partial g_j(x^*)}{\partial x_i} + \sum_{k=1}^{q} \rho_k \frac{\partial h_k(x^*)}{\partial x_i}, \quad i = 1, 2, \cdots, n, \tag{A.8.5}$$

or in more concise notation

$$\nabla f(x^*) = \sum_{j=1}^{m} \rho_j \nabla g_j(x^*) + \sum_{k=1}^{q} \rho_k \nabla h_k(x^*);$$

(ii) for every $k = 1, \cdots, q$

$$\rho_k (h_k(x) - c_k) = 0,$$

i.e., $\rho_k = 0$ for every constraint k that does not hold with equality.

The above theorem is also applicable to local minimizers, with the only change being that the sign restriction on all of the multipliers is now $\rho_k \leq 0$, $k = 1, \cdots, q$.

Theorem A.8.3. Suppose that there are no equality constraints (i.e., $m = 0$) and that every inequality constraint k is given by a quasiconvex function h_k. Suppose also that the objective function satisfies

$$\nabla f(x)(x' - x) > 0 \quad \text{for any } x \text{ and } x' \text{ with } f(x') > f(x),$$

then if $x^* \in \Omega$ satisfies the Kuhn-Tucker conditions (conditions (i) and (ii) in theorem A.8.2), and if the constraint qualification holds at x^*, it follows that x^* is a global maximizer.

If we have

$$\nabla f(x)(x' - x) < 0,$$

for any x and x' with $f(x') < f(x)$ and the multipliers have the nonpositive sign that corresponds to a minimization problem, then x^* is a global minimizer.

A.9 Dynamical optimization[4]

Consider a state variable $x(t)$ which belongs to an interval I of R; its motion through time is governed by

$$x(t+1) = f(x(t), C(t)), \tag{A.9.1}$$

where

$$C(t) = (c_1(t), c_2(t), \cdots, c_n(t))$$

is the vector of control variables. Assume that the initial stock x_0 is given and that the set of feasible decisions in t, denoted by $\Psi(x(t))$, depends on the level of the stock $x(t)$, $C(t) \in \Psi(x(t))$. A sequence

$$(x(t), C(t)), \quad t = 0, 1, \cdots,$$

[4] Appendix A.9 is mainly referred to appendix A.5 in de la Croix and Michel (2002).

which satisfies (A.9.1), the initial condition, and $C(t) \in \Psi(x(t))$ is called a *feasible trajectory starting from* x_0.

Consider a payoff function, depending on the stock variable and the decisions at each date as

$$U(x(t),\, C(t)) = U(x(t),\, c_1(t),\, c_2(t),\, \cdots,\, c_n(t)).$$

Given a positive discount δ, the objective is to maximize the discounted flow of payoff

$$Max \sum_{t=0}^{\infty} \delta^t U(x(t), C(t)), \qquad \text{(A.9.2)}$$

on the set of feasible trajectories starting from x_0.

Assumption A.9.1. For all elements, x, of I, $\Psi(x)$ is non-empty, and for all vectors C in the subset $\Psi(x)$ of R^n, $f(x, C)$ is defined and belongs to I, i.e., the dynamics are defined on I.

Assumption A.9.2. For C given, the functions $f(x, C)$ and $U(x, C)$ are non-decreasing with respect to x. The upper bound of $f(x, C)$ with respect to C in $\Psi(x)$ is finite

$$\phi(x) = \sup_{C \in \Psi(x)} f(x, C) \in I.$$

The upper bound of $U(x, C)$ with respect to C in $\Psi(x)$ is finite

$$u(x) = \sup_{C \in \Psi(x)} U(x, C) \in I.$$

For any initial condition x_0, there exists a constant b_0 and a scalar $\delta_0 > \delta$ such that the sequence $\bar{x}(t)$ defined by

$$\bar{x}(t+1) = \phi(\bar{x}(t)), \quad \forall t \geq 0,$$

satisfies

$$\delta_0^t u(\bar{x}(t)) \le b_0, \quad \forall t \ge 0.$$

Proposition A.9.1. (the convergence of the infinite sum) Under assumptions A.9.1 and A.9.2, every discounted sum of feasible payoff is defined and has values in $R \cup \{-\infty\}$

$$\sum_{t=0}^{\infty} \delta^t U(x(t), C(t)) = \lim_{T \to +\infty} \sum_{t=0}^{T} \delta^t U(x(t), C(t)) \in R \cup \{-\infty\}.$$

Assumption A.9.3. For all $x_0 \in I$, there exists a feasible path

$$(x(t), C(t))_{t \ge 0}$$

starting at x_0 such that the sequence

$$\sum_{t=0}^{T} \delta^t U(x(t), C(t))$$

is bounded below when $T \to +\infty$.

Proposition A.9.2. (the value function) Under assumptions A.9.1 - A.9.3, the function

$$V(x_0) = \sup\left\{ \sum_{t=0}^{\infty} \delta^t U(x(t), C(t)); \ (x(t), C(t)) \text{ feasible from } x_0 \right\},$$

is defined on I and satisfies $\forall x \in I$

$$V(x) = \sup\{g(x, C) + \delta V(f(x, C)); C \in \Psi(x)\}.$$

Proposition A.9.3. (the characteristics of optimal trajectories) Under assumptions A.9.1 - A.9.3, a feasible path $(x^*(t), C^*(t))$ starting from $x_0^* = x_0$ is optimal if and only if we have for all t

$$V(x^*(t)) = U(x^*(t), C^*(t)) + \delta V(x^*(t+1)). \tag{A.9.3}$$

We now examine the same problem in this section in a different way. We maximize (A.9.2) such that (A.9.1) and the initial condition are satisfied and

$$C(t) \in \Psi(x(t))$$

by introducing the Lagrangian.

Definition A.9.4.1. The *Lagrangian* $L_t(x(t), C(t))$ of period t is obtained as the sum of the payoff $U(x, C)$ with the increase in the value of the stock

$$L_t(x(t), C(t)) = U(x(t), C(t)) + \delta q(t+1) f(x(t), C(t)) - q(t)x(t).$$

A *feasible trajectory* $(x^*(t), C^*(t))$ is supported by a sequence of shadow prices $(q(t))$ if, for every integer $t \geq 0$, the Lagrangian $L_t(x(t), C(t))$ attains its maximum at $(x^*(t), C^*(t))$ on the set of vectors $(x(t), C(t))$ which verify

$$x(t) \in I, \quad C(t) \in \Psi(x(t)).$$

We then have

$$L_t(x^*(t), C^*(t)) = \underset{x(t) \in I,\, C(t) \in \Psi(x)}{Max} L_t(x(t), C(t)),$$

for all $t \geq 0$.

Assumption A.9.4. The set A of the feasible triplets of payoff, current stock, and resulting stock is convex. Formally, A is the set of elements (a, x, y) of $R \times I \times I$ for which there exists $C \in \Psi(x)$ such that

$$a \leq U(x, C), \quad y = f(x, C).$$

Proposition A.9.4 (the characteristics of optimal trajectories)[5]
Let us consider a feasible trajectory $\left(x^*(t), C^*(t)\right)$ starting from x_0 for which $x^*(t)$ is interior to I for all $t \geq 0$. Under assumptions A.9.1 - A.9.4, the trajectory $\left(x^*(t), C^*(t)\right)$ is optimal if and only if there exists a sequence of shadow prices $(q(t))$ such that: (i) the trajectory $\left(x^*(t), C^*(t)\right)$ is supported by the sequence of shadow prices $(q(t))$; and (ii) for any other feasible trajectory $(x(t), C(t))$ starting from x_0 such that

$$\sum_{t=0}^{\infty} \delta^t U(x(t), C(t)),$$

is finite, we have

$$\lim_{t \to \infty} \delta^t q(t)\left(x(t) - x^*(t)\right) \geq 0. \tag{A.9.4}$$

The condition (A.9.4) is the transversality condition. It means that the discounted value of the optimal capital is exhausted in the long run and that the value of any of other feasible stock should be greater than or equal to that of the optimal stock.

[5] The necessary condition is given by Michel (1990) and the sufficient condition is standard.

Bibliography

Abe, N., 1995, Poverty trap and growth with public goods, Economic Letters 47, 361-366.

Abell, M.L. and J.P. Braselton, 2004, Differential Equations with Mathematica. Elsevier, Amsterdam.

Aghion, P. and P. Bolton, 1997, A theory of trickle-down growth and development, The Review of Economic Studies 64, 151-172.

Aghion, P. and P. Howitt, 1992, A model of growth through creative destruction, Econometrica 60, 323-51.

Alligood, K.T., T. Sauer, and J.A. Yorke, 1997, Chaos: An Introduction to Dynamical Systems. Springer-Verlag, New York.

Arrow, K.J., 1962, The economic implications of learning by doing, The Review of Economic Studies 29, 155-173.

Azariadis, C., 1993, Intertemporal Macroeconomics. Blackwell, Oxford.

Barro, R.J. and G.S. Becker, 1989, Fertility choce in a model of economic growth, Econometrica 57, 481-501.

Barro, R.J. and X. Sala-i-Martin, 2004, Economic Growth. McGraw-Hill, New York.

Becker, G, K. Murphy, and R. Tamura, 1990, Human capital, fertility, and economic growth, Journal of Political Economy 98, s12-37.

Becker, R.A., 1980, On the long-run steady state in a simple dynamic model of equilibrium with heterogeneous households, The Quarterly Journal of Economics 95, 375-82.

Benabou, R., 1996, Equity and efficiency in human capital investment: the local connection, The Review of Economic Studies 63, 267-274.

Benedicks, M. and L. Carleson, 1991, The dynamics of the Hénon map, Ann. Math. 133, 73-169.

Benhabib, J. and R. H. Day, 1981, Rational choice and erratic behavior, The Review of Economic Studies 48, 459-471.

Berman, A. and R.J. Plemons, 1979, Nonnegative Matrices in Mathematical Sciences. Academic Press, New York.

Bernstein, D.J. and J.C. Lagarias, 1996, The $3x + 1$ conjugacy map, Canadian J. Math. 48, 1154-1169.

Bertinelli, L. and D. Black, 2004, Urbanization and growth, Journal of Urban Economics 56, 80-96.

Bhaduri, A., 2002, Chaotic implications of the natural rate of unemployment, Structural Change and Economic Dynamics 13, 357-366.

Blanchard, O., 1985, Debt, deficits, and finite horizons, Journal of Political Economy 93, 223-247.

Block, L.S. and W.L. Coppel, 1992, Dynamics in One Dimension. Springer, New York.

Burmeister, E. and A.R. Dobell, 1970, Mathematical Theories of Economic Growth. Collier Macmillan Publishers, London.

Carr, J., 1981, Applications of Center Manifold Theory. Springer-Verlag, New York.

Carvalho, L.A.V., 1998, On a method to investigating bifurcation of periodic solutions in retarded differential equations, Journal of Differential Equations and Applications 94, 17-27.

Champ, B. and S. Freeman, 2001, Modeling Monetary Economies. Cambridge University Press, Cambridge.

Chakrabarti, R., 1999, Endogenous fertility and growth in a model with old age support, Economic Theory 13, 393-416.

Chang, W.W. and D.J. Smyth, 1971, The existence and persistence of cycles in a non-linear model: Kaldor's 1940 model re-examined, The Review of Economic Studies 38, 37-44.

Chiang, A.C., 1984, Fundamental Methods of Mathematical Economics. McGraw-Hill Book Company, London.

Chiarella, C. and P. Flaschel, 2000, The Dynamics of Keynesian Monetary Growth: Macro Foundations. Cambridge University Press, Cambridge.

Coppel, W.A., 1955, The solution of equations by iteration, Proceedings of the Cambridge Philosophical Society 51, 41-43.

Dawkins, R., 1986, The Blind Watchmaker. Longman, London.

Day, R.H., 1982, Irregular growth cycles, The American Economic Review 72, 406-414.

De la Croix, D. and P. Michel, 2002, A Theory of Economic Growth: Dynamics and Policy in Overlapping Generations. Cambridge University Press, Cambridge.

De Melo, W. and S. van Strien, 1993, One-Dimensional Dynamics. Springer-Verlag, Berlin.

De Vault, R., V. Kocic and G. Ladas, 1992, Global stability of a recursive sequence, Dynam. Systems Appl. 1, 13-21.

Dendrinos, P.S. and M. Sonis, 1990, Chaos and Socio-Spatial Dynamics. Springer-Verlag, Berlin.

Denecker, R. and K. Judd, 1992, Cyclical and Chaotic Behavior in a Dynamic Equilibrium Model, 1992, Cycles and Chaos in Economic Equilibrium, Ed. J. Benhabib, Princeton University Press, Princeton.

Devaney, R., 1989, An Introduction to Chaotic Dynamical Systems, 2nd edition. Addison-Wesley, Reading.

Devereux, M.B., A.C. Head and B.J. Lapham, 1996, Monopolistic competition, increasing returns, and the effects of government spending, Journal of Money, Credit, and Banking 28, 233-254.

Devereux, M.B., A.C. Head and B.J. Lapham, 2000, Government spending and welfare with returns to specification, Scandinavian Journal of Economics 102, 547-561.

Diamond, P., 1965, National debt in a neoclassical growth model, American Economic Review 55, 1125-1150.

Dornbusch, R., 1976, Expectations and exchange rate dynamics, Journal of Political Economy 84, 1161-76.

Durlauf, S., 1991, Path dependence in economics: the invisible hand in the grip of the past, American Economic Review 81, 7-74.

Durlauf, S., 1993, Nonergodic economic growth, The Review of Economic Studies 60, 349-366.

Durlauf, S., 1996, A theory of persistent income inequality, Journal of Economic Growth 1, 75-93.

Either, T.S. and C. García-Peñalosa, 2001, Inequality and growth: the dual role of human capital in development, Journal of Development Economics 66, 173-197.

Elaydi, S.N., 1996, A converse to Sharkovsky's theorem, Amer. Math. Month. 103, 386-392.

Elaydi, S.N., 1999, An Introduction to Difference Equations. Springer, Berlin.

Elaydi, S.N., 2000, Discrete Chaos. Chapman & Hall/CRC, London.

Enns, R.H. and G.C. McGuire, 2001, Nonlinear Physics with Mathematica for Scientists and Engineers. Birkhäuser, Boston.

Ezekiel, M., 1938, The cobweb theorem, Quarterly Journal of Economics 52, 255-280.

Farmer, R.E.A., 1986, Deficits and cycles, Journal of Economic Theory 40, 77-88.

Feigenbaum, M.J., 1978, Quantitative universality for a class of nonlinear transformations, J. Stat. Phys. 19, 25-52.

Ferguson, B.S. and G.C. Lim, 1998, Introduction to Dynamic Economic Models. Manchester University Press, Manchester.

Flaschel, P., R. Franke and W. Semmler, 1997, Dynamic Macroeconomics. The MIT Press, Mass., Cambridge.

Friedman, M., 1968, The role of monetary policy, American Economic Review 58, 1-17.

Gabisch, G. and H.W. Lorenz, 1989, Business Cycle Theory, 2nd edition. Springer-Verlag, Berlin.

Gale, D., 1983, Money: In Disequilibrium. Cambridge University Press, Cambridge.

Gale, D., 1996, Delay and cycle, The Review of Economic Studies 63, 169-198.

Gallas, J.A.C. and H.E. Nusse, 1996, Periodicity versus chaos in the dynamics of cobweb models, Journal of Economic Behavior and Organization 29, 447-464.

Galor, O., 1992, Two-sector overlapping-generations model: a global characterization of the dynamical system, Econometrica 60, 1351-86.

Galor, O. and D.N. Weil, 2000, Population, technology, and growth: from Malthusian stagnation to the demographic transition and beyond, The American Economic Review 90, 806-828.

Galor, O. and J. Zeira, 1993, Income distribution and macroeconomics, The Review of Economic Studies 60, 35-52.

Gandolfo, G., 1996, Economic Dynamics, third edition. Springer, Berlin.

Ghiglino, C. and G. Sorger, 2002, Poverty traps, indeterminacy, and the wealth distribution, Journal of Economic Theory 105, 120-39.

Gilbert, J. and L. Gilbert, 1995, Linear Algebra and Matrix Theory. Academic Press, New York.

Goeree, J.K. and C.H. Hommes, 2000, Heterogeneous beliefs and the nonlinear cobweb model, Journal of Economic Behavior & Organization 24, 761-798.

Goodwin, R.M., 1967, A Growth Model, 1967, Socialism, Capitalism and Economic Growth, Ed. C.H. Feinstein, Cambridge University Press, Cambridge.

Grandmont, J.M., 1985, On endogenous competitive business cycles, Econometrica 53, 995-1045.

Guo, J.T., 2004, Increasing returns, capital utilization, and the effects of government spending, Journal of Economic Dynamics & Control 28, 1059-1078.

Haavelmo, T., 1954, A Study in the Theory of Economic Evolution. North-Holland, Amsterdam.

Hahn, F.H. and R.M. Solow, 1985, A Critical Essay on Modern Macroeconomic Theory. Blackwell, Oxford.

Hénon, M., 1976, A two-dimensional mapping with a strange attractor, Comm. Math. Phys. 50, 69-77.

Hess, G. and S. Iwata, 1997, Asymmetric persistence in GDP? a deeper look at depth, Journal of Monetary Economics 40, 535-554.

Hommes, C.H., 1991, Chaotic Dynamics in Economic Models: Some Simple Case-Studies. Wolters-Noordhoff, Groningen.

Hommes, C.H., 1994, Dynamics of the cobweb model with adaptive expectations and nonlinear supply and demand, Journal of Economic Behavior and Organization 24, 315-335.

Ishida, J. and M. Yokoo, 2004, Threshold nonlinearities and asymmetric endogenous business cycles, Journal of Economic Behavior & Organization 54, 175-189.

Jovanovic, B. and R. Rob, 1990, Long waves and short waves: growth through intensive and extensive research, Econometrica 58, 1391-1409.

Jullien, B., 1988, Competitive business cycles in an overlapping generations economy with productive investment, Journal of Economic Theory 46, 45-65.

Jury, E., 1964, Theory and Applications of the Z-transform. Wiley, New York.

Kaldor, N., 1940, A model of the trade cycle, Economic Journal 50, 78-92.

Kalecki, M., 1937, A theory of the business cycle, The Review of Economic Studies 4, 77-97.

Kalecki, M., 1939, A Theory of the Business Cycle, 1939, Essays in the Theory of Economic Fluctuation, Ed. M. Kalecki, Allen-Unwin, London.

Kocic, V.L. and G. Ladas, 1993, Global Behavior of Nonlinear Difference Equations of Higher Order with Applications. Kluwer Academic Publishers, Boston.

Kulenovic, M.R.S. and G. Ladas, 2001, Dynamics of Second Order Rational Difference Equations. Chapman & Hall, London.

Lagarias, J.C., 1985, The $3x + 1$ problem and its generalizations, Amer. Math. Monthly 92, 3-23.

Li, T.Y. and J.A. Yorke, 1975, Period 3 implies chaos, American Mathematical Monthly 82, 985-992.

Lorenz, H.W., 1993, Nonlinear Dynamic Economics and Chaotic Motion. Springer-Verlag, Berlin.

Martelli, M., 1999, Introduction to Discrete Dynamical Systems and Chaos. John Wiley & Sons, Inc., New York.

Matsuyama, K., 1991, Endogenous price fluctuations in an optimizing model of a monetary economy, Econometrica 59, 1617-32.

Matsuyama, K., 1999, Growing through cycles, Econometrica 67, 335-347.

Matsuyama, K., 2000, Endogenous inequality, The Review of Economic Studies 67, 743-759.

Matsuyama, K., 2001, Growing through cycles in an infinitely lived agent economy, Journal of Economic Theory 100, 220-234.

Metzler, L.A., 1941, The nature and stability of inventory cycles, The Review of Economics and Statistics 23, 113-129.

Michel, P., 1990, Some clarifications on the transversality conditions, Econometrica 58, 705-723.

Mitra, T., 2001, A sufficient condition for topological chaos with an application to a model of endogenous growth, Journal of Economic Theory 96, 133-152.

Moav, O., 2002, Income distribution and macroeconomics: the persistence of inequality in a convex technology framework, Economics Letters 187-192.

Myrberg, P.J., 1958, Iteration der reellen polynome zweiten grades, I, Ann. Acad. Sci. Fennicae 256, 1-10.

Myrberg, P.J., 1959, Iteration der reellen polynome zweiten grades, II, Ann. Acad. Sci. Fennicae 268, 1-18.

Myrberg, P.J., 1963, Iteration der reellen polynome zweiten grades, III, Ann. Acad. Sci. Fennicae 336, 1-10.

Nagomo, J. and S. Sato, 1972, On a response characteristic of a mathematical neuron model, Kybernetic 10, 155-164.

Nerlove, M., 1958, Adaptive expectations and cobweb phenomena, Quarterly Journal of Economics 72, 227-240.

Neugart, M., 2000, Nonlinear Labor Market Dynamics. Springer, New York.

Neugart, M., 2004, Complicated dynamics in a flow model of the labor market, Journal of Economic Behavior & Organization 53, 193-213.

Nishimura, K. and K. Shimomura, 2002, Trade and indeterminacy in a dynamic general equilibrium model, Journal of Economic Theory 105, 244-60.

Nishimura, K. and M. Yano, 1995, Nonlinear dynamics and chaos in optimal growth: an example, Econometrica 63, 981-1001.

Okuguchi, K. and K. Irie, 1990, The Schur and Samuelson conditions for a cubic equation, Manchester School 58, 414-18.

Onozaki, T., G. Sieg and M. Yokoo, 2000, Complex dynamics in a cobweb model with adaptive production adjustment, Journal of Economic Behavior & Organization 41, 101-115.

Onozaki, T., G. Sieg and M. Yokoo, 2003, Stability, chaos, and multiple attractors: a single agent makes a difference, Journal of Economic Dynamics & Control 27, 1917-38.

Panico, C. and N. Salvadori, Eds., 1993, Post Keynesian Theory of Growth and Distribution. Elward Elgar, Vermont.

Pasinetti, L.L., 1960, A mathematical formation of the Ricardian system, The Review of Economic Studies 27, 78-98.

Pasinetti, L.L., 1974, Growth and Income Distribution - Essays in Economic Theory. Cambridge University Press, Cambridge.

Peitgen, H.O., H. Jürgens and D. Saupe, 1992, Chaos and Fractals - New Frontiers of Science. Springer-Verlag, Heidelberg.

Peterson, G.L. and J.S. Sochacki, 2002, Linear Algebra and Differential Equations. Addison-Wesley, Boston.

Phillips, A.W., 1954, Stabilisation policy in a closed economy, Economic Journal 64, 290-323.

Piketty, T., 1997, The dynamics of the wealth distribution and interest rate with credit-rationing, The Review of Economic Studies 64, 173-189.

Puu, T., 1989, Nonlinear Economic Dynamics. Springer-Verlag, Berlin.

Ralf, K., 2001, Do complementary factors lead to economic fluctuations? Economics Letters 71, 97-103.

Rasband, S.N., 1990, Chaotic Dynamics of Nonlinear Systems. John Wiley & Sons, New York.

Raut, L.K. and T.N. Srinivasan, 1994, Dynamics of endogenous growth, Economic Theory 4, 770-790.

Rivera-Batiz, L.A. and P.M. Romer, 1991, Economic integration and endogenous growth, Quarterly Journal of Economics 106, 531-555.

Rosser, J.B.Jr., 1991, From Catastrophe to Chaos: A General Theory of Economic Discontinuities. Kluwer Academic Publishers, Boston.

Samuelson, P.A., 1958, An exact consumption-loan model of interest with or without the social contrivance of money, Journal of Political Economy 66, 467-82.

Samuelson, P.A. and F. Modigliani, 1966, The Pasinetti paradox in neo-classical and more general models, The Review of Economic Studies 33, 269-301.

Salvadori, N., 1991, Post-Keynesian Theory of Distribution in the Long Run, 1991, Nicholas Kaldor and Mainstream Economics - Confrontation or Convergence?, Eds. E.J. Nell and W. Semmler, Macmillan, London.

Sandefur, J.T., 1990, Discrete Dynamical Systems: Theory and Applications. Clarendon Press, New York.

Sato, K., 1966, The neoclassical theorem and distribution of income and wealth, The Review of Economic Studies 33, 331-36.

Shone, R., 2002, Economic Dynamics – Phase Diagrams and Their Economic Application. Cambridge University Press, Cambridge.

Simon, H.A., 1973, The Organization of Complex Systems, 1973, Hierarchy Theory: The Challenge of Complex Systems, Ed. H. Pattee, G. Braziller, New York.

Singer, D., 1978, Stable orbits and bifurcation of maps of the interval, SIAM J. Appl. Math. 45, 260-67.

Solow, R., 2000, Growth Theory – An Exposition. Oxford University Press, New York.

Sorger, G., 2000, Income and wealth distribution in a simple model of growth, Economic Theory 16, 23-42.

Sorger, G., 2002, On the long-run distribution of capital in the Ramsey model, Journal of Economic Theory 105, 226-43.

Stutzer, M., 1980, Chaotic dynamics and bifurcation in a macro economics, Journal of Economic Dynamics and Control 2, 253-273.

Thom, R., 1977, Structural Stability and Morphogenesis. Addison Wesley, New York.

Tirole, J., 1985, Asset bubbles and overlapping generations, Econometrica 53, 1499-528.

Tobin, J., 1965, Money and economic growth, Econometrica 33, 671-84.

Tobin, J., 1969, A general equilibrium approach to monetary theory, Journal of Money, Credit and Banking 1, 15-29.

Uzawa, H., 1965, Optimal technical change in an aggregative model of economic growth, International Economic Review 6, 18-31.

Von Thadden, L., 1999, Money, Inflation, and Capital Formation. Springer, Berlin.

Waugh, F.V., 1964, Cobweb models, Journal of Farm Economics 46, 732-750.

Yokoo, M., 2000, Chaotic dynamics in a two-dimensional overlapping generations model, Journal of Economic Development and Control 24, 909-934.

Zhang, F.Z., 1999fz, Matrix Theory: Basic Results and Techniques. Springer, Berlin.

Zhang, W.B., 1991, Synergetic Economics. Springer, Berlin.

Zhang, W.B., 1996, Knowledge and Value - Economic Structures with Time and Space. Umeå Economic Studies, Umeå.

Zhang, W.B., 1999, Capital and Knowledge - Dynamics of Economic Structures with Non-constant Returns. Springer-Verlag, Berlin.

Zhang, W.B., 2000, A Theory of International Trade – Capital, Knowledge and Economic Structures. Springer, Berlin.

Zhang, W.B., 2002, An Economic Theory of Cities – Spatial Models with Capital, Knowledge, and Structures. Springer, Berlin.

Zhang, W.B., 2003a, Taiwan's Modernization. World Scientific, Singapore.

Zhang, W.B., 2003b, A Theory of Interregional Dynamics – Spatial Models with Capital, Knowledge, and Structures. Springer, Berlin.

Zhang, W.B., 2005a, Differential Equations, Bifurcations and Chaos in Economics. World Scientific, Singapore.

Zhang, W.B., 2005b, Economic Growth Theory. Ashgate, Hampshire.

Zhang, W.B., 2005c, A discrete economic growth model with endogenous labor, Discrete Dynamics in Nature and Society, 2, 101-109.

Zhang, W.B., 2005d, Path-dependent economic evolution with capital accumulation and education, Nonlinear Dynamics, Psychology, and Life Sciences. (to appear).

Index

Mathematics in Science and Engineering
Edited by C.K. Chui, Stanford University

Recent titles:
I. Podlubny, *Fractional Differential Equations*
E. Castillo, A. Iglesias, R. Ruíz-Cobo, *Functional Equations in Applied Sciences*
V. Hutson, J.S. Pym, M.J. Cloud, *Applications of Functional Analysis and Operator Theory (Second Edition)*
V. Lakshmikantham and S.K. Sen, *Computational Error and Complexity in Science and Engineering*
T.A. Burton, V*olterra Integral and Differential Equations (Second Edition)*
E.N. Chukwu, *A Mathematical Treatment of Economic Cooperation and Competition Among Nations: with Nigeria, USA, UK, China and Middle East Examples*

Printed in the United States
128777LV00002B/124-126/A

9 780444 521972